DINOSAUR TRACKS
AND OTHER FOSSIL
FOOTPRINTS OF EUROPE

DINOSAUR TRACKS AND OTHER FOSSIL FOOTPRINTS OF EUROPE

■

by *Martin Lockley and Christian Meyer*

COLUMBIA UNIVERSITY PRESS NEW YORK

Columbia University Press
Publishers Since 1893
New York   Chichester, West Sussex

Library of Congress Cataloging-in-Publication Data
Lockley, M. G.
Dinosaur tracks and other fossil footprints of Europe / by Martin Lockley and Christian Meyer.
p.   cm.
Includes bibliographical references and index.
ISBN 0-231-10710-2 (alk. paper)
1. Footprints, Fossil—Europe. I. Meyer, Christian, 1953–
II. Title.
QE845.L622 1999
560′.94—dc21   99-31154

∞

Casebound editions of Columbia University Press books
are printed on permanent and durable acid-free paper.
Printed in the United States of America
c 10 9 8 7 6 5 4 3 2 1

*To Georges Demathieu, Paul Ellenberger, and William ("Bill") Sarjeant, who have all made important contributions to the vertebrate ichnology of Europe**

* This book is the second in a regional series on vertebrate ichnology by Columbia University Press. The previous volume on *Dinosaur Tracks and Other Fossil Footprints of the Western United States* was dedicated to two other European ichnologists—Hartmut Haubold and Giuseppe Leonardi—for their work on global synthesis of vertebrate ichnofaunas.

CONTENTS

PREFACE
THE DINOSAUR-TRACKING RENAISSANCE IN EUROPE

There has been a long and distinguished tradition of tracking dinosaurs and other extinct animals in Europe. This tradition dates back to the first scientific report on a set of Permian fossil footprints from Scotland in 1828, which were at first incorrectly interpreted as the spoor of tortoises. This tradition of early track discoveries, and confusion over the trackmakers, continued in the 1830s with the discovery of the controversial Triassic "hand animal" footprint named *Chirotherium.* Although this was the first fossil footprint ever given a formal Latin name, it was not attributed to the proper trackmaker for almost a century. Even so, some of the leading paleontologists and geologists of the day paid considerable and serious attention to fossil footprints, knowing full well that they shed new light on the history of vertebrates that could not be obtained from the sparse record of bones then available.

Between the 1840s and 1860s, large Cretaceous dinosaur tracks from England were discovered and attributed, by some observers, to the well-known herbivorous dinosaur *Iguanodon.* By then the first fossil footprints from North America had been described and attributed to giant birds. Thus, American trackers also had trouble identifying trackmakers. It appears, however, that the identification of *Iguanodon* footprints was correct, and so it was the first dinosaur correctly correlated with its spoor. Early in the twentieth century Sir Arthur Conan Doyle incorporated information on *Iguanodon* and its tracks into his famous story "The Lost World."

Despite ongoing discoveries and a steady stream of detailed reports, including some encyclopedic compilations, on dinosaur tracks and other fossil footprints from many parts of Europe during the twentieth century, the magnitude and scope of discovery and documentation that we have witnessed in the past few years are unprecedented.

The current dinosaur track renaissance in Europe is part of a worldwide dinosaur-tracking revolution that, during the past decade, has thrust the study of fossil footprints into a prominent position in the world of paleontology and sedimentary geology. In Europe, an area that is traditionally regarded as having been well studied, important discoveries have been made in many areas where few would have predicted that such a wealth of new material could remain unrecognized for so long. For those of us involved in the study of tracks, or "vertebrate ichnology," the discovery of so much new material comes as somewhat of a surprise, but, after some reflection, we realize that paleontologists and geologists mostly find what they are looking for. Thus, the rate of discovery has increased remarkably as more trackers have dedicated themselves to serious research in this field.

The rate of new discovery has been such that it is hard to pick only a few examples that are representative of the great progress made by European trackers in the past decade. By way of introduction, therefore, we have chosen just a few representative Triassic, Jurassic, and Cretaceous examples to whet the appetite (figure 0.1) and refer the reader to the main text for a fuller account of the ongoing dinosaur tracks renaissance in Europe.

The Triassic of Europe has long been known as a source of abundant archosaur footprints, and substantial documentation exists in the form of short articles and more extensive monographs, many in German and French. Many of these contributions, however, deal with individual specimens and specific small sites that were studied before vertebrate ichnology came of age. In the past few years, however, this puzzling collage of information has begun to be organized into a coherent picture. This modern synthesis provides us with a global view of the distribution of trackmakers in the Triassic world. For example, it is now abundantly clear that track assemblages in Europe, North America, and elsewhere are frequently very similar in all Triassic epochs.

Such observations are perhaps not surprising when we consider that Europe and North America were united with other continents at this time as part of the Pangaean supercontinent. Archosaurs and other vertebrates, therefore, were able to range widely across the Triassic world, establish themselves in their preferred habitats, and leave their distinctive tracks and traces alongside other distinctive plant and animal remains. Thus, we find that many of the distinctive "hand-shaped" *Chirotherium* tracks known from the Early Triassic of Germany and England are essentially the same as those found in North American rocks of the same age. The same correlations are evident in a comparison of Late Triassic dinosaur and "dinosauroid" footprints between Wales, Switzerland, North America, and elsewhere. Such observations point to the similarity of trackmakers, climates, and environments throughout much of the Triassic world. Moreover, these observations help to demonstrate that the composition and distribution of vertebrate

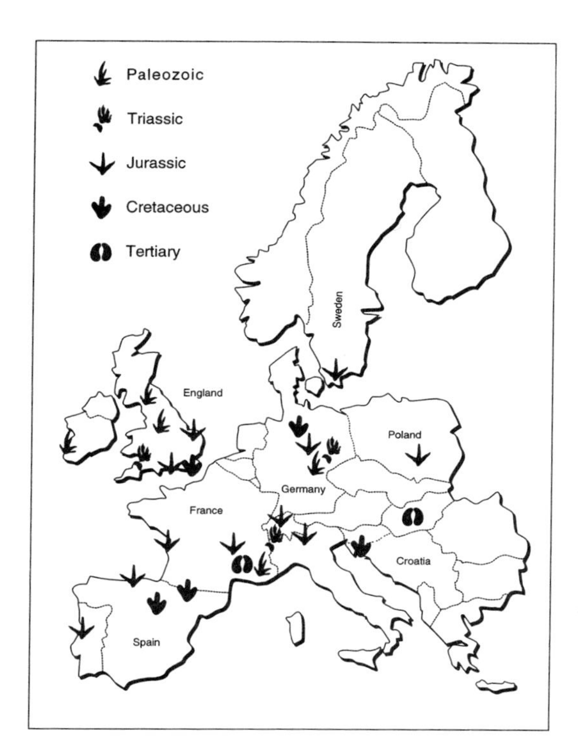

FIGURE 0.1. Simplified map of fossil footprint sites in Europe, showing regions covered in this book (Spitzbergen and Greenland not shown).

trackmakers are less complicated than previously supposed. We note that such obvious correlations are now being widely recognized, not just in the Triassic, but throughout the Mesozoic. Such similarities in track assemblages over wide areas led us to conclude that footprint correlations demonstrate a useful applied component in the field of vertebrate ichnology, leading to simple and often elegant "track" correlations. Such correlations, which can be based on comparisons of any fossil material, form the basis of the field of "biostratigraphy," a well-established branch of the earth sciences that allows us to use fossils to subdivide the geologic time scale into eras, periods, epochs, and ages. Discussion of such important correlations and comparisons of track assemblages in Europe and around the world are developed as one of the recurrent themes of this book and remind us that Europe is a modern geographic construct that had no significance for Mesozoic animals.

Moving on into the middle Mesozoic, perhaps the simplest way to view the track record of Europe is to consider the three Jurassic epochs (Early, Middle, and Late) in sequence. The Early Jurassic record, especially in Sweden, Poland, Italy, and France, is characterized mainly by tridactyl tracks, probably of carnivorous theropods but with a few tracks attributed to ornithopods, early ancestors of the duckbills. Prosauropod or early sauropod tracks are known from a large site in Italy, which was discovered only recently. This pattern is continued in the Middle Jurassic track record, which, although poorly known worldwide, is represented in Europe by an interesting track record in England and a remarkable site in Portugal that boasts the longest complete brontosaur trackways in the world.

Tracks from the Late Jurassic provide abundant information on the culmination of Jurassic sauropod evolution in the so-called age of brontosaurs by revealing fascinating and widespread evidence of sauropods, especially in Iberia, Germany, and Switzerland. In Portugal, for example, we have evidence of gregarious behavior among juvenile sauropods, and in Switzerland a recently discovered "Megatrack-site" or "dinosaur motorway" provides intriguing insights into the wandering of large sauropods over hundreds of kilometers along the northern shores of the ancient Tethys seaway. It is not just sauropods and other well-known dinosaurs that left a track record at this time. The footprint record for this epoch also provides tantalizing glimpses of other reptiles, including giant turtles and various pterosaurs.

The Cretaceous track record of Europe is characterized by a considerable abundance of tracks in Spain, Germany, Croatia, and England. Most tracks are attributable to duckbilled herbivores such as *Iguanodon* and its relatives, with footprints of theropods and a few sauropods still in evidence at various sites. Again, however, there are many intriguing traces of other dinosaurian activity, including herding or social activity among small herbivorous bipeds such as *Hypsilophodon.* Among the nondinosaurian groups we see further evidence of pterosaur and turtle activity along with traces of crocodilians.

The track record of the Late Cretaceous, the final Mesozoic epoch, provides us with some interesting scientific and political challenges. On the one hand, the track record is sparse and difficult to interpret; on the other hand, it includes an important site in the outskirts of Lisbon about which an entire book has been written. This now famous Carenque site boasts one of the longest continuous trackways in Europe and has been the subject of a national conservation campaign and high-level political decisions. Thus, our journey through the Mesozoic tracks of Europe reminds us that these ancient fossil footprints are still part of the modern scientific and cultural landscape of Europe.

This point is perhaps best illustrated by the recent discovery of tracks of Ice Age mammals, hunted, painted, and, in some cases, domesticated by our own ancestors. These animals are not merely remote curiosities from a former dinosaurian era; rather, they are creatures on which our own species depended for survival and whose domestication led to irreversible changes in the structure of human society. The challenge for all of us, but trackers especially, is to study, interpret, and conserve this fascinating ancient resource and make it accessible to the public. To this end, many local conservation and public education efforts have been launched to establish updated museum exhibits and outdoor interpretive trails based on tracks, and we list such sites in the appendix so that the reader may be encouraged to explore and appreciate them. As we enter the twenty-first century, this ancient and invaluable facet of Earth history remains an essential part of Europe's prehistoric heritage.

This book is the second in a series of books that address dinosaur tracks and other fossil footprints of particular geographic regions. The first, *Dinosaur Tracks and Other Fossil Footprints of the Western United States,*[1] attempted to synthesize the vast amount of new data discovered and described in that region in the past decade. Owing to the fact that much of these American data were new, they were easier to summarize, since they involved a much smaller legacy of historical literature than we find in Europe. For this reason, we were able to address the entire spectrum of fossil footprints from Paleozoic through Cenozoic. Even so, one reviewer noted that our publisher had placed too much emphasis on dinosaur tracks in our title. We defend this emphasis because dinosaur tracks do make up the majority of fossil footprints found to date, not just in the western United States, but in many regions including Europe. This is because dinosaurs made tracks for at least half of the entire history of tetrapod trackmaking on land.

In *Dinosaur Tracks and Other Fossil Footprints of Europe* we have been a little less ambitious in our coverage of footprints from the pre- and post-dinosaurian eras. This is partly because we have not been directly involved in new research in the field and so do not feel qualified to write at great length or with great authority on the subject. Even so, we have attempted to summarize the important sites and the insights they offer into vertebrate activity and evolution in Europe. Mesozoic tracks, however, are an entirely different story. Not only have we been involved in active research at many new European sites, some of which we have found ourselves, but we have also been witness to the emergence of the vast amount of new data that have been reported in recent years. The potential subject matter and scope of the book are growing as we write. We therefore offer no apologies for focusing mainly on the subject of dinosaur tracks, so keeping the length of the book manageable.

BIBLIOGRAPHIC NOTE

1. Lockley, M. G. and A. P. Hunt. 1995.

ACKNOWLEDGMENTS

We have had the opportunity to work on Mesozoic fossil footprints in Great Britain, France, Switzerland, Italy, Spain, and Portugal, and have many colleagues to thank for their kindness and generosity. In particular, Martin Lockley thanks the University of Colorado at Denver, Dinosaur Trackers Research Group, and the Earth Science Department, Cambridge University, England, for resources and support provided during spring 1998, when the writing of this manuscript was completed. Christian Meyer thanks the offices of the Naturmuseum Solothurn and the University of Basel. While this book was being written, the Tracking Dinosaurs exhibit toured in Europe (Switzerland and Great Britain), and we particularly thank Köbi Siber of the Aathal Museum, the Naturmuseum Solothurn, Olten, and the National Museum of Wales (particularly Stephen Howe and Tom Sharp) for their effort to display fossil footprint exhibits to the general public.

We shall not attempt to list everyone who has assisted us in our studies; many colleagues are mentioned by name in the text. We select, however, a few names of those with whom we have co-authored several manuscripts and who have made significant contributions to the vertebrate ichnology of Europe. First, from Portugal and Spain, Vanda Faria dos Santos and Joaquin Moratalla; from Italy, Giuseppe Leonardi and Fabio Dalla Vecchia; from Germany, Hartmut Haubold; from France, George Demathieu, Jean Le Loueff, and Paul Ellenberger, to whom the book is dedicated; from England and Canada, Bill Sarjeant, to whom the book is also dedicated;

and two persons who have recently completed Ph.D.s in vertebrate ichnology, Joanna Wright and Mike King.

Last but not least, we thank Mike Parrish of the University of Northern Illinois and Spencer Lucas of the New Mexico Museum of Natural History and Science for their reviews of the manuscript and many constructive suggestions. Also in this regard, we thank Ed Lugenbeel of Columbia University Press and our copyeditors, Ron Harris and John Curley, Jr. Thanks also to Chris Chippendale, Cambridge University, for reading an early draft of chapter 10. We also thank John Sibbick, Paul Koroshetz, Mauricio Anton, Kaspar Graf, and Annemarie Burzynski for providing artwork used herein.

CHAPTER I

INTRODUCTION TO TRACKING

THE IMPORTANCE OF FOSSIL FOOTPRINTS

Anyone who studies footprints can be referred to as a "tracker." In the specialized jargon of paleontology and geology, the study of footprints and other traces, such as worm trails, is known as "ichnology." The word derives from the Greek *ichnos* (meaning "trace"), and paleontologists always make a clear distinction between the study of "trace fossils" and the study of "body fossils," which are the fossilized remains of parts of actual organisms such as bones and other skeletal remains. Because trace fossils are shadowy evidence of creatures that passed by in the dim and distant past, body fossils are often considered more important. This is because they are the actual tangible remains of creatures. When complete skeletons are found, they give us the best possible picture of what the whole creature looked like. Although fossil footprints caused much excitement when first discovered in the early nineteenth century, many researchers since have considered fossil footprints to be of secondary importance. Consequently, the field was neglected for an entire century. But as we shall soon see, this view has changed considerably in recent years, and the scientific study of tracks is enjoying a healthy revival.

Fossil footprints were made by living creatures. Unlike bones, which are the product of death and decay, tracks represent the dynamic behavior of animals walking, running, and going about their day-to-day activities. They are the nearest things we have to movies of extinct animals.[1] Without tracks, we have to reconstruct extinct animals from their lifeless bones, and as experience has taught us, such reconstructions are not always correct. In addition, tracks are always made exactly where animals passed by—what geologists call *in situ.* By contrast, bones may be washed long distances before they finally come to rest in the sedimentary layers in which we find them. Dinosaurs may have floated far out to sea. Thus, tracks are an integral part of the ancient ecosystems in which they are found. As we shall see, this special relationship between tracks and ancient landscapes is one of the central themes of ichnology.

Footprints can be made by anything with feet. This includes many invertebrates, notably the arthropods, ranging from insects to crabs and spiders, but excludes fish and other vertebrates such as snakes. Because this book is not about the traces made by arthropods or fish, we are essentially concerned with vertebrates with feet, which are know as "tetrapods." We can talk therefore about vertebrate ichnology, by which we generally mean tetrapod ichnology.

Owing to the lack of widespread interest in vertebrate footprints, prior to the 1980s many ichnologists focused their attention on the traces left by worms, molluscs, and other lowly invertebrates. Although not considered a newsworthy branch of paleontology by members of the general public, these invertebrate ichnologists were nevertheless always regarded as making important contributions to paleontology and to the interpretation of ancient environments. Indeed they were considered a part of the paleontologic mainstream, whereas vertebrate ichnologists were regarded as the esoteric lunatic fringe. Following the dinosaur renaissance of the 1970s, however, interest in dinosaur tracks and other fossil footprints was dramatically revived. Suddenly dinosaur biology and dinosaur metabolism became a hot topic and everyone wanted to know how dinosaurs lived, how fast they ran, who attacked who, and whether they traveled in herds. Suddenly the obscure subject of vertebrate ichnology moved rapidly into the paleontologic mainstream, a position it had not enjoyed since the early days of the nineteenth century.

Much of what has been called the dinosaur footprint renaissance[2] can be attributed to the fact that ichnologists have demonstrated, in the past decade or so, that footprints are far more common than had been realized. The more data one has, the harder they are to ignore. As we shall see, one theme of this book is the investigation of the relative importance of tracks in comparison with bones. In contrast to previous misconceptions which held that tracks were rare, we find that they are incredibly abundant. In many areas, and in many sequences of strata, tracks are turning out to be many orders of magnitude more abundant than bones. Traditionally paleontologists spoke of the "fossil record" in reference to the sum total of the evidence of past life. It is now time to expand our vocabulary and also speak of the "track record"—a record that has assumed far greater importance than paleontologists ever suspected.

HOW TO STUDY ANCIENT TRACKS

In recent years there have been a growing number of new books published on the subject of fossil footprints. These are cited below, so it is not necessary to go into great detail on the subject of how to study or interpret fossil footprints in our in-

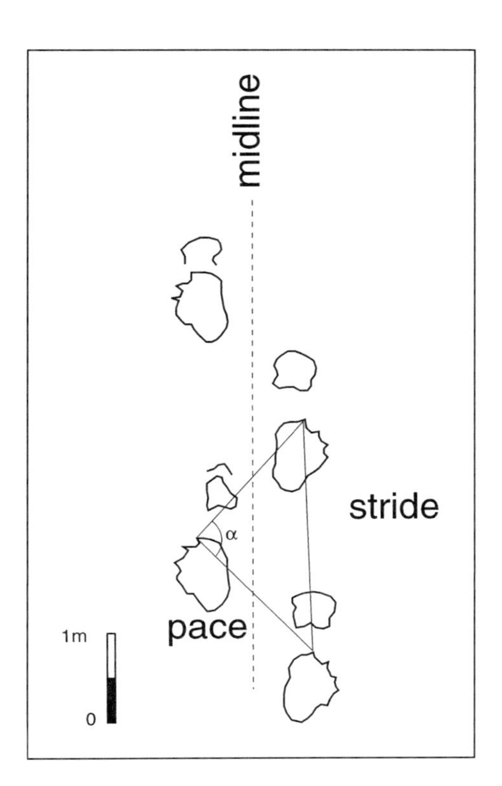

FIGURE 1.1. The anatomy of a trackway, showing essential features, including step (pace), stride and pace angulation (α).

troduction. If we were to do this, we would duplicate much of what is written in the introductory chapters of such books.[3] We also hope our descriptions and interpretations of the many tracksites discussed in this book will be self-explanatory, so that the reader should be able to understand any section without first acquiring a specialized knowledge and vocabulary.

It is nevertheless necessary to present some basic terms. Most of these are also self-explanatory and easy to understand because they use familiar vocabulary. For example, animals are either bipeds or quadrupeds and—except in the case of snakes and other very short-legged, long-bodied animals such as newts or lizards—rarely leave tail drag marks. Thus, trackers deal primarily with footprints. As we walk along we take steps (also known as "paces"). Our steps alternate from left to right, right to left, and so on. Two steps are referred to as a stride (figure 1.1). Consecutive footprints form a trackway that can be wide or narrow, depending on the type of trackmaker and the speed at which it is traveling. The trackway width (or straddle) is therefore, in part, a measure of the extent to which an animal sprawls or walks erect. This can also be deduced by the angle between steps, known as "pace angulation." Thus, it is a rule of thumb among trackers that three consecutive footprints, comprising a full stride, give us the basic information we need to analyze a trackway.

Having established the basic trackway configuration, we can look closely at the anatomy (or morphology) of the front and hind feet as expressed in the footprints. The most obvious characteristics of footprints are the number of toes, ranging from one to five (figure 1.2). Single-toed feet, typical of horses, are referred to as

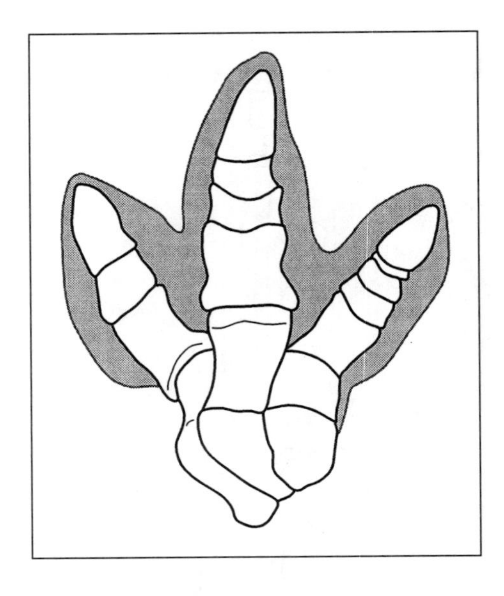

FIGURE 1.2. The number of toe impressions observed in footprints provides one of our most basic clues to the anatomy of the feet of the trackmaker. Example based on *Iguanodon* (see chapter 8).

"monodactytl" (literally "singled-toed"). Cloven-hoofed animals, such as deer, sheep, bison, pigs, and so forth, leave "didactyl" footprints. The ostrich also has a didactyl footprint. Three-toed or "tridactyl" footprints are typical of birds and many bipedal dinosaurs. Four-toed or "tetradactyl" footprints are typical of cats and dogs and are also characteristic of such diverse animals as the hippopotamus and horned dinosaurs. Five-toed or "pentadactyl" footprints are characteristic of creatures ranging from human beings to lizards and brontosaurs. Some creatures, including the occasional human, are "polydactyl," meaning that they have more than five digits on their feet or hands. Currently, however, there are no convincing examples of fossil footprints that reveal this condition of polydactyly.

The number of toes or digits expressed in a footprint is only one aspect of anatomy, for, as examples as distinct as humans lizards and brontosaurs indicate, the footprint shape is fundamentally different, despite a numerical similarity. Thus, the tracker has to become familiar with other aspects of morphology, which boil down to simple observation and knowledge of foot anatomy. For example, it takes no special skill or extraordinary terminology to describe footprint features that reflect the length and width of digits, the shape of individual pad impressions, the marks made by claws, or the impressions left by the skin.

INDIVIDUAL BEHAVIOR

The standard methods of track analysis just outlined deal with examination of the footprints of individuals in motion, which we can refer to as "individual activity" or "behavior." Such behavior can be very simple, such as walking, running, speeding up, or slowing down. Using various mathematical formulas, we can easily estimate the speed of animals from their fossil trackways. The most widely used formula, proposed by the British zoologist R. McNeil Alexander,[4] is as follows:

$$\text{Velocity } (V) \text{ in meters per second} = 0.25\ g^{0.5} \cdot SL^{1.67} \cdot h^{-1.17}$$

This formula only requires that we have a trackway in which we can measure the length of the stride (*SL*) and the length of the footprint (*FL*), which is considered to be one fourth of the hip height (*h*) of the trackmaker (so $FL \times 4 = h$).*

It is important to stress that these are only estimates of speed. The determination of exact or "absolute" speed is impossible from trackways because we do not know how much time elapsed between footfalls. One way to visualize this problem is to imagine two world-class sprinters both running 100 m in 10.0 seconds. We will assume that both have feet (footprints) that are identical in size, but one is taller than the other and takes slightly longer strides. The result of an analysis of their trackways, using this formula, is that the longer-striding individual would appear faster. It is also possible that much slower runners could run the same 100-m course, placing their feet exactly in the places where the olympian feet had fallen, yet they would take much longer to cover the course. The rate or timing of footfalls, known as "cadence," would be slower, yet the formula would estimate the same speed as the olympians. This is the only way that most of us could match an olympic champion stride for stride. We would be cheating, of course, because we would be ignoring the important time factor.

Having said this, however, if we consistently use the same formula on fossil footprints, we are going to get a fair indication of the relative speed of various animals. Work on this subject has revealed that some carnivorous dinosaurs were moving at speeds estimated to be in the range of 40 km per hour, or almost exactly 100 m in 10.0 seconds. The trackways of most other dinosaurs, and other extinct vertebrates, indicate slower, walking speeds. This does not mean that these animals could not run or attain faster maximum speeds. It simply means that, because most animals spend most of the time walking, not running, trackways seldom capture high-speed activity.

Other aspects of individual behavior that are easily determined from trackways include turning, or sudden changes in direction, and various abnormal behaviors such as limping (figure 1.3).[3,5] It is worth pointing out that we cannot usually determine why an animal changed direction or demonstrated abnormal locomotion. It is tempting to attribute a motive to such behaviors, but such interpretation is usually completely speculative and unscientific. The fossil footprint literature contains many examples of unwarranted speculation on this subject. It is important to maintain a clear distinction between behavior and motive. For example, running is a behavior, but if we say an animal was doing it because it was chasing another animal or was itself being chased, we are reading in motives. Such interpretations are only valid if there is some compelling evidence to support such a scenario.

*All measurements should be in meters, giving a result in meters per second.

FIGURE 1.3. Example of a trackway attributed to a limping dinosaur from the Cretaceous of Spain (see chapter 8).

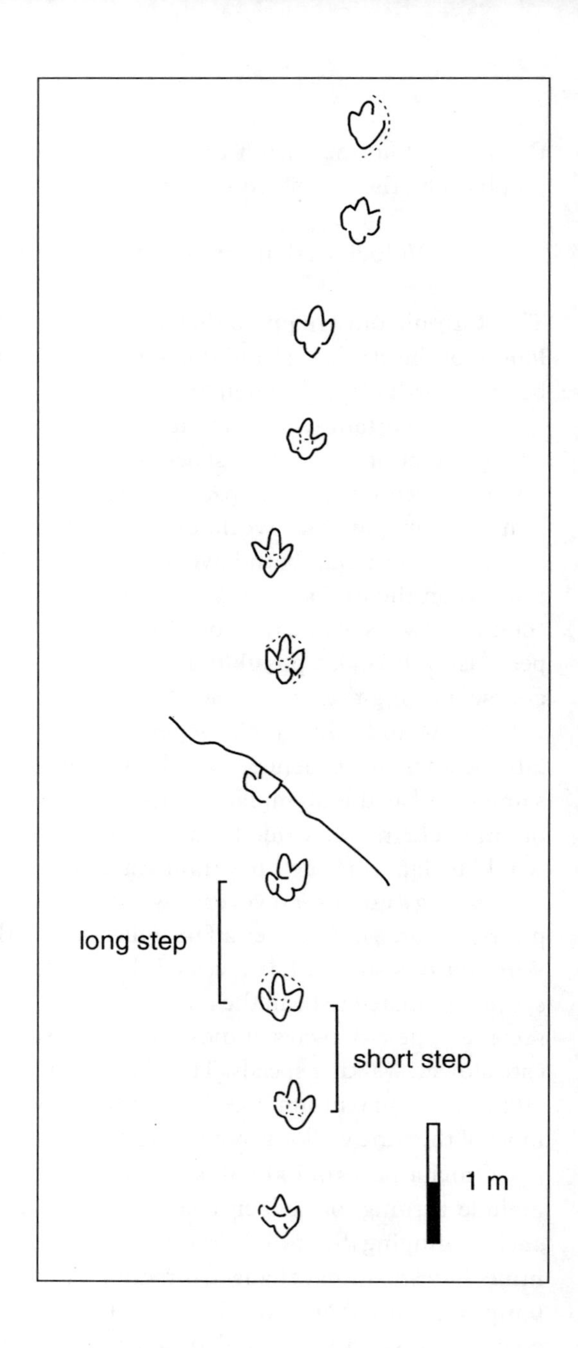

WHERE TO GO TRACKING

Now that we have spent a few minutes focusing on how we describe fossil footprints, it is only natural to ask where we might find interesting examples. From the outset it is important to realize that fossil footprints are not just lying around for anyone to collect. This is especially true in Europe, where almost all known sites are on private land or land owned or administered by various municipalities, counties, or government agencies. We therefore divide our advice on where to go tracking into two categories: the first is aimed at professional paleontologists, and those with an understanding of field geology, who could potentially find new sites; the second is aimed at those who are happy to visit existing sites, exhibits, or museums that are open to the public.

The professional or serious tracker, with a good grasp of paleontology, will already know that fossil footprints can only be found in strata of the right age and right type. Tetrapods first began to walk on land in the Devonian, becoming relatively abundant by Carboniferous times and continuing to be represented in the rock record until the present time. Most footprints were made in deposits that represent ancient shorelines of lakes, lagoons, rivers, and other bodies of water, including the sea itself. Thus, a knowledge of the local geology is essential if one is to improve one's chances of finding tracks. Simply put, one must look in strata of the right age that represent the likely habitat of potential trackmakers. The reader who has not had experience interpreting the ancient environments represented by various sedimentary rock layers will soon learn something of how paleontologists interpret the strata in which footprints typically occur, for an important theme of this book is explaining the relationship between fossil footprints and the ancient habitats in which they were made.

It is always possible that those without a working knowledge of geology might find tracks, perhaps more by luck than by design. Important tracksites have been found by keen-eyed observers who were out hiking or mountaineering in suitable terrain. Tracks occur on surfaces that represent ancient beaches, mud flats, lake shorelines, and other similar settings. Such surfaces may reveal ripple marks, raindrop impressions, or mud cracks of the type we observe in similar modern environments. In order to get preserved in the rock record, such surfaces must first be buried, but in order to be exposed again so that we can see them, they must be later exhumed by erosion. Moreover, it is necessary for the process of erosion to actually expose the surfaces of sedimentary layers. The best analogy is that we cannot read a book by looking at the edges of its pages when it is closed. We must open the book or, more accurately, find where mother nature has opened the book for us, and look at the surface of its pages. This is where the footprint message is inscribed. So, unlike many paleontologists, who actually break open layers of strata and look "inside" for buried fossils, when trackers go in the field, they look "on" these sedimentary surfaces. Such surfaces have been called "paleosurfaces" (literally "ancient surfaces") and represent where the ancient landscape has once more been exhumed and coincides with Earth's present topographic surface. It is always a source of amazement to the imaginative tracker that a surface that is 100, 200, or even 300 million years old can be exhumed and become part of the modern landscape.

For those who are unlikely to go in search of new fossil footprint sites that still await discovery, there are many known sites that are accessible to the public. From the Iberian Peninsula, to the alpine regions of Italy and Switzerland, to the lowlands of France and Germany, and to Britain's coastal exposures, there are sites where one may view dinosaur tracks and other fossil footprints in situ. Many of these sites have been developed into outdoor interpretive centers, and many are adjacent to natural history museums, where important specimens are also on display. We have listed these sites and museums in the appendix.

NAMING FOOTPRINTS

The cynic might say that scientists cannot resist inventing lengthy and hard-to-pronounce names for the objects of their study. Paleontologists are sometimes regarded as particularly notorious in this regard. But there are reasons for what appear to be paleontologic long-windedness. In the modern world, we can speak of dogs, cats, blackbirds, and earthworms, and everyone knows what we mean without us having to resort to the Latin names. However, all these creatures do have Latin names: *Canis familiaris, Felix domesticus, Turdus merula,* and *Lumbricus terrestris,* respectively, are the scientific names of the four examples just cited. Most fossil species do not have common names. We do not speak of black pterosaurs, earth dinosaurs, or even old men to describe particular species. Instead we speak of *Tyrannosaurus rex* or *Homo erectus.* This system, which applies to both living and extinct plants and animals, is known as the "binomial" (two names) system of classification. The whole field of naming and classification of organisms is known as "systematics" or "taxonomy," or in normal language simply as classification or nomenclature. Any formally named individual species or label applied to a family, class, or group of species can therefore be called a "taxon" (plural "taxa").

When dealing with fossil footprints, we may be able to match tracks with a particular species known from bones or body fossils. But we could equally well be looking at the tracks of extinct animals that are not even known from corresponding skeletal remains. This latter case makes it quite impossible to match tracks with bones. So how do we describe or label the footprints? The simplest solution is to have a different but parallel system of nomenclature, formally called a "parataxonomy." This parallel system of taxonomy for ichnologists is called "ichnotaxonomy," with corresponding "ichnospecies," "ichnogenera," and "ichnotaxa." Some people object to the naming of tracks, but these dissenters are usually not ichnologists. The problem is that if we do not have labels for tracks, we have no formal or

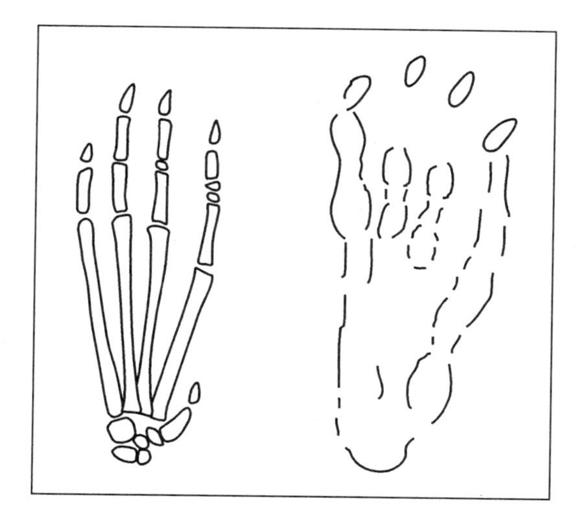

FIGURE 1.4. Matching a pterosaur track with a foot skeleton resulted in the naming of the footprint type *Pteraichnus.*

systematic way of describing them other than referring to them as "big, three-toed Jurassic footprint" or "small Cretaceous footprint from Spain." Such an unsystematic alternative is clearly of little or no utility. So ichnotaxonomy, which was first introduced in the 1830s, is here to stay.

Many footprint names have an ending such as *-pus, -podus,* or *-ichnus,* which immediately identifies them as track names rather than the names of body fossils. For example, the ichnogenus *Pteraichnus* refers to footprints attributed to pterosaurs and quite literally means "pterosaur trace" (figure 1.4). Throughout this book, when we first refer to a particular track type we may use the double-barreled name. Subsequently, for convenience and brevity, we may use only the ichnogenus name. So, if we wished to refer to a specific track that has been attributed to *Tyrannosaurus,* we would use the formal name *Tyrannosauripus.*

WHO DUN IT? IDENTIFYING ANCIENT TRACKMAKERS

If we label a footprint as *Tyrannosauripus,* the obvious question arises as to how we know it was made by a *Tyrannosaurus,* as the name implies. Obviously determining which animals were responsible for making particular footprints is a prime objective of tracking. In the modern world we can correlate all footprints with trackmakers simply by observation and by being patient enough to track down the animals responsible. As explained in the previous section, we cannot always do this with extinct animals, but in many cases we can. The task of tracking down "who dun it" can be a process of elimination. By going through all the possible candidates and examining their feet we may be lucky enough to arrive at the foot that is the perfect fit for the footprint we are trying to match. This reminds us of the closing scene of an Agatha Christie drama, when the detective eliminates all but one suspect and finally reveals the culprit "who dun it." In many cases, however, we do not have the luxury of a complete menu of choices. The classic case of *Chirotherium,* the footprint that was not matched with a trackmaker for more than a century, would be better characterized as a process of "accumulation" because at first no potential trackmaking candidates were known from the fossil record, but little by little the skeletal record improved until finally a suitable candidate was discovered (see chapter 2).

Therefore, we cannot determine what animals made particular fossil footprints without some knowledge of the skeletal or body fossil record. We know that large, elongate footprints with three-digit impressions are probably those of carnivorous dinosaurs because foot skeletons of these creatures fit the footprints very well. Similarly, the tracks of pterosaurs are an indisputably compelling, one might say "perfect" match for pterosaur feet. In short we are guided by the anatomy of feet when trying to determine what animal made the tracks.

TIME AND TIME AGAIN

It is also important to consider the age and geographic location of footprints when trying to determine probable trackmakers. If we find the track of a very large carnivorous dinosaur in 150-million-year-old Upper Jurassic strata in Europe (as reported in chapter 6), it does not make much sense to attribute the track to *Tyrannosaurus rex,* which lived 67 million years ago in North America.

Time is an important component of geologic and paleontologic research and so is equally important in the study of fossil footprints. The geologic time scale has been constructed largely through the study of the evolutionary history of fossils. As each new species appears and disappears, it leaves its remains in particular units of strata, thus allowing us to identify a stratigraphic zone corresponding to the creature's history on earth (figure 1.5). Time intervals defined by fossils, known to geologists as "biostratigraphic zones," are the geologic equivalent of the reigns and dynasties of kings, queens, and emperors on the historical time scale. In fact the science of biostratigraphy is the geologic equivalent of history, and is literally how we have reconstructed Earth history.

By looking at the appearance and disappearance of dinosaurs, for example, we have been able to estimate the average longevity of various dinosaur species as about 7 or 8 million years.[6] (This is not to be confused with the life span of individual animals, which we presume to be, like comparable modern animals, on the order of a few decades.) If new dinosaur species replaced old ones every few million years, as the fossil record shows, then we would expect there to be changes in the footprint record every 7 or 8 million years. This is essentially what we find. There are various footprint zones or ichnologic zones characterized by distinctive tracks. These were first identified by European trackers who coined the term "palichno-

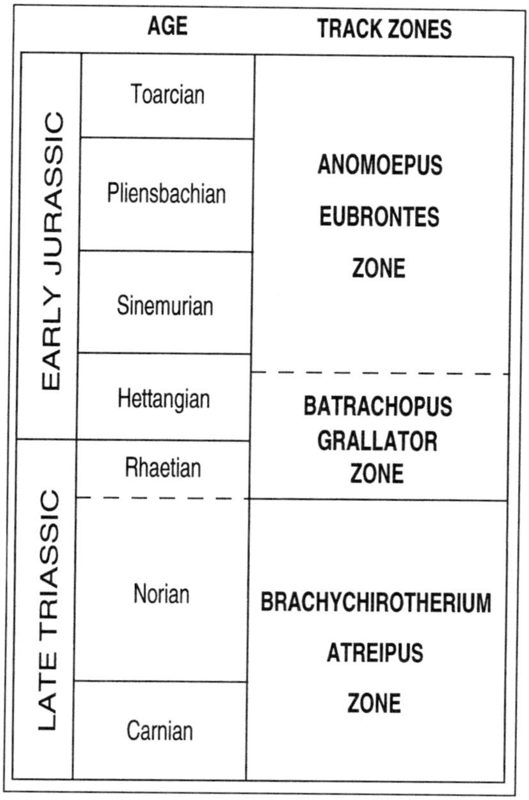

	AGE	TRACK ZONES
EARLY JURASSIC	Toarcian	ANOMOEPUS EUBRONTES ZONE
	Pliensbachian	
	Sinemurian	
	Hettangian	BATRACHOPUS GRALLATOR ZONE
LATE TRIASSIC	Rhaetian	
	Norian	BRACHYCHIROTHERIUM ATREIPUS ZONE
	Carnian	

FIGURE 1.5. Track zones reflect the turnover of vertebrate species and communities through time. Modified after Haubold.[14]

stratigraphy,"[7] literally meaning "ancient track stratigraphy." In fact, the whole concept of a sequence or chronology of events through time is so fundamental to paleontology and evolution that almost every paleontologic book adheres to a chronological or historical structure. This book is no exception. Although we prefer to view it as a "trail through time" it is, technically speaking, also a "palichnostratigraphic journey."

COLLECTING FOOTPRINT DATA

In the section on "How to Study Ancient Tracks" we outlined details of how to measure step and stride and count the number of digit impressions found in footprints. But there is more to the study of a site than just obtaining measurements. Large sites may be the size of several football fields and can contain hundreds of different trackways. In such cases, a complete map of the site is useful and forms a basis for a census of all the animals that crossed the ancient surface or landscape

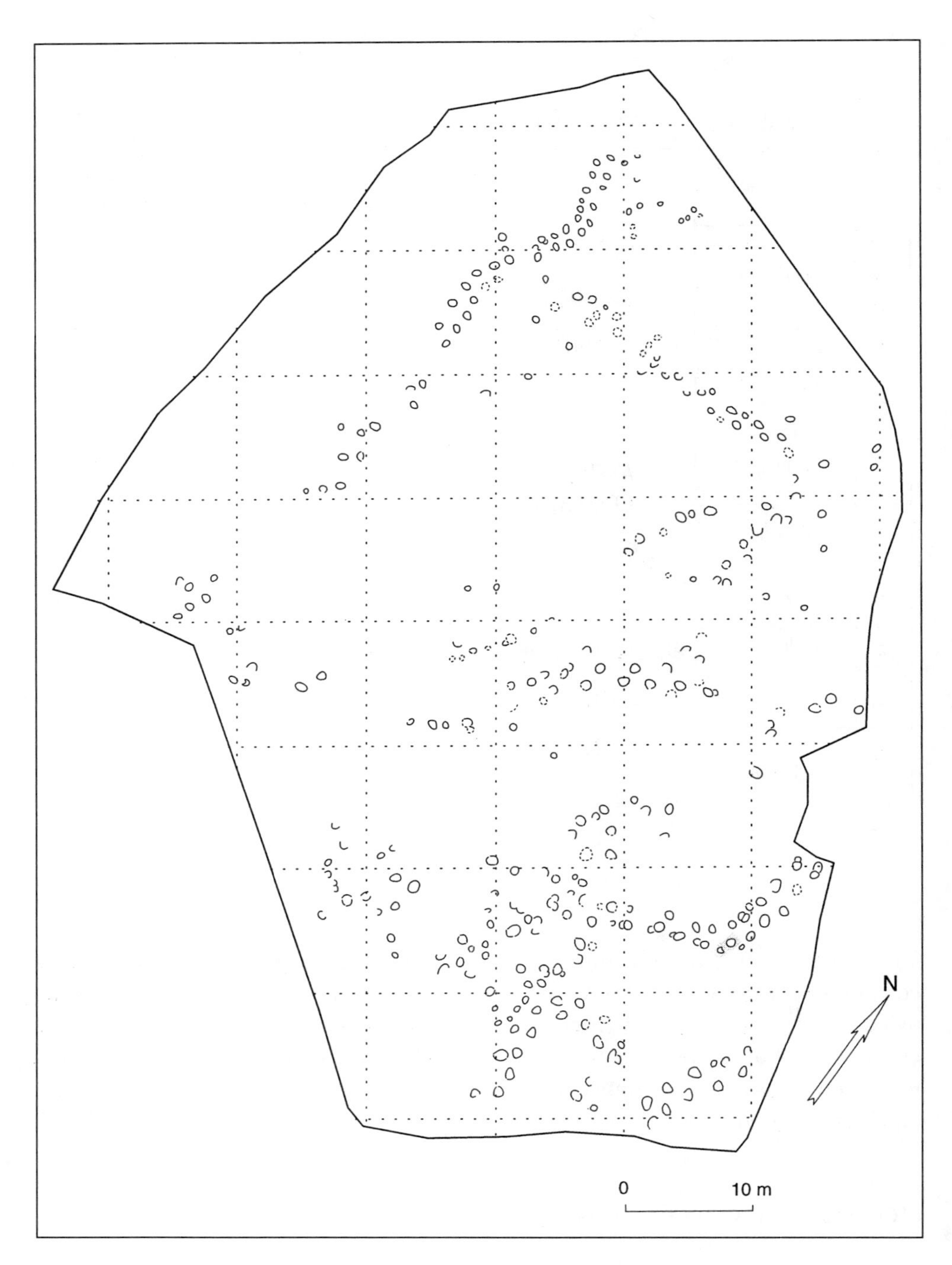

FIGURE 1.6. A simple tracksite map forms the basis for any further analysis of footprints and trackways, based on the Lommiswil tracksite, Switzerland (see chapter 7).

represented by a particular site (figure 1.6). In the past, trackers would use a tape measure and compass to lay out a grid and map all the tracks onto graph paper. Modern surveying technology now allows more sophisticated techniques to be used. These techniques, however, can be more expensive. In recent years at least one site in Europe has been mapped using state-of-the-art photogrammetry methods. Such techniques require investigators to photograph the site from a helicopter, or at least from some high vantage point, in order to produce a three-dimensional image of the surface, which can then be turned into a map through computer plotting.

Once a map is made, trackers can determine the number of trackways that were registered on the surface, as well as the direction in which the trackmakers were going. These trackway orientation data are useful for determining preferred trends or directions of movement and may indicate whether different types of animals moved in different directions, whether they showed gregarious tendencies or not, or whether they moved in particular directions dictated by the shoreline or other features of the ancient topography.

SOCIAL BEHAVIOR

The subject of dinosaur social behavior has become popular in recent decades during the so-called dinosaur renaissance. Ideas about social behavior are linked in various ways with debates about dinosaur metabolism and warm-bloodedness. The old, pre-renaissance view of dinosaurs is often presented as a scenario in which dinosaurs were cold-blooded, solitary creatures. This view is in line with our characterization of many modern reptiles, such as lizards, snakes, and crocodiles, as antisocial loners. By contrast, many mammals are viewed as gregarious creatures that often travel in herds and engage in much more sophisticated social behavior than reptiles. These views can be borne out by observation in some cases, but they are not necessarily correct, and correlations between social behavior and metabolism are rather tenuous.

The study of tracks and nesting sites has provided considerable persuasive evidence in support of the idea that some dinosaurs were gregarious. The work of Jack Horner and his colleagues,[8] for example, indicates that dinosaurs nested in colonies to which they returned year after year. In addition, Horner and colleagues presented evidence to suggest that dinosaurs were good parents that attended their young, in the nest, long after hatching. Such behavior is reminiscent of that of birds—dinosaur descendants that happen to be warm-blooded. But such behavior is also observed in crocodiles, which protect their nests fiercely and also attend to their young after hatching.

Trackway evidence also provides compelling evidence that some dinosaurs traveled in herds or large groups. Based on various tracksites where we find dozens of parallel trackways, it is now quite well established that Jurassic and Cretaceous

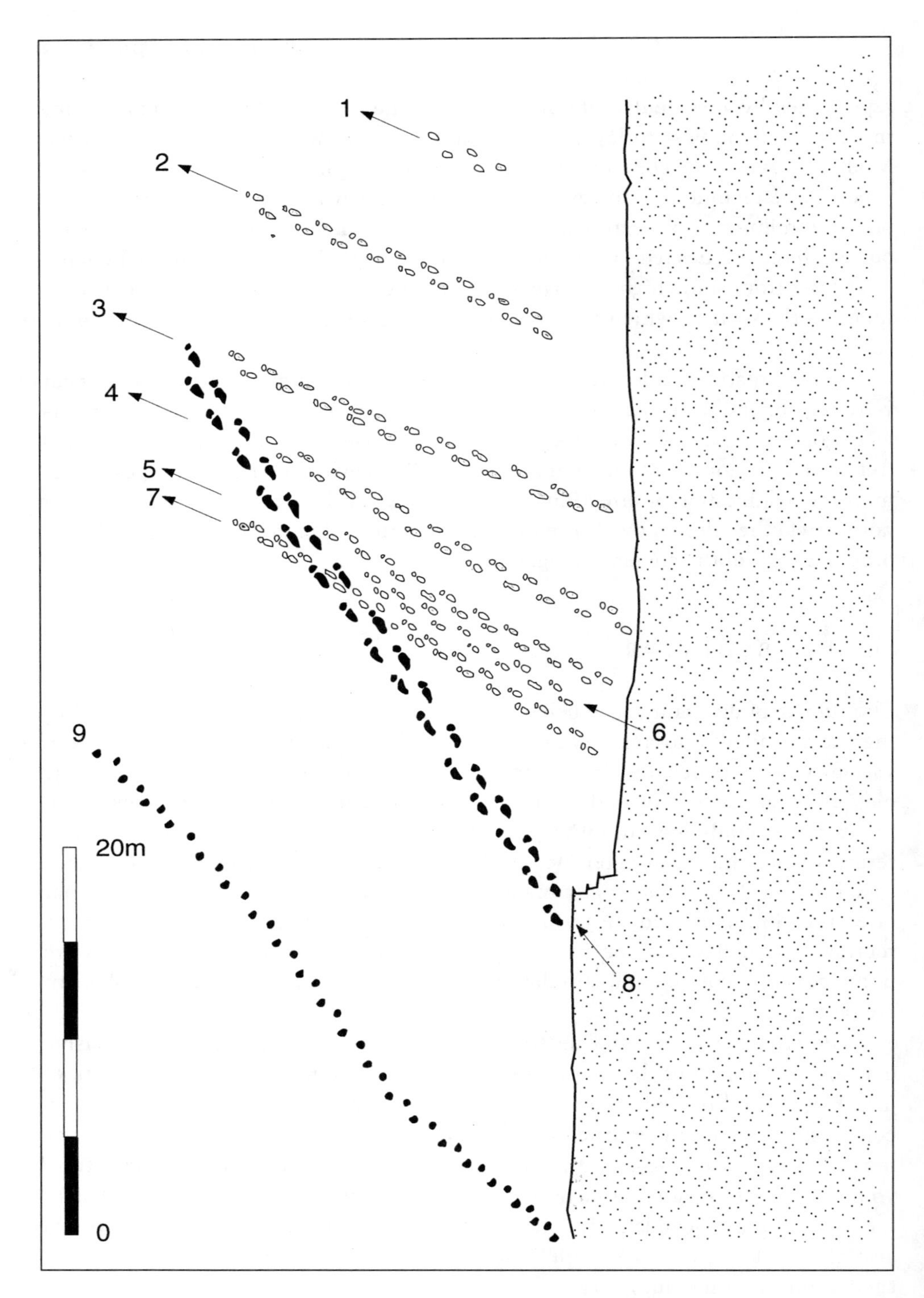

FIGURE 1.7. Parallel trackways from the Late Jurassic Cabo Espichel tracksite, Portugal (see chapter 7).

brontosaurs and ornithopods often traveled in herds (figure 1.7). The first discoveries of such sites were made in North America, but in recent years, several such sites have come to light in Europe. Present indications suggest that the habit of gregarious behavior was well established early in the Mesozoic. Such evidence suggests that dinosaurs may have evolved from gregarious ancestors; indeed, it is possible to find trackway evidence from before the age of dinosaurs that suggests social behavior among certain nondinosaurian reptiles.[9]

Because modern mammals are often gregarious, it is tempting to attribute the same degree of social sophistication to dinosaurs that we do to a herd of buffalo, and then extend the comparison to note that such mammals are warm-blooded. Thus, by implication dinosaurs also might be warm-blooded and socially sophisticated. But such assumptions, although intriguing, are tenuous. Fish, for example, are gregarious and swim in shoals. Amphibians and reptiles are also gregarious in some instances. So social or herding tendencies do not provide proof of a particular metabolic condition or social sophistication. However, it is interesting to be able to use tracks to demonstrate gregarious or herding behavior among various groups of dinosaurs, and other tetrapods, and report when and where such activity occurred.

USING TRACKWAYS TO READ ANCIENT ECOLOGY: THE ICHNOFACIES CONCEPT

We have now explained why it is important to make maps of tracksites and build up a complete picture of how many animals of different types were present in a particular area and how the orientation of their trackways might reveal interesting behavioral patterns. We can treat each site as a sample or census of animals in a given region, just the way we would a collection of bones from a particular paleontologic excavation. The complete sample of tracks from a site can be called the "footprint assemblage," just as a sample of body fossils might be called a "bone" or "skeletal assemblage." Ichnologists who collect information on different invertebrate traces from various sites would refer to this assemblage of trace fossils as an "ichnocoenosis" (literally "the trace of a living community").

In the case of bones, the assemblage of remains may represent indigenous animals, or they could represent a mixture of exotic, nonindigenous animals that were washed into the area. In the case of tracks, the assemblage is always indigenous. This means that if we can interpret the ancient sedimentary environment in which the footprints are found, we really are looking at evidence of animals in some part of their original habitat. Through years of observation, ichnologists have noted that a particular assemblage of trace fossils (say, types A, B, and C) often show up repeatedly in the same type of sedimentary rocks. This means that we have a recurrence of a particular assemblage in a particular environment (figure 1.8). This result is scientifically significant, for it demonstrates that by sampling the same ancient environment repeatedly we can obtain the same reproducible results.

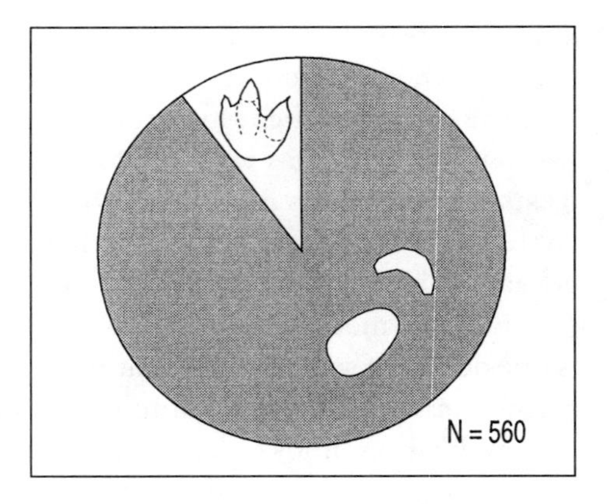

FIGURE 1.8. Pie diagram representing a census of trackmakers in a given area.

Ichnologists recognize the recurrence of trace fossil assemblages (or ichnocoenoses) in the same type of deposit as "ichnofacies." The simplest translation or definition of this term is a distinctive sedimentary deposit representing a particular ancient habitat in which the same tracks and traces recur repeatedly. We cannot go so far as to assume that the trace fossils that we find represent the whole community, but in a more general sense they represent everything that is preserved at a given site and so are the best census of tracks available.

Because worms and various other invertebrates are small and live in the sediment, it is obvious that they are unlikely to travel very far from the environment in which we find them making traces. Thus, it is not surprising that they repeatedly make the same traces in the same area. Vertebrates, however, have the potential to wander far and wide, and one might expect that they would leave their tracks in many different environments. If they did this, however, we would not find the repeated relationship between tracks and environment that we call vertebrate ichnofacies. So geology and nature are our guides, showing us that tetrapod ichnofacies really do exist, despite theoretical arguments to the contrary.[10] This encouraging result allows us to use ichnofacies to reconstruct ancient animal communities and so gain insight into the ancient ecosystems in which the trackmaking species were active.

TRACKS AND BONES: TWO PIECES OF THE INCOMPLETE PUZZLE

As the study of ichnology has matured in recent years, trackers have begun to show the extent to which tracks add to the paleontologic picture provided by bones. Traditionally we relied almost exclusively on bones for reconstructing the history of vertebrates. It was assumed that tracks were relatively rare, and that when found they usually added nothing new to the picture. Thanks to the landslide of new footprint information that has emerged in recent years, this view has proved incorrect. The most obvious examples of footprints adding substantive information to an otherwise incomplete picture is in the case of rock formations that contain no bones, but abundant footprints. Again, the traditional view was that this was a rare case in which footprints proved useful.

More recent work has begun to demonstrate that even in deposits where bones of particular animals are known, we often find tracks of entirely different creatures. There are two significant conclusions to be drawn from such observations. First, we must recognize that we cannot list, describe, or analyze the fauna from such deposits without studying both the footprints and the bones and, second, that the track and bone records are "selective" in sampling different components of the ancient fauna. In relation to the second point, we might find that in a particular deposit the bones represent mainly the aquatic fauna that lived in ponds, rivers, and lakes (for example, turtle and crocodile remains are common in many such deposits). By contrast, the footprint record may represent various terrestrial species that typically do not get preserved as skeletal remains.

THE MYSTERIES OF TRACK PRESERVATION

It is not too difficult to picture a large, solid dinosaur bone being buried and preserved for posterity. Tracks, however, are often regarded as ephemeral, and easily washed away by the ebb and flood of tides or the action of rains. One of the questions most often asked of trackers is "How can tracks possibly be preserved for so long and still, in some cases, look as fresh as if they were made yesterday?" The answer is not always obvious or easily explained. Traditionally geologists have resorted to what we call the "cover-up" explanation.[3,11] The scenario goes something like this. An animal walks on a wet beach or mud flat, leaving well-preserved footprints. Then the sun bakes the sediment as hard as cement. This way the tracks are not washed away by the next tide or flood that washes into the area. Certainly such a scenario is possible, but how probable? Does every preserved footprint call for a hot, sunny spell to bake the sediment just after the track was made? Many of the best tracks we find were made on layers of sediment freshly deposited after a flood, and it is often cool and humid after the wet weather that produces such floods.

An alternative cover-up explanation calls for very gentle water action that produces little or no erosion. For example, a dry lake or empty lagoon might fill up gradually and gently without scouring away the tracks. Once submerged, the tracks would then be gradually covered by sediment settling out of the water. Ultimately such scenarios can be tested by looking at the rocks in which we find the tracks. Signs of sun baking or the gradual and gentle accumulation of sediment should be relatively easy for a good geologist to identify.

In trying to address these questions, it pays to think of tracks in three dimensions. Rather than thinking of them as features associated with flat, two-

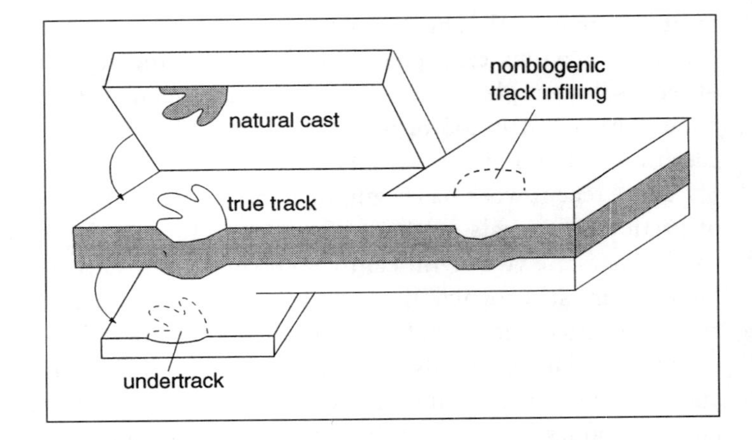

FIGURE 1.9. Undertracks, true tracks, and fillings. *Modified after Lockley.*[2]

dimensional surfaces, we should regard them as features that penetrate through these surfaces and leave their mark below the surface that the animal actually passed over. By definition, in order to make an indentation, a footprint must be below the surface. Anything below the surface is already buried or on the way to being buried. Given this perspective, animals that make footprints are making direct deposits into subsurface layers. They are leaving their tracks in the perfect place for them to become fossilized. Imagine a dinosaur walking over a mud flat and sinking in up to its ankles so that its feet come to rest 20 cm below the surface. In some cases, if the mud is soft enough, it will slop back into the track, instantly burying and preserving it. Even if the mud is not sloppy, the track becomes a trap for the gradual accumulation of sediment. The footprint cannot be scoured away because the footprint indentation is a protected pocket. Such a track preservation scenario stresses the importance of underlayers below the surface.

When animals make footprints, especially in sediments that consist of thin layers, their weight transmits the footprint into underlayers (figure 1.9). Such footprints are referred to as "underprints," "undertracks," or "ghost prints." Strictly speaking, they represent the deformation of layers at some point below the actual level where the foot finally came to rest (i.e., below the actual footprint). In some thinly layered sediment, there may be several underprints stacked on top of each other. We can also say that if the footprint is deeper than the thickness of layers in the sediment, the foot will, by definition, disturb underlayers and so in most cases create underprints. A moment's thought will tell us that underprints are a peculiar phenomenon for at least two reasons. First, there is really no other body fossil or paleontologic equivalent such as an underbone or undercoprolite that can be dynamically transmitted into another layer of strata. Second, underprints create traces in layers that are older than the surfaces over which the animals walked. In this sense, animals are capable of leaving their mark during the time of their ancestors. Usually it is the other way around—creatures may make a mark that carries over into their children's or grandchildren's generation, rather than traveling back in time to the days of their parents or grandparents.

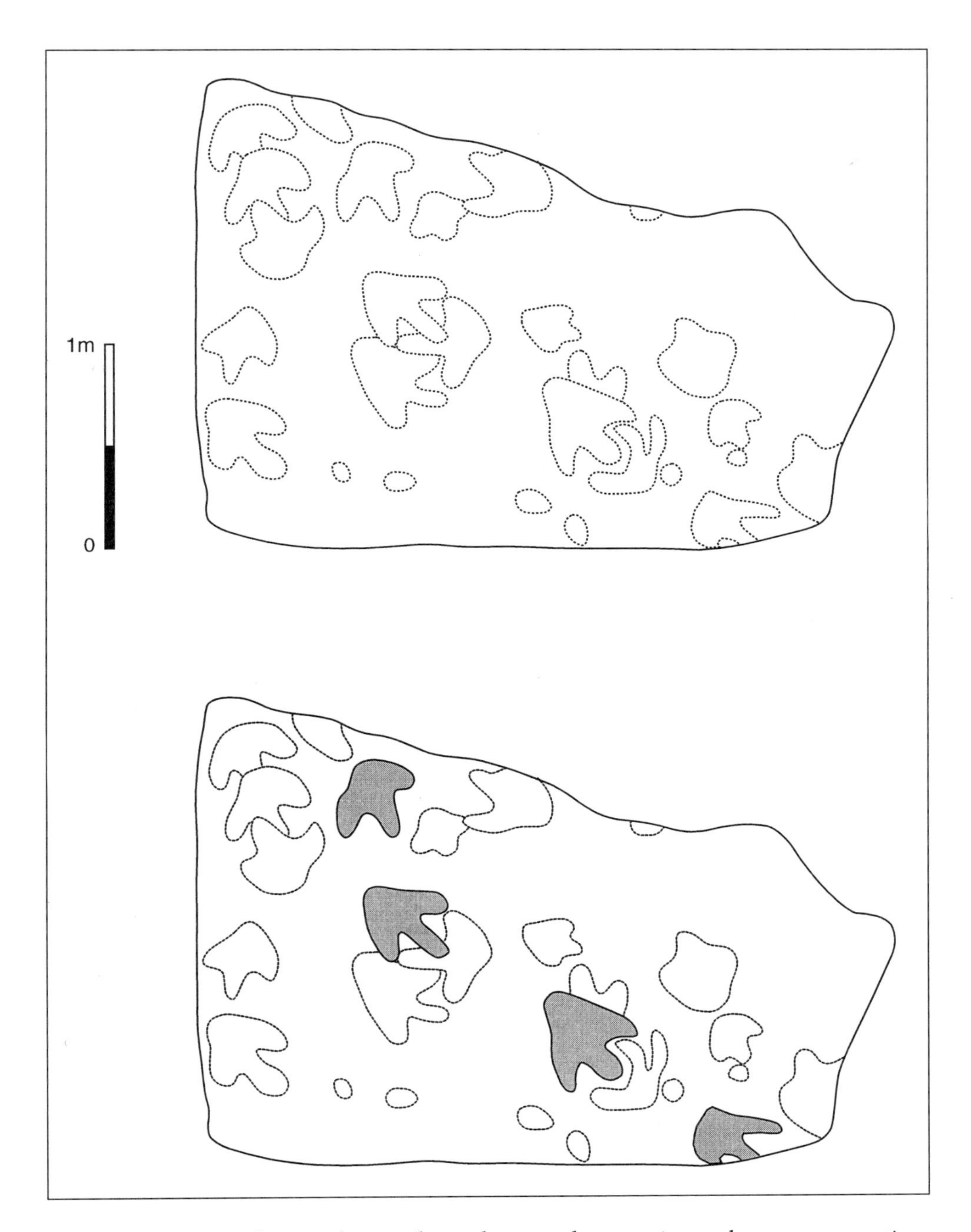

FIGURE 1.10 Elite tracks are those that stand out owing to better preservation. *Modified after Haubold.*[14]

Underprints can usually be distinguished from true tracks because they are less clearly demarcated than real footprints. Real footprints, when well preserved, can show skin impressions and other details of the fleshy tissue such as digital pad impressions. Undertracks, however, are generally less distinct in outline and detail. Obviously such loss of detailed information can be a problem for trackers, creating a situation analogous to trying to identify something through a veil or a mist. For this reason, it is important not to read too much into such tracks. Several significant blunders have been made by trackers who allowed themselves to be misled by underprints. The most striking example is the swimming sauropod scenario proposed by the American tracker Roland T. Bird in 1944.[12] He found what appeared to be a sauropod trackway in which all but one of the hind footprints were missing. He postulated that it had been swimming along touching the bottom only with its front feet. It later became apparent that the footprints were underprints. The animal had simply been walking along on the next layer of sediment, without its broad hind feet sinking in so deeply.[3]

Such a cautionary tale serves to emphasize the need to search for well-preserved tracks. It is particularly important to focus attention on such well-preserved tracks, sometimes referred to as "elite" tracks when naming new tracks.[11,13] We illustrate this principle using an example from the Lower Cretaceous of Germany.[14] Here we see a trackway which can be separated from the others owing to better or elite preservation. The information it provides is more useful than that seen in the "messy," poorly preserved background region (figure 1.10). This is simply the commonsense approach. Nevertheless, many poorly preserved tracks have been given names that are of little or no utility. Even though a proliferation of names for such poorly preserved tracks has no scientific benefit, in the past some trackers even defended this practice. The majority of trackers, however, recognize that only well-preserved footprints give a clear indication of the morphology of the trackmaker's foot. Such footprints provide the maximum amount of useful information, and when tracking extinct animals we are glad to make use of every clue available.

BIBLIOGRAPHIC NOTES

1. Paul, G. 1988.
2. Lockley, M. G. 1991a.
3. Lockley, M. G. 1991b.
4. Alexander, R. McN. 1976.
5. Lockley, M. G., A. P. Hunt, J. J. Moratalla, and M. Matsukawa. 1994.
6. Dodson, P. 1990.
7. Haubold, H. and G. Katzung. 1978.
8. Horner, J. R. and G. Gorman. 1990.
9. Lockley, M. G. 1995.

10. Lockley, M. G., A. P. Hunt, and C. Meyer. 1994.
11. Lockley, M. G., and A. P. Hunt. 1995.
12. Bird, R. T. 1944.
13. Lockley, M. G. 1993.
14. Haubold, H. 1984.

Reconstruction of the giant myriapod *Arthropleura* making tracks among standing plants during the Carboniferous period. *Courtesy of Annemarie Burzynski.*

CHAPTER 2

THE TRADITION OF TRACKING DINOSAURS IN EUROPE

The variety and number of these impressions have created a new science, and Ichnology has taken a definite place as a branch of paleontological research.

Elizabeth Gordon (daughter of William Buckland), 1894, p. 216

EARLIEST DISCOVERIES

The tradition of tracking extinct animals began in Europe in the 1820s when the Reverend Henry Duncan of Dumfrieshire, Scotland, obtained a slab of Permian sandstone from a quarry at a place called Corncockle Muir. The slab revealed 24 small footprints, and it is reported that Duncan built the rock into the wall of a summerhouse in the garden. Duncan, however, was intrigued by the discovery and soon visited the quarry, where he obtained additional specimens (figure 2.1). At that time nothing was known about fossil footprints, and so it is not surprising that Duncan sought the help of a paleontologic expert. The man he chose to consult—the reverend William Buckland of Oxford University—has become a legend in the annals of European paleontologic and geologic history. Among his many claims to fame was his historic 1824 paper on *Megalosaurus,* the first dinosaur ever described (see chapter 7).

Buckland had a sharp mind, intense curiosity, and unflagging energy. He almost immediately concluded that the tracks must be those of some form of reptile, for although the age of the strata was not precisely known, it was recognized as part of the so-called New Red Sandstone. Even then this Permo-Triassic sequence was known to be ancient. Buckland's daughter and biographer reported that at first "he was greatly puzzled [by the tracks]; but at last, one night, or rather between two and three in the morning, when according to his wont, he was busy writing, it suddenly occurred to him that these impressions were those of a species of tor-

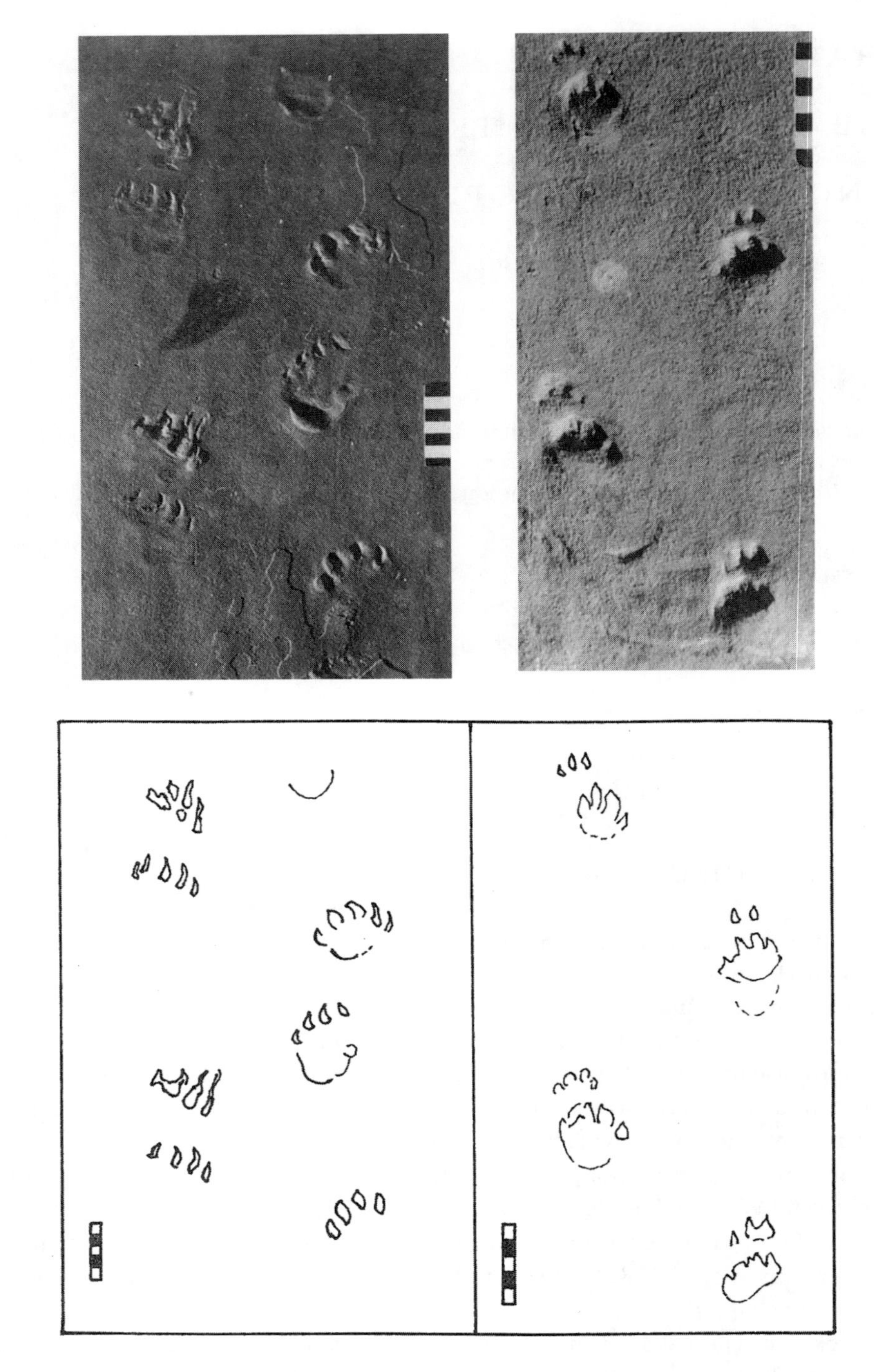

FIGURE 2.1. Permian tracks named *Chelichnus* from the Permian of Scotland were the first fossil footprints ever reported in the scientific literature. They were at first attributed to a tortoise by William Buckland. They are now known to be those of mammal-like reptiles.[7,16,19] *Photographs courtesy of Patrick McKeever.*

toise. He therefore called his wife to come down and make some paste, while he went to fetch a tortoise from the garden. . . . The delight of this scientific couple may be imagined when they found that the footmarks of the tortoise on the paste were identical with those on the sandstone slab."[1]

As British tracker William (Bill) Sarjeant has pointed out, these were the first paleoecologic experiments made with the objective of trying to identify the makers of fossil footprints.[2] Buckland became quite famous for his experiments, which he reported to Duncan and others in letters penned in the late 1820s.[3] In these he reports making crocodiles walk on "pye-crusts" which showed "decidedly that [the] sandstone foot-marks are not crocodiles." Buckland tried these pie crust experiments with three "distinct species" of tortoise, and also had them walk on wet sand and soft clay, leaving him in "little or no doubt" that they were the trackmakers responsible for making the Scottish footprints. On a subsequent occasion, Buckland acted as "master of the ceremonies" while a gathering of distinguished geologists watched the experiment repeated. On this occasion, however, the paste was too sticky and the animals "stuck so fast to the pie crust as only to be removed with half a pound of dough sticking to each foot. The philosophers, flour besmeared, worked away with tucked-up sleeves so that a proper consistency of paste was attained, and the animals walked over the course in a very satisfactory manner; insomuch that many who came to scoff returned rather better disposed towards believing."[1]

With such a weight of experimental evidence furnished by such a scientific giant as Buckland, who in turn performed his experiments in front of the leading intellectuals of the day, it is hard to imagine not being convinced that tortoises were responsible for making the footprints. Thus, it is not surprising that the Reverend Duncan liberally quoted these letters in his report to the Royal Society of Edinburgh entitled "An Account of the Tracks and Footprints of Animals Found Impressed on Sandstone in the Quarry of Corncockle Muir." This paper, reiterating Buckland's interpretations, was published by the society in 1831 and is the first account of fossil footprints ever to appear in the scientific literature.[4] Historical sticklers, including Bill Sarjeant, have pointed out that the first written reports appeared in 1828. One of these, a newspaper article which would have been preserved only in library archives, was believed lost to wartime bombing, but was rediscovered in a Parisian library and republished in 1996.[5]

As we shall soon see, the tracks are not those of tortoises, although it must be admitted, in Buckland's defense, that they resemble tortoise tracks in many details. The great British anatomist Sir Richard Owen, best known for coining the name "Dinosauria" in 1841, also supported Buckland's interpretation, and in the following year (1842) suggested the tracks be named *Testudo duncani.* Because *Testudo* is a generic name applied to living turtles it is inappropriate for a track, and, in 1850, the alternate name *Chelichnus duncani* (meaning turtle or tortoise trace) was proposed by Sir William Jardine.[6] Jardine also coined various other names for footprints that, in his opinion, were different from those he labeled as *C. duncani.*

It is outside the scope of this book to go into great detail on the subject of the validity of these various names, and what they might indicate about differences between one trackway and the next. Suffice it to say that we currently view many of these names as redundant because they are simply synonyms of *Chelichnus.*

New Permian sites revealing similar *Chelichnus*-like tracks were discovered in 1851 in Permian sandstones near Cummingstone in the Scottish Highlands. As a result, these deposits, which also contained bones, became known as the "Reptiliferous beds." When these tracks were examined in 1877 by Thomas Henry Huxley, he disagreed with the interpretations of Buckland and Owen and stated that "There is really no ground for ascribing these tracks to chelonians," and further doubted that the tracks could "be safely referred to any known form of Reptile or Amphibian."[7] It is worth noting that Huxley was known for his strong antitheology views and iconoclast tendencies. Thus, it is perhaps not surprising that he attacked Richard Owen, who believed in a "transcendental creator." In fact, Huxley was quick to dismiss the views of anyone who even introduced any hint of theological principles into a discussion of evolution or paleontology. Despite his possible motivations for disagreeing with the "Reverend" Buckland and Owen, Huxley was evidently correct, for we now know that true turtles did not exist in the Permian. Moreover the tracks are found in deposits that represent the steep avalanche faces of desert sand dunes, an unlikely habitat for turtles or tortoises. As noted below, current opinion holds that *Chelichnus* represents the spoor of some form of mammal-like reptile.

THE OLDEST EUROPEAN TRACKMAKERS

We need to make a clear distinction between the oldest known animal tracks, which are those of invertebrates, and the oldest known vertebrate tracks. We also need to distinguish between invertebrate traces made in marine or subaqueous settings, such as lakes, and those made by animals that ventured on land. Marine invertebrate traces date back to the Precambrian, about 600 million years ago, whereas the oldest terrestrial traces are around 450 million years old, from Ordovician rocks, as described below. Such distinctions also hold for vertebrates. Although their traces are rare and hard to identify, fish from marine or lake habitats began making trails and traces long before four-legged tetrapods ventured on land to register the first footprints.

As any student of evolution knows, the most successful invertebrates are the arthropods, including insects, spiders, crabs, lobsters, and countless others. They were among the first animals to crawl onto land. The oldest trackway of any terrestrial animal with legs appears to be that reported from 450-million-year-old Ordovician strata in the English Lake District.[8] Europe cannot confidently claim the world's oldest vertebrate trackway, which at present appears to come from Australia. However, it can at least claim the oldest trackway of a land animal reported

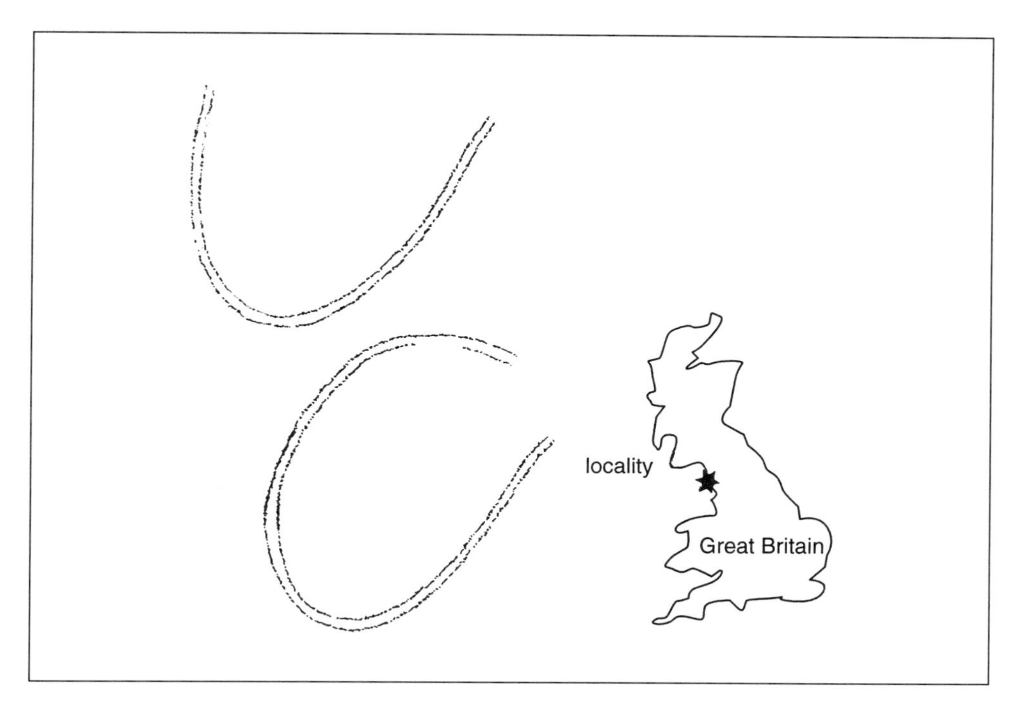

FIGURE 2.2. Ordovician tracks of creatures resembling wood lice were the first animals to walk on land. *After Johnson et al.,*[8] *figure 8.*

to date. The trails in question appear to be those of a creature that resembled a wood louse (genus *Oniscus*). These creatures left distinctive narrow trackways consisting of two rows of footprints situated only about 0.5 cm apart (figure 2.2). The trails, which meander, loop around, and crisscross randomly, have been named *Diplichnites* (meaning "double trace").

In describing the European track record from a geologic or chronologic perspective, we also need to make a clear distinction between the first fossil footprint discoveries made in Europe and the oldest known trackways. As we learned in the previous section, the first discoveries represent Permian footprints dating back 245 to 285 million years. But much older footprints have been reported from Carboniferous strata (dating from 285 to as much as 360 million years) and from Devonian deposits dating back 360 to as much as 400 million years.

Part I: Dragging Through the Devonian

The oldest convincing examples of vertebrate footprints from Europe were reported by Iwan Stössel of the Geological Institute of Zurich in a paper published in 1995.[9] Stössel was working on rocks of Devonian age on Valentia Island in a remote part of southwestern Ireland when he discovered the "meandering path" of a tetrapod consisting of an 8-m-long trackway segment consisting of 150 footprints. The trackway is "superbly exposed" on a surface washed clean by powerful Atlantic storms.

The trackway was found in a deposit know as the Valentia Slate Formation, which has been deformed by late Paleozoic mountain-building processes. This tectonic activity caused the compression or shortening of the trackway by about 20%. When the trackway is stretched out again to its original shape, it is about 10 m long (figure 2.3). The individual footprints in the trackway are essentially rounded depressions, 5 to 10 cm in diameter and about 2 cm deep. Stössel estimated that the animal, which had a step and stride of about 35 cm, had a trunk length, from shoulder to hip, of about 35 to 40 cm.

The tracks could be as old as Mid-Devonian (385 to 375 million years) but might be Late Devonian (375 to 360 million years). At present, we know of no tetrapod body fossils that are older than Late Devonian. Thus, if the tracks are Mid-Devonian, as further study might show, they are older than any known skeletal remains.

The Valentia Island tracks illustrate the importance of fossil footprints in providing information that is not available from skeletal remains. There have been various reports of Devonian trackways from around the world, although not all have been convincing. Probably the only other convincing report comes from Victoria in southeastern Australia, where the tracks are evidently even older than those reported from Ireland. Reports of a single footprint from Brazil have been reinterpreted as a starfish trace,[10] and reports of traces from Scotland, described below, are less than convincing.

Possible vertebrate trails ("tracks left by some creature") were reported from the Devonian "Old Red Sandstone" exposed in the Orkney Islands, Scotland.[2] The possibility that these are vertebrate footprints has been suggested by William Sarjeant, who noted that lobe-finned rhipidistian fish were venturing onto land at this time. Apparently the specimens were never studied seriously or preserved in a museum, so little could be said in support of this hypothesis. Recently trails fitting this description were rediscovered, and it appears they may be of vertebrate origin.[11] Similar trails were reported from Devonian rocks in Ayrshire on the Scottish mainland, but, again, they have not been studied in detail.

Paleontologists tend to have a stereotyped image of the first lobe-finned fish crawling onto land, an event that, at least symbolically, represents a great moment in evolution. It is nice to think that fossil footprints might record this great step forward at the dawn of the age of tetrapods, but a reevaluation of the evidence has recently suggested a new slant on this favorite paleontologic story. Jenny Clack, from Cambridge University, made a careful study of all the known Devonian amphibian occurrences and all reports of trackways. She concluded that there is little compelling evidence that any of the reported trackways from the early and middle Devonian are reliably dated.[12] In other words they could all be no older than Late Devonian. She and her colleagues have also suggested an entirely new scenario for

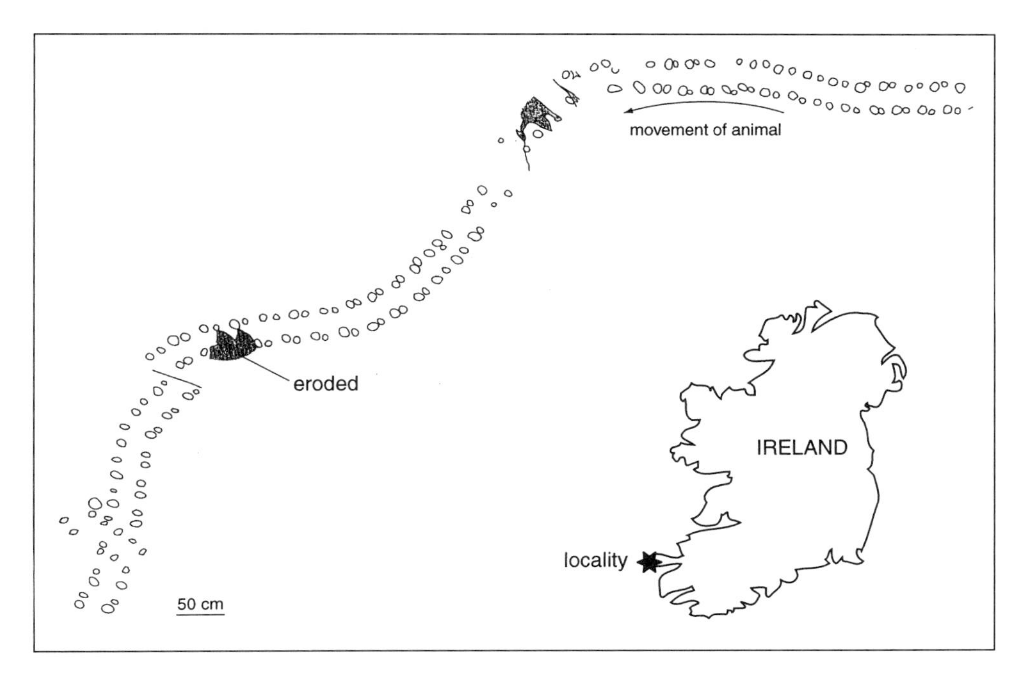

FIGURE 2.3. Carboniferous tetrapod (amphibian) tracks from Valentia Island, Eire, are Europe's oldest vertebrate tracks. *After Stössel,[9] figure 2.*

the origin of tetrapod locomotion. They suggest that the early tetrapod limb was not designed for walking on land because their limbs were not stout enough to support their weight. So limbs did not evolve as a result of fish with stout fins trying to scurry from one shallow pool to another in times of drought. It appears, instead, that early amphibians evolved limbs to "walk" subaqueously along the bottom of the sea and other water bodies.[13] Such a scenario, which is now gaining acceptance, has implications for ichnology. It means, at least in theory, that tracks may exist that represents tetrapods that walked on sea or lake beds before they had the capability of walking on land. It has not yet been ascertained if any of the Devonian trackways so far reported fit this category.

Part II: Cruising the Carboniferous

The Carboniferous Period (360 to 285 million years ago) is well known as the age of coal swamps, as the name suggests, and is also known as the age of amphibians. Having established that amphibian footprints first appear in the Devonian, along with the first records of their skeletal remains, it is not surprising to find that their tracks are also reported, quite frequently, from Carboniferous deposits. But the Carboniferous is also known as the age of giant insects and other arthropods. Although this book is about vertebrate footprints, we cannot resist the temptation to mention a few of the more spectacular invertebrate trails that have been reported. In fact, it is helpful to make this digression because such trails have been confused with those of vertebrates. So herein lies a cautionary tale.

We have already mentioned the trackways of minute multi-legged Ordovician arthropods that resembled wood lice. Such creatures can be referred to as "myriapods" and are related to millipedes and centipedes. Fossil evidence clearly reveals that myriapods were well established in the Carboniferous, where they formed part of the forest litter community that evolved as extensive forests first colonized the Carboniferous lowlands and began shedding litter onto previously barren ground. Such ancient coal swamp habitats have dictated the location of Europe's economically important coal fields, around which the industrial revolution was built. It is in just such an area, associated with the Scottish coalfields, that we find some of the most remarkable of all known arthropod trackways.

The most striking feature of these trackways is their remarkable size. A trackway reported in 1979 from Middle Carboniferous rocks on the geologically famous island of Arran is 36 cm wide and is preserved in a segment more than 6 m long.[14] It is attributed to the giant millipede-like arthropod *Arthropleura,* known from body parts with equally gigantic proportions. Additional trackways in Scotland were found in Lower Carboniferous rocks near Fife.[15] At this site a dozen trackways were reported, ranging in width from 18 to 46 cm (figure 2.4). Such an abundance of trackways suggests that *Arthropleura* was a common genus of arthropod during the Carboniferous.

It is hard to imagine a millipede, or any terrestrial arthropod, with a body width of 46 cm. If we were to try and visualize such a creature walking among the tree stumps in the Carboniferous fossil forest display in Glasgow, we would have to concede that it would occupy most of the space between adjacent trees. It has been suggested that arthropods were able to grow so large during the Carboniferous because of high levels of oxygen in the atmosphere, exceeding present concentrations. Indeed some science journalists have referred to this period as the "oxygeniferous." There can be no doubt that conditions favored gigantism among some arthropods and that such forms as *Arthropleura* were actually larger than most vertebrates. For example, the amphibian described in the previous section is narrower than the largest Fife trackway by a good 10 cm, and *Arthropleura* trails from North America have been mistakenly identified as vertebrate trackways.

Giant *Arthropleura* trackways are only known from the Carboniferous, where paleontologists have named them *Diplichnites,* reflecting their similarity to much smaller trails, such as the Ordovician example just cited. (Although the ichnogenus name is the same, the ichnospecies names are different.) Whatever the reasons for the success of *Arthropleura,* whether it was high oxygen levels, an abundant new organic litter food supply, or the absence of serious threats from predators, it is evident that they left their mark. It is often hard to identify the genus or species responsible for making tracks, especially when studying invertebrates. In this case, however, it appears we can identify *Arthropleura* trails with

FIGURE 2.4. Reconstruction of a giant myriapod (*Arthropleura*) making a trackway. Courtesy of Annemarie Burzynski.

confidence—a happy circumstance for trackers of extinct animals, and one that we will encounter more frequently as our knowledge and tracking skills improve.

Now let us turn to the trackways of Carboniferous vertebrates, many of which represent animals that were dwarfed by the monster myriapods just described. The best sites for finding Carboniferous footprints are again the coal measures, where conditions were always wet underfoot, creating substrates in which it was easy to register tracks. The best known of such sites are in England, France, and Germany.

It is outside the scope of this book to give a comprehensive account of the variety of Paleozoic (Carboniferous and Permian) footprints known from Europe. The interested tracker should consult *Saurierfährten.* This specialist volume, by the leading German ichnologist Hartmut Haubold, was, until recently, the only available handbook for the identification of fossil footprints and is still available only in

German.[16] Although much has been published on dinosaur tracks that supersedes this book, it is still valuable as a source of information on nondinosaurian footprints.

Carboniferous vertebrates are not well known outside the specialist ranks of vertebrate paleontology, and even in such select circles, their footprints are almost unknown. Haubold's efforts to correlate Carboniferous tracks with potential trackmakers are therefore of considerable help in trying to make sense of the array of footprints extracted from coal mines and various other sites. For the nonspecialists, which in the case of Carboniferous tracks includes almost everyone, the first step is to recognize that Carboniferous tetrapods were either amphibians or reptiles (although modern cladistic classification makes this traditional distinction somewhat blurred). There is no foolproof way to distinguish the tracks of these two groups. Although amphibians are sometimes said to lack sharp or elongate claws, there are also many footprints with blunt toe impressions that have been attributed to reptiles.

Late Paleozoic amphibians have traditionally been divided into about 20 families that show a considerable diversity of form, ranging from the snake-like aistopods, to hammerhead nectridians, to lizard-like microsaurs, to alligator-sized temnospondyls such as *Eryops.* Obviously, lack of feet or very diminutive limbs will result in trackways and trails that are significantly different from those made by species that have stout limbs and large feet. For example, long-bodied short-legged creatures are more likely to sprawl and leave tail traces than are long-legged erect species. However, differences in the anatomy of the rest of the body, which are of significance to paleontologists, may not help distinguish trackmakers when all we have to go on is feet. Nevertheless, we can compare foot skeletons with footprints in an effort to narrow down potential groups of trackmakers, or at least eliminate some from consideration (see chapter 1). To date, Haubold's encyclopedic work, *Saurierfährten,* is still the main summary of this type of information for Paleozoic tracks. Similarly, reptiles fall into a number of different groups, known only to specialists. These various types may also be distinguished to some degree by their foot anatomy. The best approach, and the one taken here, is to simply describe representative trackways and comment on potential trackmakers.

One of the largest Carboniferous trackways reported from Europe was recently discovered on the east coast of Northumberland, England.[17] The trackway, which is 65 to 70 cm wide, reveals a short, stumpy, five-toed foot about 18 cm long and a tail or belly drag mark (figure 2.5). David Scarboro and Maurice Tucker, the authors of this 1995 study, suggested that the trackmaker might have been a temnospondyl amphibian that was about 1.5 m in length and lived in a deltaic environment during the Middle Carboniferous. The trackmaker was larger than anything known from the bone record in Great Britain at this time, although larger amphibians became more common later in the Carboniferous. Here again we see the utility of footprints in adding to information obtained from the skeletal record.

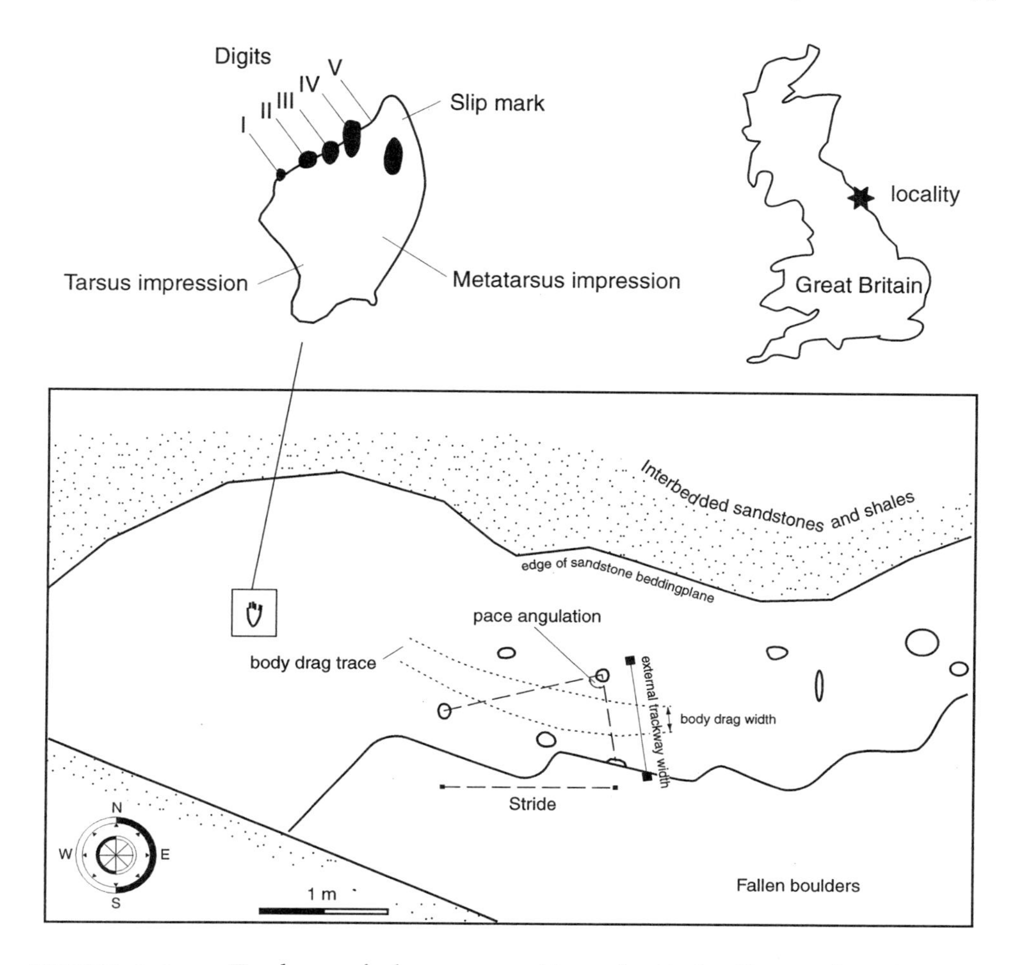

FIGURE 2.5. Trackway of a large tetrapod from the Carboniferous of northern England. *After Scarboro and Tucker,*[17] *figures 7a and 11.*

OF DESERTS AND SWAMPS

The typical, stereotyped reconstruction of Carboniferous landscapes is a vista of swamps. After all, this is the age of coal swamps and the age of amphibians. By comparison, our stereotype of the Permian world typically revolves around arid landscapes in which only tough, desiccation-resistant reptiles were able to survive and pioneer the colonization of the dry interiors of continents. There is some truth in these stereotyped views. But a better perspective on Carboniferous and Permian habitats is gained if we realize that there were wet and dry habitats in different areas during both time periods. Such differences in environment and climate obviously exerted a significant influence on the distribution of plants and animals.

Thus, as we might expect, the flora and fauna vary from habitat to habitat. If we compare the tracks found in desert sand dune deposits and those associated with wetlands, we usually find very different types of footprints even in the same time periods. On the other hand, tracks from the same types of environments often show notable similarities, despite differences in the age of the strata. These habitat preferences among a wide range of fossil vertebrates are reflected in the track record by what we have referred to as ichnofacies (see chapter 1). In short, the tracks appear to represent some portion of the indigenous animal communities of the particular ancient environment in which the footprints are found. We can only make the claim that footprints reflect particular animal communities in particular habitats if we find a "repeated association" of footprints with a particular type of sedimentary deposit, or facies.

As recent studies have shown, this is exactly what we find. Possibly the best example is the repeated associations of various Permian mammal-like reptile tracks such as those of *Chelichnus* with desert sand dune deposits. Such associations are not confined to Europe. Recent work in North America shows that the track of the mammal-like reptile known as *Laoporus* occurs at various sites in Arizona, Utah, and Colorado in fossil sand dune deposits assigned to four differently named geologic formations. In all areas the footprints are associated with tracks and traces of spiders, scorpions, and other arthropods. For this reason we named the *Laoporus* ichnofacies[18] and consider it representative of Permian desert faunas.

Work by Hartmut Haubold and Patrick McKeever has shown that *Laoporus* is in fact similar to *Chelichnus*, which, as mentioned above, is also found in desert sand dune deposits[19] and was probably made by a mammal-like reptile (see figure 2.1). Thus, there can be little doubt that similar track types occur repeatedly in the same type of sedimentary deposits (at approximately the same time) in both Europe and North America. In short, not only should our concept of *Chelichnus* be expanded to include serious consideration of *Laoporus* (and vice versa), but our concept of the *Laoporus* ichnofacies should be expanded to include consideration of repeat *Chelichnus* assemblages in dune deposits (what we might call the *Chelichnus* ichnofacies). Such an approach is integrative and coherent, suggesting that tracks are an expression of ancient ecologic systems. In this regard we are indebted to Hartmut Haubold for suggesting that the *Laoporus* and *Chelichnus* ichnofacies, which are more or less two different geographic expressions of the same type of ecosystem, might be called the "dune ichnofacies." There is nothing wrong with this alternative, although it is generalized and so could also include similar ichnofacies that contain different track types in Triassic or Jurassic deposits. In short, the terms *Laoporus* and *Chelichnus* ichnofacies refer to a subset of the dune ichnofacies. In reality, even if the trackmakers of *Laoporus* and *Chelichnus* were members of the same species, they presumably represented different populations, with different distributions and different dynamics.

Another pertinent issue related to this topic is the subject of palichnostratigraphy or track correlation. The same track types may occur in the same types of deposits simply because they are the same age (in this case Permian). But if the tracks (and trackmakers) bore no relationship to the sedimentary deposits, we would expect

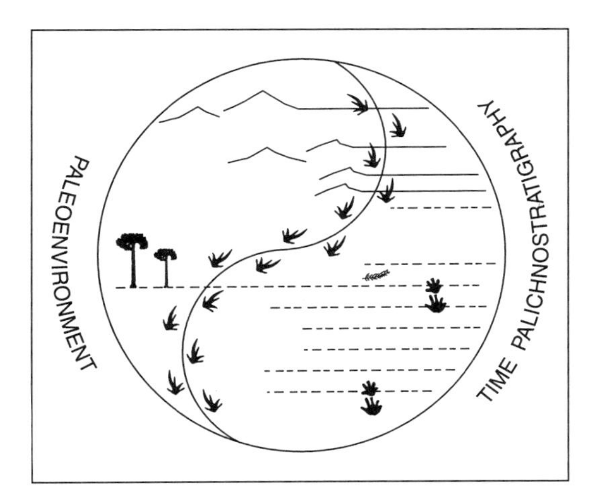

FIGURE 2.6. The distribution of track assemblages and ichnofacies is controlled by both the age of the fauna and the ecological range of species in ancient ecosystems. This integrated relationship can be represented using the yin-yang motif.

them to occur more widely and not be confined to one sedimentary facies, such as dunes. This is sometimes the case, presumably for animals with a wide range of habitat preferences. Thus, we must be aware of both ichnofacies patterns and palichnostratigraphy. Both are integrative themes, of ecology and time, respectively, in the broader paleontologic picture. Neither can operate or be understood entirely without reference to the other; like yin and yang, they are part of an integrated whole in which one component is ecologic space and the other is ecologic time (figure 2.6).

Moving from desert to swamp settings, or at least from ancient environments that were predominantly arid to those that were wetter, we can observe similar ichnofacies relationships between track assemblages and particular sedimentary deposits. One of the more common Carboniferous and Permian track types is *Limnopus,* which loosely translates as "lake foot."[20] Clearly, the association with lakes (or limnology) provides a significant contrast to dune settings. This track type (figure 2.7) is commonly regarded as amphibian in origin and is frequently associated with much smaller footprints such as those of *Anthichnium* (also known as *Batrachichnus,* denoting an amphibian origin). The association is sufficiently common in Europe to suggest what we have called an *Anthichnium-Limnopus* assemblage.

It is beyond the scope of this book to delve too deeply into the composition of such assemblages, although a summary is given in the sections that follow. Such an endeavor reveals a plethora of names that require explanation and detailed reevaluations. As we have already seen, even the ichnogenera that give their names to ichnofacies labels are, like any paleontologic names, potentially subject to change. The important point is simply that the repeated occurrence of certain suites or assemblages of tracks in certain environments is a clear indication of the preference of certain fauna for such settings. This does not mean that such fauna are not representative of any other environments (although the ichnofacies principle shows that this is usually not the case). It also does not mean that the tracks will necessarily match the skeletal record of rocks of the same age, although varying degrees of correspondence may be established. We can state, however, that tetrapod ichnofacies are a real phe-

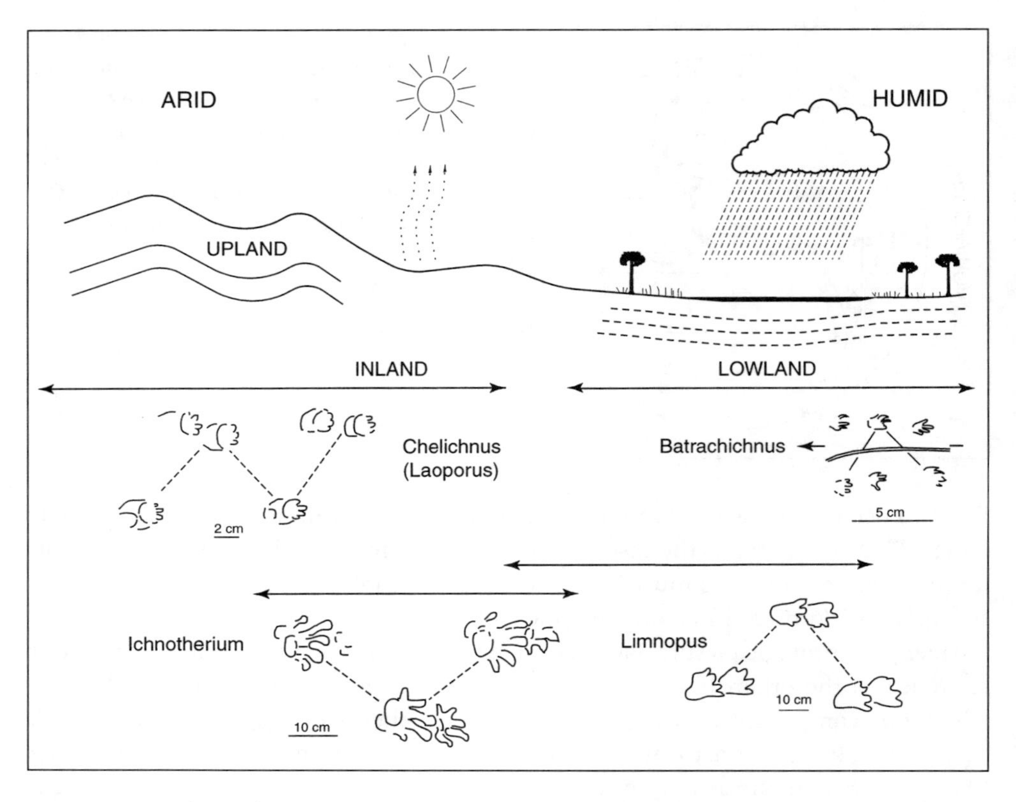

FIGURE 2.7. Of swamps and deserts: comparison of different track types from humid and arid habitats.

nomenon, not just in Paleozoic rocks, but throughout the tetrapod track record. They represent some facet of ancient ecology, and it is up to paleontologists to determine what that is and to what extent it substantiates or adds to the bone record.

ROTLIEGENDES: PERMIAN TRACKWAY HEAVEN

Rotliegendes means "red layers," and among geologists the concept of "red beds" is well known and usually implies terrestrial or continental sediments that contain a lot of iron oxide (rusted iron minerals). In a historical context, red Devonian sediments of this type were labeled the "Old Red Sandstone" and Permian and Triassic sandstones were known as the "New Red Sandstone," with the nonred or less red sediments of the Carboniferous in between. The track-bearing sediments of Scotland fall into the new red category. All such red beds in Europe have proved hard to date accurately owing to a lack of fossils. For this reason, tracks are potentially useful and have been used enthusiastically, sometimes too enthusiastically, to help estimate the age of the rocks.

FIGURE 2.8. *Ichnotherium* trackway (left) and *Dimetropus* trackway (right) from the Gotha Natural History Museum. Individual footprints are 5 to 6 cm in width.

In the Thuringian forest region of central Germany, the new red Rotliegendes sediments are more than a kilometer in thickness, and are, in most places, rich in tracks. Thus, Thuringia has become the cradle of European Permian ichnology. Tracks were first reported in a letter by a Mr. Cotta in 1847 and later named *Saurichnites cottae* (meaning "Cotta's reptile tracks"). Since that time, countless tracks have been collected and described from dozens of sites, and many central European museums contain substantial collections. For example, long well-preserved trackways known as *Ichnotherium* (meaning "trace of a beast") are known from the Tambach quarries near Gotha, and many fine examples are preserved in the Natural History Museum at Gotha (figure 2.8). This particular track type is thought to be attributable to a sphenacodontid pelycosaur. Another track type found in this area is *Dimetropus* (meaning "*Dimetrodon* tracks"). Together these track types suggest the activity of this famous large fin-backed reptiles and its synapsid relatives in a conti-

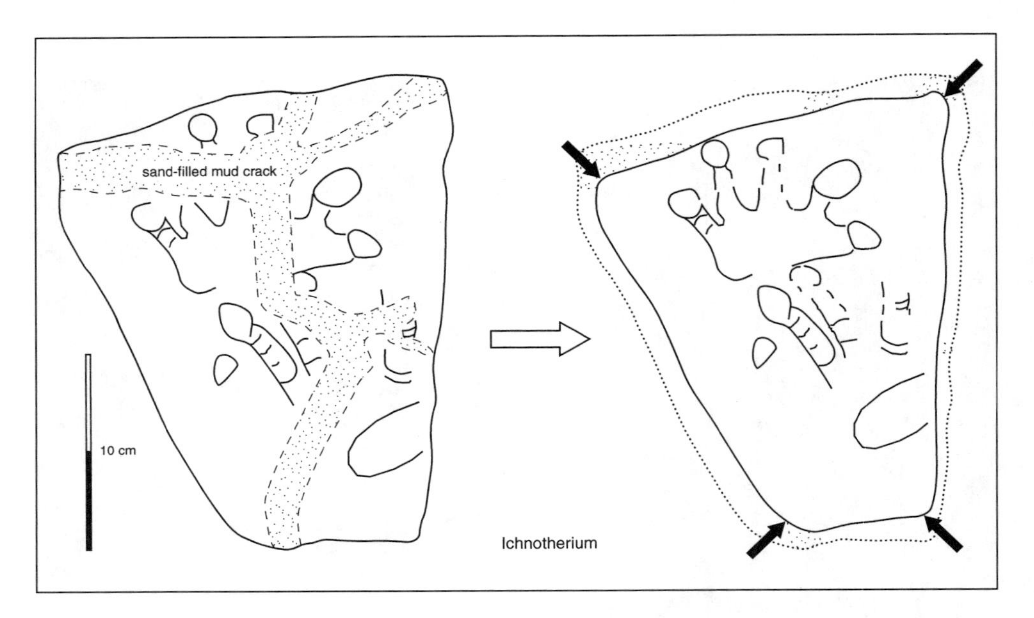

FIGURE 2.9. Reconstructing *Ichnotherium* tracks by removing the mud cracks.

nental location. Synapsids were a group of primitive mammals sometimes referred to as "protomammals" or "mammal-like reptiles," especially in the older literature.

Although beautifully preserved, the *Ichnotherium* trackways are associated with many mud cracks, which were made after the tracks had been registered. The mud cracks are sufficiently wide to cause the overall shape and size of the footprints to be exaggerated. This distortion of the true track shape, however, can be rectified by "collapsing" the footprint back to its original size and shape, as it would have appeared before the mud cracking occurred (figure 2.9). This simple technique was demonstrated by the well-known prize-winning German ichnologist Adolf Seilacher.

Other red bed (Rotliegendes) sites reveal a variety of smaller track types, including those of *Batrachichnus, Limnopus,* and *Dromopus,* attributed to a variety of early reptiles of different sizes.[21] Yet other sites, particularly those associated with the Cornberg Sandstone, which is a sand dune deposit, yield tracks such as *Chelichnus,* similar to those found in Scotland.

Owing to the abundance of tracks at many different locations, there was a natural tendency for them to be examined by many different paleontologists and amateur enthusiasts. The result was a proliferation of detailed publications about local discoveries and a corresponding proliferation of names. This approach reflects the nineteenth-century tradition of classifying all flora and fauna. Many track names, however, have proved to be ill conceived and the result of overexuberance, which has left us with a legacy of far too many names. We shall review this problem and the latest solution in the section that follows.

Before we move on to other perspectives, however, we should not give the impression that all Rotliegende or Rotliegende-like track-rich sediments are confined

FIGURE 2.10. Trackways of *Limnopus* (large) and *Anthichnium* (small) from the Carboniferous of Alveley, near Birmingham, represent a typical assemblage. *Courtesy of the National Museum of Wales and Birmingham University.*

FIGURE 2.11. Trackway of *Ichnotherium* crossing trackways of smaller tetrapods, from red beds near Birmingham. *Courtesy of the National Museum of Wales and Birmingham University.*

FIGURE 2.12. *Anthichnium* trackway, showing a clear tail drag trace, from Permian red beds near Birmingham. *Courtesy of the National Museum of Wales and Birmingham University.*

to the Thuringian Forest region. Similar deposits with similar tracks are found to the west in the Saar-Nahe basin, not to mention the Autun region of France, northeast of the Massif Central, midway between Lyon and Paris. Autun, the medieval capital of an ancient mining district, gives its name to the Autunian stage. Many of the red beds have been mined to produce uranium. Tracks not only occur here but also in the Lodève region close to Montpellier and the shores of the Mediterranean. Tracks from these areas have been studied by French ichnologists Paul Ellenberger, George Demathieu, and George Gand, who together have been responsible for a large number of publications[21,22] and the introduction of a bewildering number of names. As discussed in a later chapter, Paul Ellenberger learned tracking from the Bushmen of southern Africa, who could identify any extant species from their tracks and even discriminate individuals. The result of this African influence and the tradition of nineteenth-century classification help explain the proliferation of names. As we repeat in a subsequent chapter, "C'est l'Afrique."

Another important collection of red bed tracks was reported from the Keele and Enville beds of the Birmingham region in England.[23] The former, which are considered Carboniferous deposits, contains *Limnopus* and *Anthichnium* (figure 2.10). The latter, which may be Permian in age, includes trackways of *Ichnotherium* (figure 2.11). The collection, housed at Birmingham University, includes a fine example of an *Anthichnium* trackway showing a tail drag impression (figure 2.12). Recent studies reveal that skin impressions are associated with tracks from these assemblages.[24]

THE GERMAN SUMMIT CONFERENCE

Although an international symposium on "Dinosaur Tracks and Traces" was held in New Mexico in 1986,[25] there had never been an international gathering to discuss Paleozoic footprints until 1997. This meeting of Paleozoic trackers convened under the title of a "workshop" on Permian ichnofacies and ichnotaxonomy. As is appropriate for trackers, the group was mobile, tracking animals in the Rotliegende from the Frankfurt region to Martin Luther University in Halle, visiting the Nierstein, Cornberg, Gotha, and Geiselthal Museums en route.

The workshop was the brainchild of Hartmut Haubold, who also acted as field trip leader and master of ceremonies. His aim was to get everyone interested in the subject to try to speak the same language. Haubold is an excellent person to orchestrate such a meeting of minds, not just because he is an expert on Paleozoic tracks, but because he was for too long restricted to working behind the iron curtain and so knows the drawbacks of enforced isolation. As noted in the previous section, there has been a tendency for workers to focus on local fossil footprints. This provincial outlook may be self-imposed by tradition or local politics, and those adopting such a view may not be aware that their vision is somewhat narrow. Haubold, of all the trackers, after being restricted by "provincial politics," was acutely aware of the need for a broader vision.

The point of this philosophical preamble is that the Paleozoic tracks led to a very confused and fragmentary model of Permian footprint diversity and distribution. Similarly the nonprovincial, holistic, or global overviews, as advocated by Haubold and like-minded colleagues, led to a radically different perspective. Just as the Dinosaur Tracks and Traces Symposium led to rejuvenation and maturing of the discipline of dinosaur ichnology, so Haubold's summit conference marked a turning point in the study of Paleozoic tracks.

In a nutshell, what Haubold has done since he emerged from the isolation of the former eastern bloc is to show that Permian tracks are much more similar from region to region than one would ever surmise from the voluminous and confusing literature. Up to the present time, there have been about 140 different track types (ichnospecies) named. In reality, there are, according to the compelling evidence presented by Haubold, perhaps no more than six or seven at any one locality. Only 10% to 15% of the 140 can be considered valid ichnotaxa.

The problem of naming too many tracks is called "taxonomic splitting," in contrast with taxonomic "lumping," where similar tracks are all given the same name. There are several reasons why such splitting occurs. First, as indicated above, the problem begins with provincial traditions, the desire to name anything interesting found in one's own backyard, without knowing what is found elsewhere or communicating fully with researchers in other areas. This problem is compounded when reports of local finds are published in obscure journals in several different languages. True dedication and much time are required to ferret out all relevant information. Second, there is the problem of incomplete preservation of tracks. A trackmaker that makes five-toed tracks under ideal circumstances may make tracks with only four-toe impressions in other circumstances, and in yet other circumstances the footprints may reveal only three- or two-toe impressions. In fact, footprints of lizard-like trackmakers often show the impression of only the two longest digits (III and IV), producing a footprint that had been named *Didactylus.*[16]

A third factor in the proliferation of names has been the positivistic philosophy[21,22] which adopts the view that any object, in this case tracks, that has a distinctive identity can be given a distinctive or separate label. This would be rather like differentiating between men, women, and children rather than regarding them as members of the same species. Both viewpoints are correct, but only if you realize the whole picture. If you believe in the three categories and do not know they are all one species, the view is incomplete and misleading.

One of the results of the historical proliferation of names was that some workers began to look for distribution patterns of track types through time. Such a goal of track zonation, or palichnostratigraphy (literally "ancient track stratigraphy") is admirable because one of the main objectives of geology and paleontology is to use fossils to calibrate time in the geologic record. In the case of the Rotliegende, the abundance of tracks and lack of other fossils encourage such an attempt at track zonation. After all, if tracks are abundant and virtually all that is available, then use them. What other options are there?

Unfortunately attempts to produce track zonations have been thwarted by the problem of oversplitting and the fact that the horizons from which the tracks originated were not well known. Without a complete sequence of strata to work through, the problem becomes difficult. One needs to know the level at which the tracks are found in the first place in order to establish or calibrate a time scale. Hartmut Haubold, in conjunction with Gerhard Katzung, pioneered the idea of track stratigraphy,[26] as noted below. Subsequent work, however, has resulted in unnecessarily complex and controversial schemes, which, because they are based on incorrect and confusing names, cannot work effectively. In short, Haubold's summit conference brought out these problems and the conferees suggested that we go back to basics.[27] It is first necessary to review the ichnotaxonomy and to jettison redundant or invalid names; then we can tackle track stratigraphy, ideally with a group of ichnologists who agree on an ichnotaxonomic classification. The fact that Permian track stratigraphy is in a mess in no way invalidates the concept of palichnostratigraphy. As we shall see in subsequent chapters, there are good examples of reliable track correlations from other time periods, where the track names are not so confused by historical circumstances.

STUCK IN THE MUD: THE COMPLETE TRACE OF A HAMMERHEAD AMPHIBIAN

In 1988 Harald von Walter and Ralf Werneberg described what appeared to be several relatively complete impressions of small amphibians, only a few centimeters long, found in the Rotliegende of the Thuringian Forest region.[28] The trace has a crescent-shaped anterior end resembling the head of a diplocaulid, a "hammerhead" amphibian. The main part of the body impression displays a reticulate pattern (figure 2.13).

FIGURE 2.13. Stuck in the mud. *Hermundurichnus fornicatus:* resting traces of a diplocaulid. *After von Walter and Werneberg,[28] figure 1.*

It is thought that these amphibians might have had lateral flaps of skin that allowed them to swim like modern skates. Regardless of the validity of this interpretation, these are the first traces to indicate the presence of this group of amphibians, and, moreover, are the first examples of relatively complete body traces from the Rotliegende.

Von Walter and Werneberg named these traces *Hermundurichnus fornicatus* (referring to the Hermunduri, a Germanic tribe that inhabited Thuringia in the first century B.C.). The species name *fornicatus* can be taken to indicate the collection or "orgy" of traces found together. The recent discoveries of skin impressions including partial body traces in England[24] hold out hope for more such finds when such track assemblages are studied in greater detail.

THE FIRST PAREIASAUR TRACKWAY

There are few families that can boast more than one vertebrate ichnologist among their members. The Leonardi family is a notable exception. The late Piero Leonardi, a distinguished Venetian geologist, may not be as famous in ichnologic circles as his son Giuseppe, but he has the distinction of having described the largest and one of the most interesting Permian tracks.[29] The track, named *Pachypes dolomiticus* (meaning "thick foot from the Dolomites"), was discovered in the Late Permian Val Gardena sandstone in the Dolomite region of northern Italy in 1975. The original discovery was of a single track cast that was insufficient to be certain if it was a hind or front footprint or to give any clear indication of the trackway pattern. Owing to the vertical exposures in the Val Gardena cliffs, an area perhaps better known for its skiing than its fossil footprints, it proved difficult to find a complete trackway. In fact it has taken more than 20 years to finally resolve the nature of the trackway (figure 2.14). Based on this new and complete picture of the trackway, the trackmaker appears to have been a large quadruped with a relatively

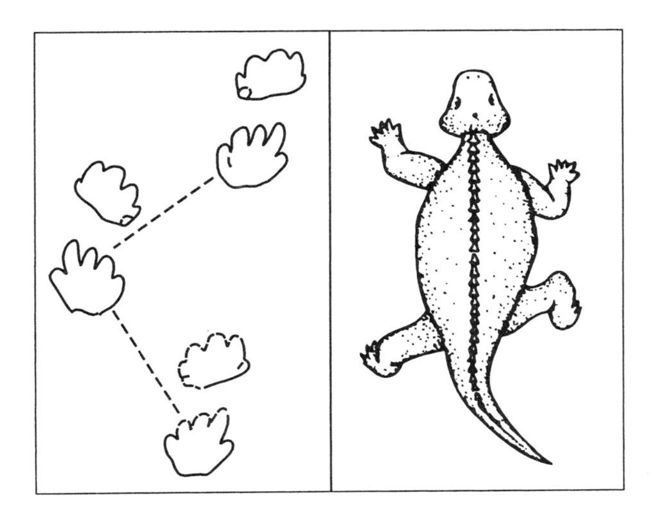

FIGURE 2.14. *Pachypes dolomiticus,* the largest footprint known from the Permian, indicates the presence of large pareiasaurs in the region that is now northern Italy. *Courtesy of G. Leonardi and colleagues.*

wide straddle.[30] The morphology of individual footprints suggests a match with the feet of pareiasaurs. The pareiasaurs are an odd group, mainly known to specialist paleontologists. Some suggest they are the ancestors of turtles.[31] They were short-bodied animals with short tails, short legs, and heavy armor.

PANGAEAN GLOBETROTTERS

Although Haubold's summit conference was held in Europe, with representatives from Germany, France, Italy, Great Britain, and Switzerland, it was also an international affair. Permian animals knew no political boundaries, and the world they inhabited was quite different geographically and ecologically. As already noted, during the Permian the world's continents were coalesced together into a supercontinent (figure 2.15). Permian footprints from the western United States closely resemble *Chelichnus* from Europe. At first the similarity was not recognized and the American tracks were named *Laoporus*, meaning "stone track."[32] We now recognize that they are very similar to *Chelichnus*,[19] obviously implying that the same types of trackmakers were present on both continents (see figure 2.7). We emphasized that this similarity was the result of the trackmakers inhabiting similar habitats, but the similarity also results from the fact that the trackmakers were of the same age and possibly belonged to the same species.

The presence of similar trackmakers in both Europe and North America is easy to explain in the context of Permian geography. At this time the world's continents were coalesced to form the supercontinent Pangaea, a continent that really did not begin to break up until well into the Mesozoic Era. Therefore, for tens of millions of years the fauna of what are now different continental plates was free to move from plate to plate. Such circumstances led to the evolution of a cosmopolitan "pandemic" or worldwide fauna, rather than the isolated or endemic faunas that typically evolve when plates are separated and faunal interchange ceases.

Hartmut Haubold and Gerhard Katzung were the first trackers to explicitly draw attention to the potential of tracks for faunal correlation and coin the term "palichnostratigraphy" (meaning "ancient track stratigraphy"). Haubold proposed a series of track zones ranging from Carboniferous to Lower Jurassic,[16] and it is no coincidence that this represents Pangaean time, when animals were free to disperse across the supercontinent, limited only by ecologic rather than geographic barriers (see figure 2.12). We shall see many other examples of Pangaean globetrotting as we continue our journey into the Triassic and Jurassic (see chapters 3 through 6). As we shall also demonstrate, these post-Permian correlations are quite precise. The identification and dating of Permian tracks is still in a state of flux after a confused, and highly provincial history of splitting. As should now be abundantly clear, the provincial perspective is highly restrictive when attempting to understand ichnofaunas that were cosmopolitan in character. The recent German summit conference was an important turning point in the reorganization and sim-

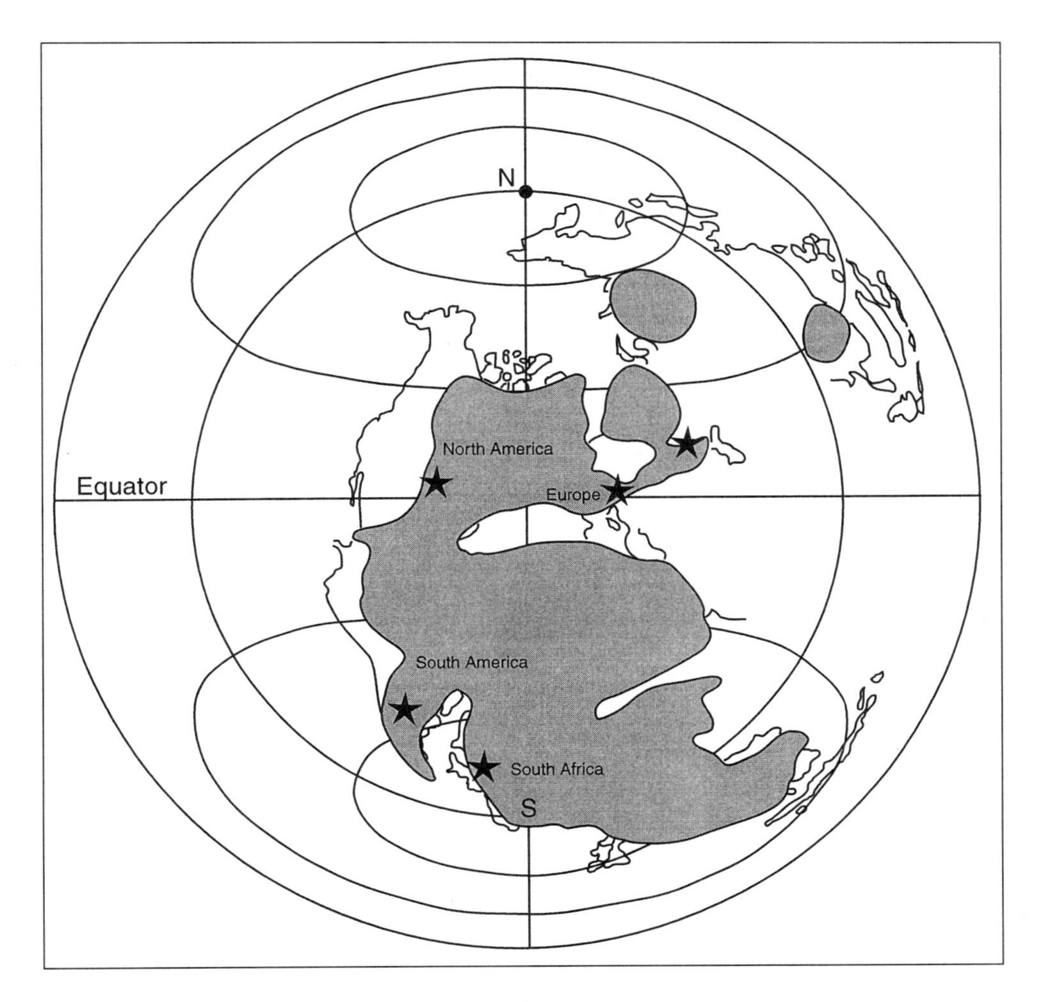

FIGURE 2.15. Permian tracksites on the supercontinent Pangaea.

plification of the world of Permian tracks. Thanks to Haubold, and to current synthetic trends in vertebrate ichnology, we can look forward to establishing a much clearer picture of the distribution of Permian tracks in the future.

BIBLIOGRAPHIC NOTES

1. Gordon, E. 1892.
2. Sarjeant, W. A. S. 1974.
3. Buckland, W. 1858.
4. Duncan, H. 1831.
5. Pemberton, S. G., W. A. S. Sarjeant, and H. S. Torrens. 1996.

6. Jardine, W. 1850.
7. Huxley, T. H. 1877.
8. Johnson, E. W. et al. 1994.
9. Stössel, I. 1995.
10. Rocek, Z. and J-C. Rage. 1995.
11. Rogers, D. A. 1990.
12. Clack, J. A. 1997.
13. Coates, M. I. and J. A. Clack. 1995.
14. Briggs, D. E. G. et al. 1979.
15. Pearson, P. N. 1992.
16. Haubold, H. 1984.
17. Scarboro, D. D. and M. E. Tucker. 1995.
18. Lockley, M. G., A. P. Hunt, and C. Meyer. 1994.
19. McKeever, P. J. and H. Haubold. 1996.
20. Lockley, M. G. and A. P. Hunt. 1995.
21. Haubold, H. 1996.
22. Haubold, H. 1997.
23. Haubold, H. and W. A. S. Sarjeant. 1974.
24. Wright, J. L. and I. J. Sansom. 1998.
25. Gillette, D. D. and M. G. Lockley. 1989.
26. Haubold, H. and G. Katzung. 1978.
27. Lockley, M. G. In press(a).
28. von Walter, H. and R. Werneberg. 1988.
29. Leonardi, P. et al. 1975.
30. Conti, M. A. et al. 1997.
31. Lee, M. 1994.
32. Lull, R. S. 1918.

The *Chirotherium* trackmaker was an erect-walking archosaur probably similar to *Tichinosuchus. Courtesy of Mauricio Anton.*

CHAPTER 3

DAWN OF THE MESOZOIC: THE EARLY AND MIDDLE TRIASSIC

THE STORY OF CHIROTHERIUM: THE DAWN OF THE ARCHOSAURS

The *Chirotherium* story is one of the most famous in the annals of fossil footprint research. It is also one of the oldest, longest, and most convoluted of paleontologic sagas. Part of the fascination of this story is that the morphology of *Chirotherium* tracks has an uncanny resemblance to the human hand, particularly that of the prominent fifth digit, then regarded as some sort of thumb (figure 3.1). Although we now know that these are the hind, not front footprints of Triassic reptiles, we must try to understand the problems our paleontologic forefathers faced in trying to decipher the Triassic track record with virtually no knowledge of the animals that existed at that time. In the late 1820s, tracks now assigned to *Chirotherium* were first reported from the vicinity of Tarporley and Storeton in Cheshire, England (figure 3.2), from Triassic strata, then known as the New Red Sandstone.[1] These tracks were not studied at that time, nor was it realized that they were fundamentally different from the Permian tracks recently discovered in Scotland (see chapter 2). This history has changed significantly over the years. Geoffrey Tresise and Bill Sarjeant[2] wrote a book on the subject of Triassic tracks from northwest England, and Mike King[3] wrote his Ph.D. thesis on the subject of Triassic tracks throughout all of England.

The story of these tracks really began in earnest in Germany, in 1834, the same year that the Triassic was named, when a man by the name of Helmut Barth no-

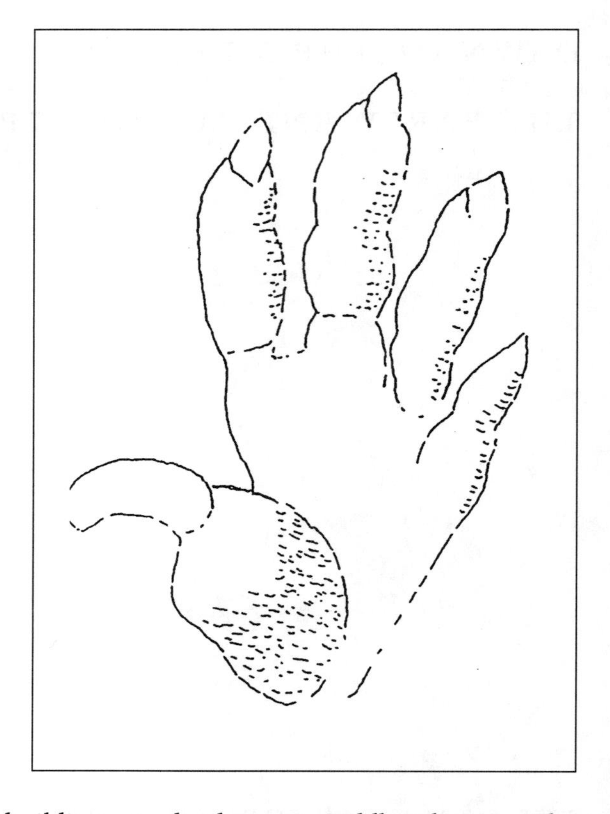

FIGURE 3.1. The classic *Chirotherium* footprint, now known to represent the hind foot of an archosaur, shows an uncanny resemblance to the print of a human hand.

ticed more of these tracks while building a garden house in Hildburghausen. The track-bearing sandstone was commonly used for building stone in all manner of famous Gothic and Renaissance structures, including the cathedral of Strassburg and the palace at Heidelberg. The new footprint finds came to the attention of various geologists and paleontologists, including Johann Jakob Kaup, who coined the name *Chirotherium*,[4] meaning "hand mammal." This mammalian interpretation was at first supported by the explorer Alexander von Humboldt, although after further collaboration with Kaup, he revised his opinion and considered the possibility that the trackmaker was a reptile. He therefore suggested the alternate name *Chirosaurus* (meaning "hand reptile"). Although one could argue that this is a better name, it has not stuck.

The story of *Chirotherium* has been told many times—for example, by Andrew Hamilton[5] and Herbert Wendt[6]—and a number of bizarre theories have been suggested to explain these petrified "hand" tracks. Friedrich Voigt believed that the tracks were of primate origin and even proposed the name *Paleopithecus* (meaning "ancient ape") for the purported trackmaker. After a change of heart he suggested that the large tracks were those of a cave bear and the small tracks were probably those of a mandrill. This was just the beginning of a series of hypotheses, most of which we now regard as ludicrous. Yet they were put forward by well-qualified and eminent scientists who appear to have given free reign to imaginative speculation.

FIGURE 3.2. *Chirotherium* track casts from original (1838) Storeton quarry. Left: Manus-pes set (left side). Right: Part of same trackway. *Courtesy of Mike King.*

William Buckland came out in favor of a mammal interpretation[7] when he suggested that the trackmaker was a marsupial, and pointed out that kangaroos have fore and hind feet of quite different sizes and a first toe "set obliquely to the others." Such bizarre and divergent hypotheses reflect a lack of knowledge of fossil vertebrates and the geologic time scale early in the nineteenth century. Like Humboldt, the British zoologist Sir Richard Owen also doubted the mammal trackmaker hypothesis and proposed that the footprints were probably those of a labyrinthodont amphibian or a reptile. The German geologist H. F. Link also suggested an amphibian trackmaker, but instead of a labyrinthodont, he proposed a species of giant toad. Jumping on the amphibian bandwagon was no less a geologic giant than the great Charles Lyell, who went so far as to create a reconstruction of the chirothere trackmaker.

Lyell's reconstruction can only be described as bizarre and unrealistic in the extreme. In order to explain the so-called "thumb" imprints on the outside of trackways, he inferred that the trackmaker crossed its legs over as it walked, setting its right foot on the left side of the trackway and its left foot on the right side

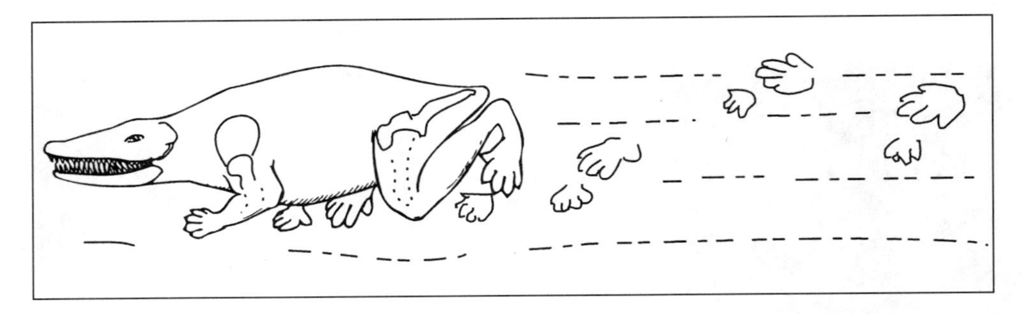

FIGURE 3.3. Charles Lyell's reconstruction of the *Chirotherium* animal shows it, in a very unflattering light, as an animal so inept that it crossed its feet over as it walked.

(figure 3.3). Lyell evidently believed that Triassic animals had been so "primitive" that they were incapable of walking normally. Such an extremely awkward gait is quite ridiculous. No modern animals are known to walk in such a clumsy fashion. Besides, *Chirotherium* trackways are very narrow, indicating an erect trackmaker. In hindsight we know that this indicates the most advanced form of locomotion exhibited in the Early and Middle Triassic.

A more reasonable interpretation of *Chirotherium* footprints was proposed by the British scientist Williamson, who in 1860 suggested that the trackmaker had been a crocodile. In the 1870s and 1890s, L. C. Miall and Albert Gaudry suggested that the trackmaker might have been a dinosaur. These interpretations put the purported trackmaker higher on the evolutionary scale and foreshadowed the continuation of the debate into the twentieth century. The young German student Karl Willruth was next to express an opinion. He concluded that the trackmaker had not walked cross-legged, but that it had possessed a fleshy appendage on the outside of the foot (digit V) that had made the thumb-like impression. He even went so far as to propose that various horseshoe-shaped markings were the isolated traces of the fleshy appendages, without any impression of the rest of the foot. We now know that they are simply current crescents caused by the scouring action of running water (figure 3.4). Willruth's professor, Johannes Walther, claimed that his clever student had solved the mystery, even though the identity of the trackmaker was not made explicit.

At this point, Baron Franz von Nopsca, one of the most interesting and brilliantly eccentric characters in the history of paleontology, enters the debate. A nobleman of Hungarian-Transylvanian descent, von Nopsca cut his paleontologic teeth by studying Cretaceous dinosaur bones discovered on the family estate, and soon became Hungary's leading geologist and a world-acknowledged expert on reptiles.[8] Von Nopsca's adventurous and restless mind took him in many different directions. He served as an intelligence agent in the Balkans and aspired to become king of Albania when the country was liberated from Turkey in 1913. To the bene-

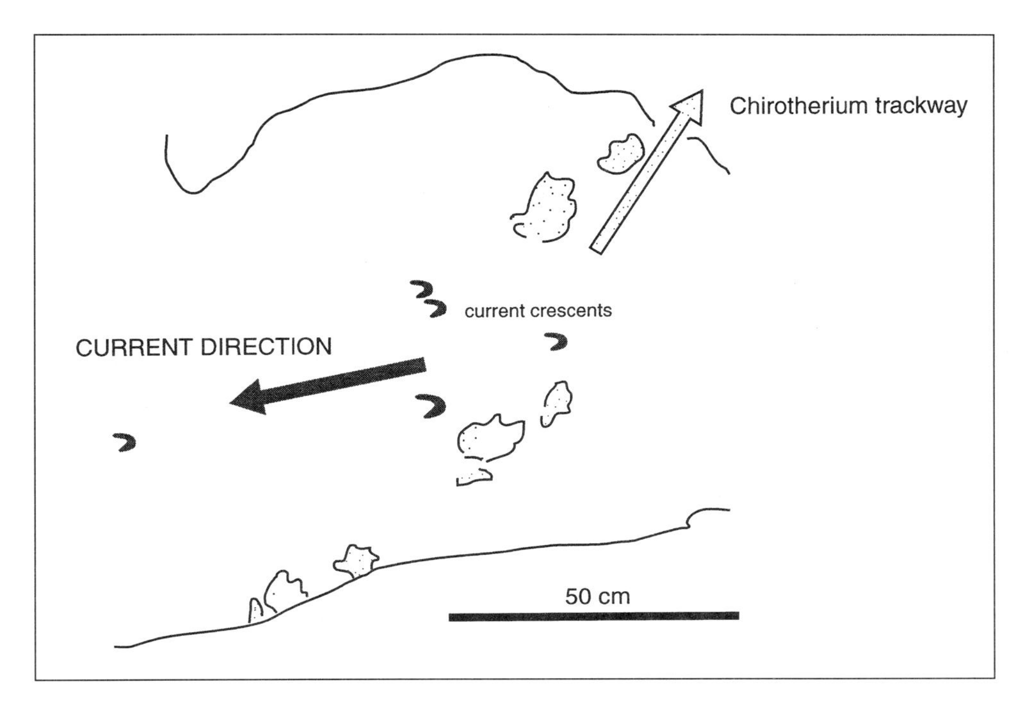

FIGURE 3.4. Crescent-shaped scour marks were once mistaken for the isolated traces of the chirotherium "thumb" and by some for a purely fleshy appendage.

fit of ichnology, he never became king and so returned to paleontology to write a seminal work on the fossil tracks of amphibians and reptiles, published in 1923. On the subject of *Chirotherium,* he proposed that the trackmaker resembled the large Triassic prosauropod *Plateosaurus,* which had recently been unearthed in Germany. Von Nopsca was evidently not too bothered by the fact that prosauropods had four toes on their hind feet, not five. Von Nopsca reconciled this inconsistency by citing Willruth's argument that the outer impression was simply that of a fleshy appendage. Despite von Nopsca's intellectual brilliance, many of his colleagues had serious reservations about this interpretation, not least because most of the German chirothere tracks were Early to Middle Triassic in age, whereas *Plateosaurus* was from the Late Triassic.

A century of debate brought a turning point in this long ichnologic saga with the publication in 1925 of a classic paper by Wolfgang Soergel,[9] "The Tracks of the Chirotheria." He made a convincing case that the "thumb" was simply the impression of a real fifth digit on the outside of the foot, not the trace of a hypothetical fleshy appendage or the trace of a digit on the inside of the foot. Moreover, Soergel identified the culprit as an archosaur allied to certain small pseudosuchian reptiles (meaning "false crocodiles") from the Triassic of South Africa. These creatures, best represented by the genus *Euparkeria,* had the right foot structure to fit chirothere

FIGURE 3.5. A reconstruction of the *Chirotherium* trackmaker by Wolfgang Sorgel. Compare with figure 3.6.

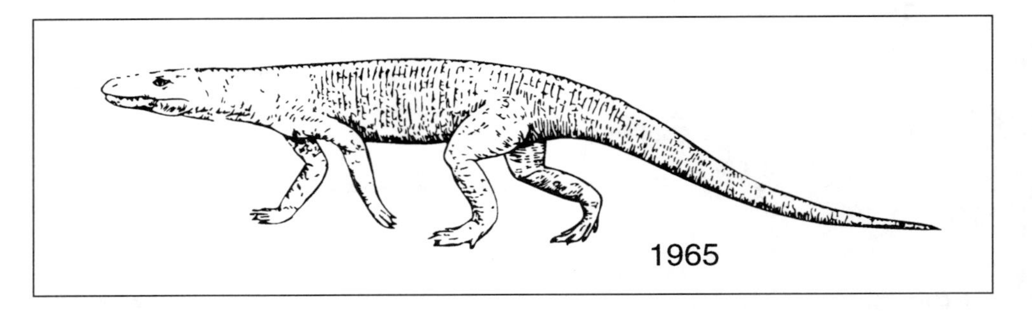

FIGURE 3.6. *Tichinosuchus* is generally regarded as the best candidate for the maker of the original *Chirotherium* (*C. barthi*) tracks. Compare with figure 3.5.

tracks, even though their feet were too small to fit the footprints of the larger chirotheres. The absence of skeletons of animals the size of the German trackmakers did not deter Soergel, and he confidently predicted that such large pseudosuchians, up to 5 m in length, must have existed. Soergel made a remarkably accurate reconstruction of the trackmaker, which was yet to be discovered (figure 3.5).

As paleontologic knowledge of Triassic reptiles improved, Soergel's interpretation appears to have been accepted, even by the eccentric von Nopsca. In the late 1920s, Friedrich von Huene, a colleague of Soergel's from Tubingen, discovered a large 5-m-long pseudosuchian in the Triassic of Brazil, which he named *Prestosuchus.* What is sometimes regarded as the final chapter in the debate over the identity of the *Chirotherium* trackmaker came in the 1960s, when another large pseudosuchian, named *Tichinosuchus,* was discovered in the Triassic of Switzerland. Here was an animal that was the right size to fit the large chirothere tracks and uncannily similar to Soergel's prophetic reconstruction (figure 3.6). After a century of debate and speculation ranging through a menagerie of possible trackmakers that included apes, toads, kangaroos, and amphibians, it finally became clear that chirothere trackmakers belong to the archosaurs, a tribe of reptiles that rose to prominence as the dominant tetrapods of the Triassic.

SEX IN THE FOOTPRINT BED

Identification of the *Chirotherium* trackmaker does not mean that the study of its tracks is complete. Indeed quite the opposite appears to be the case. *Chirotherium* tracks of many different varieties have been discovered all over Europe and on other continents, giving us a wealth of information to study. As we shall see in the following section, the number and diversity of chirotheroid tracks are such that several ichnologists have seriously proposed that it is easier to study the evolution of Triassic archosaurs through their abundant tracks than through their sparse skeletal remains.

It has also been suggested that some of the variety seen in chirotheroid tracks may be due to sexual dimorphism. In a provocatively titled article, "Sex in the Footprint Bed," published in 1995, Geoffrey Tresise proposed that the *Chirotherium* tracks at certain sites often fall into two distinct categories, slender and stout.[10] The slender tracks have been named *Chirotherium stortonense* (after Storeton on Merseyside, in England) and are tentatively attributed to the female, whereas the stouter tracks, referred to as *Chirotherium barthi* (after the German paleontologist Barth), are attributed to the male. Such interpretations are largely conjectural, though quite intriguing.

It would have been nice if Tresise had introduced further discussion on sexual dimorphism in modern reptiles before he began to talk so confidently about the possibility of detecting such differences in the fossil record. It is worth noting, however, that until the publication of this article, no one had seriously raised this intriguing idea in the realm of vertebrate ichnology. There has been the casual suggestion that small dinosaur tracks were made by females and larger ones by males. Tresise's hypothesis is subtly different. His suggestion is based on track shape, not just size, and specifically on morphologies that have already been assigned to two different species. In this regard we should note that not everyone agrees. The German ichnologist Hartmut Haubold had previously suggested that the English *Chirotherium stortonense* is simply a variety of the German *Chirotherium barthi*. Thus, the subject remains controversial. We believe that Tresise has at least pointed in the direction of potentially interesting research. Tied in with discussion of narrowness and breadth is the question of the size of the manus. It appears that in *Chirotherium*, as in some of their descendants, among quadrupedal dinosaur trackmakers there were both large and small manus varieties.

TRACKS AS KEYS TO EVOLUTION AND LOCOMOTION

The Triassic Period represents a span of approximately 40 million years. Allowing for a generation time of between 10 and 20 years, we can estimate that between 2 and 4 million generations of archosaurs came and went during this time. How many species existed during this long period is not precisely known, but allowing for the appearance of a new species or genus every few million years, we can predict that evolutionary change led to some change in the morphology of archosaurian feet. This is clearly seen in the track record, and although the literal tracking of evolution, using footprints, is difficult in detail, in a general sense it is possible to follow trends in the improvement of archosaur locomotion. By "improvement" we mean a general progression to more erect and digitigrade posture, as seen in increasingly narrow trackways and a reduced number of digits in contact with the substrate. The number of footprints made by individuals of many dozen different species must have been astronomical. Even though more than 99% of these footprints were never preserved, the remaining fraction of a percentage still represents a substantial sample available for study.

This is precisely what we see in the Triassic of Europe. Tracks are sufficiently abundant, and bones sufficiently rare, to allow for investigation of the track record as a source of new insights into archosaur evolution at this time. This problem particularly intrigued George Demathieu, who wrote a short paper entitled "*Contribution de l'ichnologie à la connaisance de l'évolution des reptiles pendant la période Triasique*" ("Contribution of Ichnology to the Understanding of the Evolution of Reptiles during the Triassic").[11] Similarly, Demathieu, in conjunction with Haubold, has written on the subject of the contribution made by footprints to our understanding of the origin of dinosaurs at this time.[12]

It is outside the scope of this book to begin to review all the available literature on Early and Middle Triassic tracks, but some general observations can be made.

Chirotherium, or *Chirotherium*-like tracks, are quite diverse in both size and morphology. This indicates a variety of archosaurian trackmakers that is greater than that represented by the skeletal record, and even without knowing which skeletal species made which track type, ichnologists have still identified and named distinctive track types and used them to compare track assemblages from site to site. For example, we know that "chirothere" tracks attributed to the ichnogenera *Chirotherium* (several ichnospecies), *Isochirotherium* (possibly *Brachychirotherium*), and the related form *Synaptichnium* are known from the Lower to the Middle Triassic of France, Germany, Switzerland, England, and Spain. Many of these tracks are the same as those found in the western United States and so are potentially useful for palichnostratigraphic correlation (see chapter 2).

We can also make further generalizations. In comparison with the trackways of Paleozoic reptiles, most *Chirotherium* trackways are narrow, indicating animals with an erect gait. Chirothere trackways are characterized by considerable differences between the size of the larger hind footprints and smaller front footprints. This phenomenon, known as "heteropody," allows us to assess the extent to which the weight was distributed over the hip and shoulder girdles. In popular parlance we can talk of the front and rear axles, or speak more technically of the pelvic and pectoral girdles. Chirothere trackways generally indicate that most of the animal's weight was distributed over the rear axle or hips, an observation made by Soergel in 1925.[9] As noted by the American ichnologist Frank Peabody, however, the chi-

FIGURE 3.7. *Isochirotherium archaem* appears to be the trackway of a bipedal animal. If so, it is the world's oldest example.

rotheres fall into a large manus group and a small manus group[13] and in this regard are similar to some dinosaur groups.

The tracks also indicate that the front foot was more fully digitigrade than the plantigrade hind foot. Such evidence suggests animals that were already on their way toward becoming bipedal. Indeed, one Early Triassic trackway from Germany, described by Demathieu and Haubold in 1982 under the name *Isochirotherium archaem,*[14] appears to be that of a bipedal animal (figure 3.7). Hartmut Haubold has warned us that this particular trackway is somewhat weathered, so it is possible that the small front footprints are missing due to poor preservation. The trackway may be the oldest evidence in the world for the origin of bipedal locomotion, although ichnologists cannot claim absolute proof of this supposition. There are other possible explanations for the absence of front footprints. For example, they may have been overprinted by the hind feet. Nevertheless, the track evidence shows that bipedal animals, probably true dinosaurs, evolved before the end of the Triassic (see chapter 4).

Another aspect of archosaur locomotion that can be investigated through the study of footprints involves the evolution of the ankle. This again is a complex though very important subject, for ankle morphology plays an important role in the classification of archosaurs. Simply put, the degree of outward rotation of the foot, as measured by the angle made between the trackway line and the long axis of the foot, along the middle toe (digit III), is an indication of ankle morphology (fig-

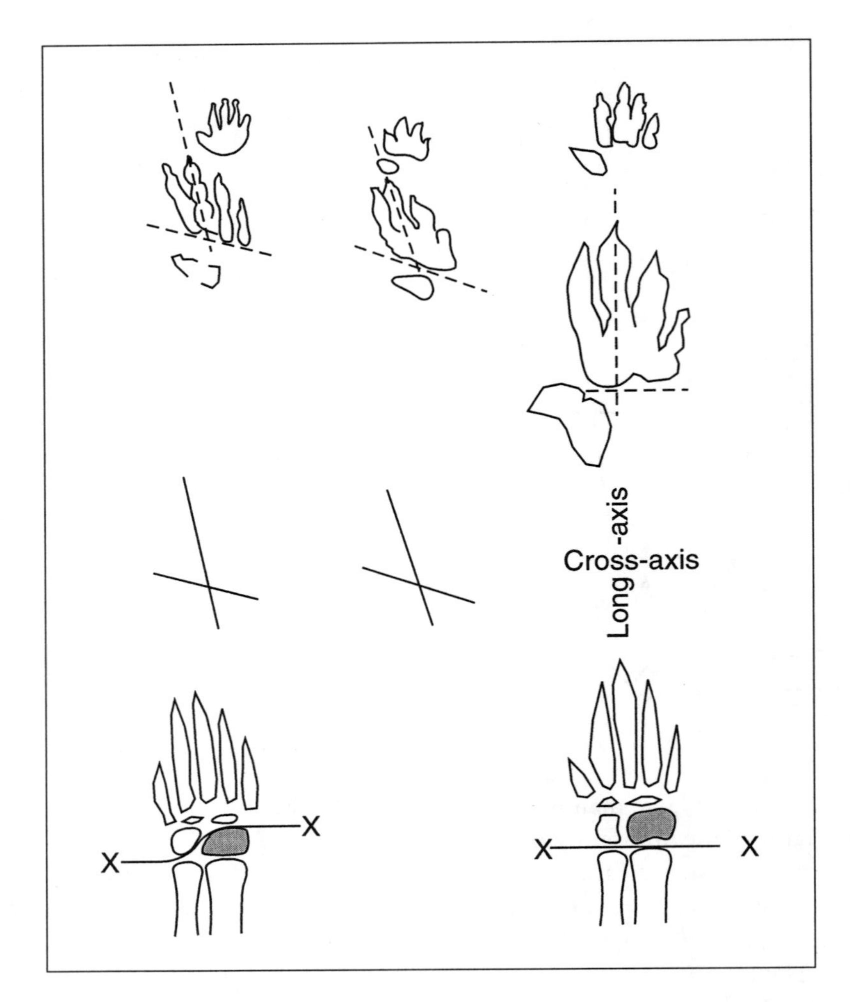

FIGURE 3.8. The long axis and cross-axis of footprints are an indication of the orientation and structure of the ankle.

ure 3.8). The foot also has a corresponding cross-axis measured along the line of the metatarsal joint (the posterior impression of digits I to IV). Trackmakers can be categorized according to the degree to which the long axis of their foot is parallel to the trackway axis or direction of travel. This, in turn, is an indication of the way in which the ankle hinged during locomotion. As has been stated another way by Michael Parrish, there is a trend in archosaur evolution toward a greater anterior alignment of the foot.[15] Along with this trend, we see the foot becoming more symmetrical. In other words, the central digit (III) becomes longer than digit IV (figure 3.8). In short, the outward splay of the more primitive foot is reflected in a more pronounced development of the outer digit (notably IV), whereas the anterior alignment, which brings the foot axis into alignment with the trackway axis, is reflected by the pleasing symmetry of the foot about the central digit (III).

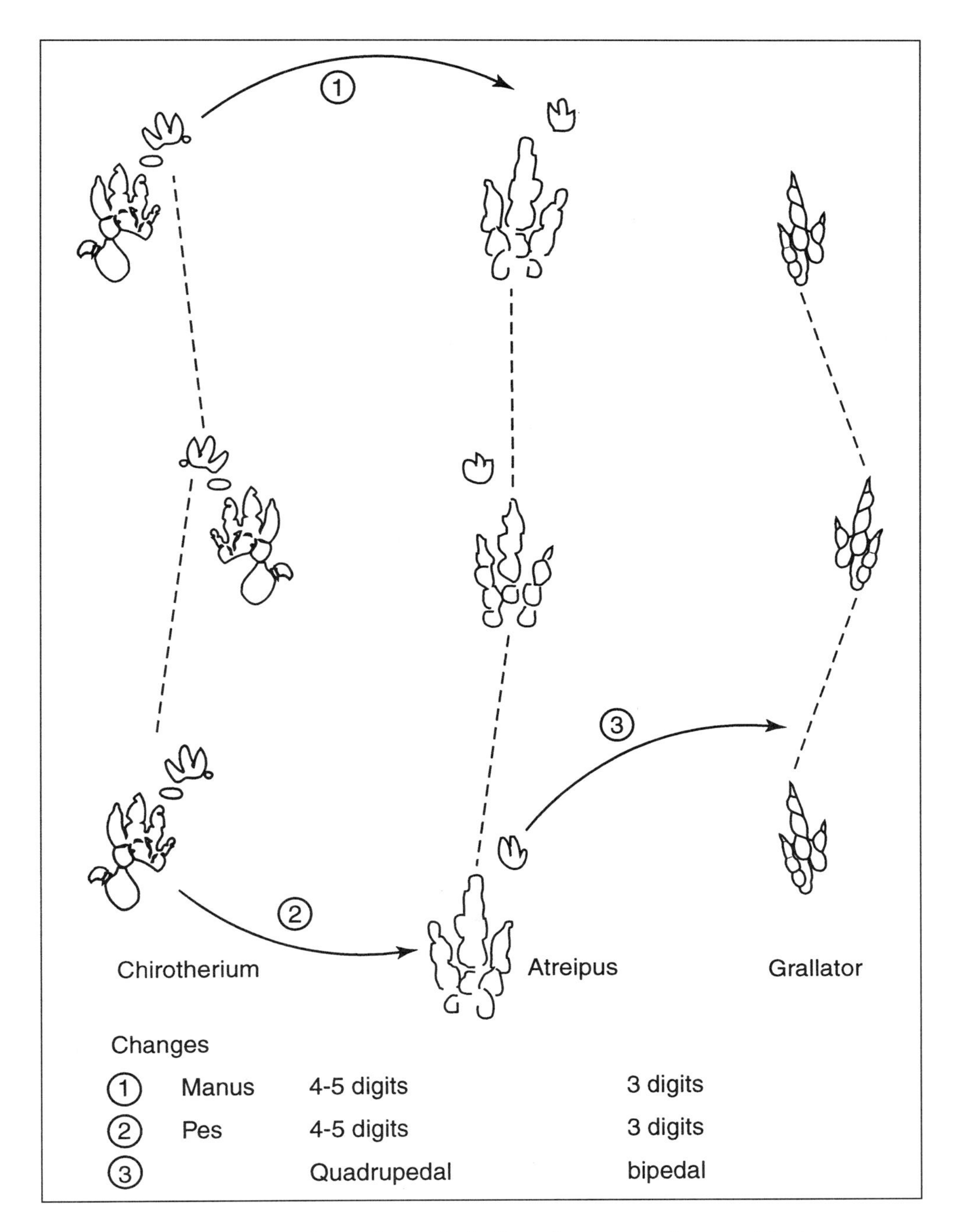

FIGURE 3.9. Comparison of a typical *Chirotherium* trackway with *Atreipus* and *Grallator* reveals a trend toward reduced plantigrady and bipedalism.

It is instructive to compare chirotheroid footprints with the tracks of other Triassic archosaurs, including both quadrupeds and bipeds. What such comparisons reveal is that during the Triassic there was a general trend in the reduction of the number of toes that made contact with the ground in both the manus and the pes. By Early Triassic times there were already chirotheroid trackmakers belonging to a small-manus group. In comparison with their large-manus contemporaries, they may have carried less weight on their front feet. By Late Triassic times we find track types such as *Atreipus,* which consist of a three-toed hind foot and a small three-toed front foot (figure 3.9). The front footprint is chirothere-like except that it reveals a reduction in functional digits from five or four down to three. The pes print is similar to dinosaur tracks such as *Grallator.* There has been debate as to whether *Atreipus* was made by a dinosaur or by some other archosaur. Regardless of the affinity of the trackmaker, such footprints are intermediate between typical chirotheroid tracks and those of three-toed dinosaurs. This does not mean that we can arrange these three track types in a sequence and say that the trackmakers formed a known evolutionary lineage. We may, however, infer that in terms of locomotor evolution *Atreipus* represents an intermediate status (figure 3.9).

LIZARD ANCESTORS AND PROTO-MAMMALS WITH HAIRY FEET

Although the Triassic has been referred to as the Age of Archosaurs by many paleontologists, we should not leave the reader with the impression that all tracks (chirotherian and dinosaurian) are attributable to this group. Indeed, there are many sites where other, quite different tracks have been reported. For example, lizard-like tracks assigned to the ichnogenus *Rhynchosauroides* (figure 3.10) are common in the Triassic and have been reported from a number of localities. These tracks are attributed to the nonarchosaurian reptiles known as "lepidosaurs," which include present-day lizards and snakes. One of the best examples of a site yielding excellent specimens of this ichnogenus is the Winterswijk locality in the eastern Netherlands,[16] where some of the tracks are so well preserved that they exhibit skin impressions. Also reported from this locality is the ichnogenus *Procolophonichnium,* which is attributed to a stem reptile.

Another nondinosaurian group that was abundant in the Early and Middle Triassic was the proto-mammals, or mammal-like reptiles or synapsids, whose tracks we encounter in abundance in Permian deposits representing arid paleoenvironments. Such tracks are less common in the Triassic, and they are generally significantly different. For example, in North America, tracks attributable to the synapsid group known as the "therapsids," and named *Therapsipus,* are the size of the footprints of a medium-sized rhinoceros. In Europe some of the most distinctive therapsid tracks have been named *Dicynodontipus* (meaning "dicynodont track") (figure 3.11) from the Bunter sandstone of the Thuringian Forest region.[17]

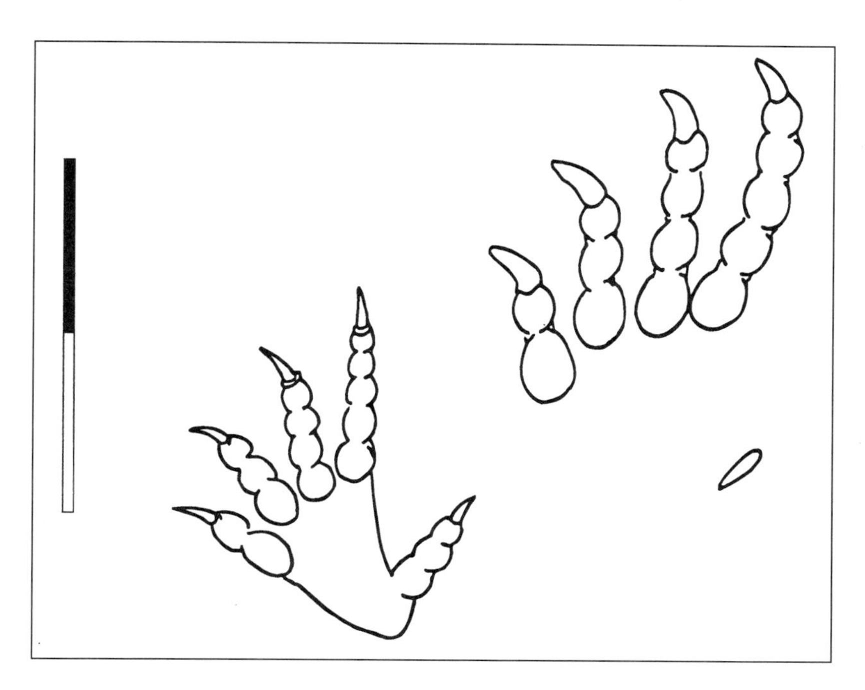

FIGURE 3.10. *Rhynchosauroides* tracks from the Middle Triassic of the Netherlands. *After Demathieu and Oosterink.*[16] Scale bar 5 cm.

According to French ichnologist Paul Ellenberger, an isolated and incomplete track from the Early to Middle Triassic of southern France revealed the footprint of another large proto-mammal, "a large but still unknown hairy vertebrate." He named the track, measuring about 20 cm in diameter, *Cynodontipus polythrix* (meaning "hairy footprint of a cynodont"),[18] and claimed it to be the left hind foot. Knowing that few mammals have hairy soles on their feet, he compared the track with that of one of the few groups that do: rabbits. Obviously a trackmaker with feet the size of a cart horse has little in common with a rabbit, but perhaps any comparison is better than none. The possibility of large proto-mammal trackmakers in the Early to Middle Triassic has been confirmed by the discovery of *Therapsipus* from North America, mentioned above. However, Ellenberger has to be credited with considerable imagination to propose this interpretation on the basis of a single incomplete footprint. Interpretation of the coarsely textured markings or striations as traces of hair is open to serious question, not least because there is no definitive evidence for hair on these proto-mammals. There are many possible causes of such markings. One possibility is horseshoe crabs, which can produce horseshoe-shaped crescentic indentations about the size of the toe impressions illustrated by Ellenberger.

One of the most enigmatic tracksites reported from the Triassic is the Vieux Emosson locality in Switzerland. This site was described, in 1982, by George Demathieu and Marc Weidmann.[19] Demathieu has considerable experience with Tri-

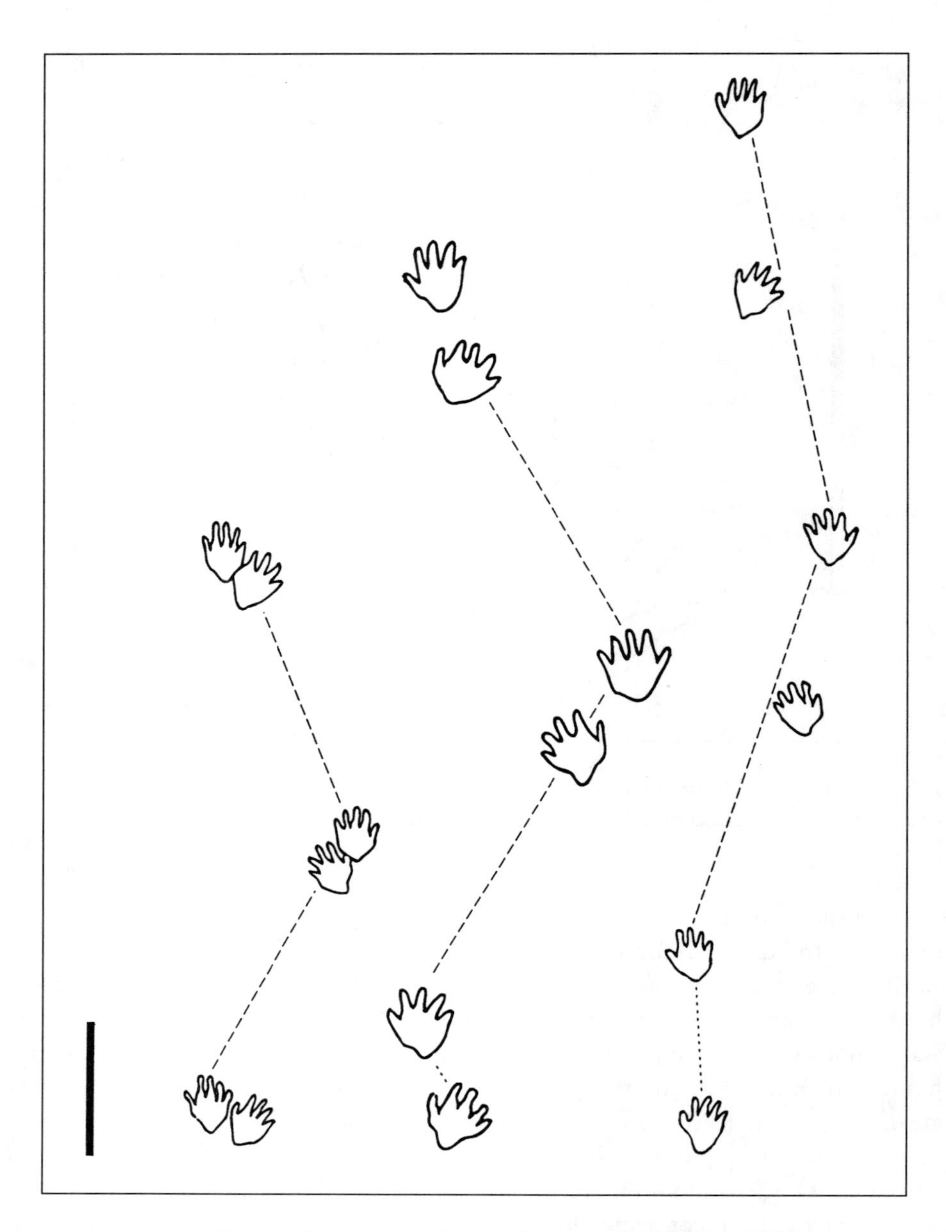

FIGURE 3.11. *Dicynodontipus,* a "dicynodont" or therapsid track from the Lower Triassic of Germany. *After Haubold.*[17] Scale bar 10 cm.

assic tracks and has published extensively on the subject, and Weidmann has done some good work on Tertiary tracks from Switzerland (see chapter 10). Their study of the Vieux Emosson site is somewhat anomalous, because nine ichnospecies are named on the basis of material that is rather dubious, at least in some cases. Some ichnogenus names such as *Brachychirotherium* and *Isochirotherium* are familiar, but several are new, and several are based on names coined by Ellenberger in his study of tracks from the Late Triassic and Early Jurassic of southern Africa.[20] Since the study was published, in 1982, these ichnogenera have not been reported from other sites, suggesting that the names have not been widely accepted as representative of distinctive morphotypes that are recognizable elsewhere. For example, *Bifidichnium ambiguum* (literally "ambiguous bifed trace") looks like the track of a cloven-hoofed ungulate (artiodactyl) and may be some sort of artifact of preservation. The authors admit that details of the tracks are no longer visible in many cases. We also know of no Early or Middle Triassic tracksites where nine ichnospecies are found in close proximity, and certainly none with these ichnospecies. We therefore regard the interpretation of this site with some skepticism.

THE WORLD'S OLDEST DINOSAUR TRACKS: FACT, FICTION, AND CONTROVERSY

As indicated in the previous section, the world's oldest dinosaur tracks appear to be attributable to bipedal animals. As such, the origin of bipedal dinosaurs is a significant event in the evolution of archosaurs and the general development of locomotion among tetrapods. Currently, skeletal evidence suggests that the oldest dinosaurs are Late Triassic in age. However, the oldest tracks that have been attributed to dinosaurs are Early and Middle Triassic in age. This raises the following multiple working hypotheses:

1. Dinosaurs existed in the Early and Middle Triassic, even though no bones have been found.
2. Dinosaurs did not exist until the Late Triassic, so dinosaur-like tracks from the Early and Middle Triassic have been misidentified.
3. Dinosaur-like tracks from the Early and Middle Triassic were made by nondinosaurs that had feet resembling those of dinosaurs.

There is also a fourth possibility, that the age of the track layers has been incorrectly assessed in some way. This is a very real possibility in the study of terrestrial deposits, which are known to be hard to date in many cases, and the possibility of Middle Triassic dinosaur bones has been proposed based on overestimates of the age of South American bone-bearing deposits now regarded as Late Triassic in age. For the purposes of this discussion, however, it does not appear that the age of the track-bearing beds is in serious dispute.

All three scenarios are possible, either individually or in any combination (1 and 2, 2 and 3, or 1 and 3). The conservative hypotheses (2 and 3) are what we might call bone-oriented viewpoints, based on the negative evidence of lack of bones older than Late Triassic age. Hypothesis 1 is a track-oriented viewpoint, supported, in principle, by other examples of tracks of major groups that predate their equivalent bone record but compromised by the lack of absolute proof that the tracks were made by dinosaurs rather than some other archosaurs (hypothesis 3).

We should look first at purported dinosaur tracks from the Early Triassic of England. These were first described, in 1970, by Leonard Wills and Bill Sarjeant from deposits in Worcestershire and Nottinghamshire at a time when the precise age of these Triassic deposits and the precise age of the oldest dinosaur were not known.[21] Among these were large, purportedly three-toed tracks assigned to the ichnogenus *Swinertonichnus*, large four-toed tracks assigned to *Otozoum*, and various three-toed tracks assigned to *Coelurosaurichnus* species. The dinosaurian affinity of all three of these track types has recently been questioned by Tony Thulborn, Mike King, and Mike Benton, and in some cases there appears to be doubt as to whether certain specimens are even tracks at all!

In 1990, Thulborn[22] was the first to question a purported Lower Triassic specimen *Coelurosaurichnus* species B from a borehole core penetrating the Lower Triassic Bunter sandstone when he suggested that it was the track of a horseshoe crab or "limulid." King and Benton[23] acknowledge this possibility but also suggest that the trace may be nothing more than an inorganic marking or sedimentary structure. They also contest the identification of other *Coelurosaurichnus* tracks from this borehole by explicitly identifying them as nonbiogenic sedimentary features such as mud flakes. Such interpretations, if correct, remove any shred of evidence of dinosaurs from the Early Triassic of England. However, this was not the end of the discussion.

In a paper published in 1996, Sarjeant[24] reclassified these tracks as follows. *Coelurosaurichnus* species A and B, respectively, are assigned to the ichnogenera *Batrachopus* and *Plectoperna*. This reinterpretation also implies that these purported tracks are no longer regarded as dinosaurian, for *Batrachopus* is usually interpreted as crocodilian in origin and *Plectoperna* is a poorly known ichnite of uncertain affinity. Both are of Jurassic age, which again makes Sarjeant's reinterpretation rather surprising. For those not intimately familiar with vertebrate ichnology, we should stress that it is unusual to find debate over whether or not tracks are really tracks or simply sedimentary features. An obvious implication of any such debate is that the material is not easily identified as a track and so should be labeled with caution. Although we have not examined this material firsthand, we have examined good-quality photographs and agree with King and Benton that the material represents sedimentary features, not tracks.

The controversy continues as we examine an assemblage of Middle Triassic tracks from Mapperley Park in England. The short version of the story is that tracks originally described by Sarjeant as *Swinertonichnus* and *Otozoum*, and assigned to dinosaurian trackmakers, have been reinterpreted by King and Benton[23] as *Chirotherium* footprints. Another *Coelurosaurichnus* specimen described by Sarjeant

FIGURE 3.12. Reinterpretation of purported tridactyl dinosaur track *Swinertonichnus* as chirotheroid removes any evidence of pre–Late Triassic dinosaurs from England. *Courtesy of Mike King.*[23]

was reinterpreted by King and Benton as mud flakes. Sarjeant countered with his own 1996 reassessment of the ichnofauna, describing *Swinertonichnus* as "crocodilian," *Otozoum* as *Paratetrasauropus,* and the purported mud flakes as *Chirotherium.*[24] (*Paratetrasauropus* was described by Ellenberger from younger, Late Triassic to Early Jurassic deposits of southern Africa, and has been variously attributed to prosauropods or crocodilians.) Again, this reinterpretation of the Mapperley material suggests an absence of compelling evidence for Middle Triassic dinosaurs and highlights the problems associated with trying to identify incomplete or poorly preserved specimens. For example, *Swinertonichnus,* according to King and Benton, is a broken *Chirotherium* track, reduced from a five-toed to a three-toed specimen (figure 3.12).

Just when ichnology is beginning to fall in line with the skeletal record and suggest a lack of compelling evidence for dinosaurs prior to the Late Triassic, we find that the evidence is not quite so convergent. Tracks reported in 1989 by Demathieu[25] from the Middle Triassic of France, in the region of the Massif Central, are extremely dinosaurian in appearance and are so well preserved that there can be no doubt that they are unequivocal vertebrate footprints. It also appears that the Middle Triassic date for the track-bearing layers is accurate.[26] Furthermore, the footprints are preserved as part of a narrow trackway that is similar to typical *Grallator* trackways from late Triassic and Jurassic deposits. Moreover, the tracks have

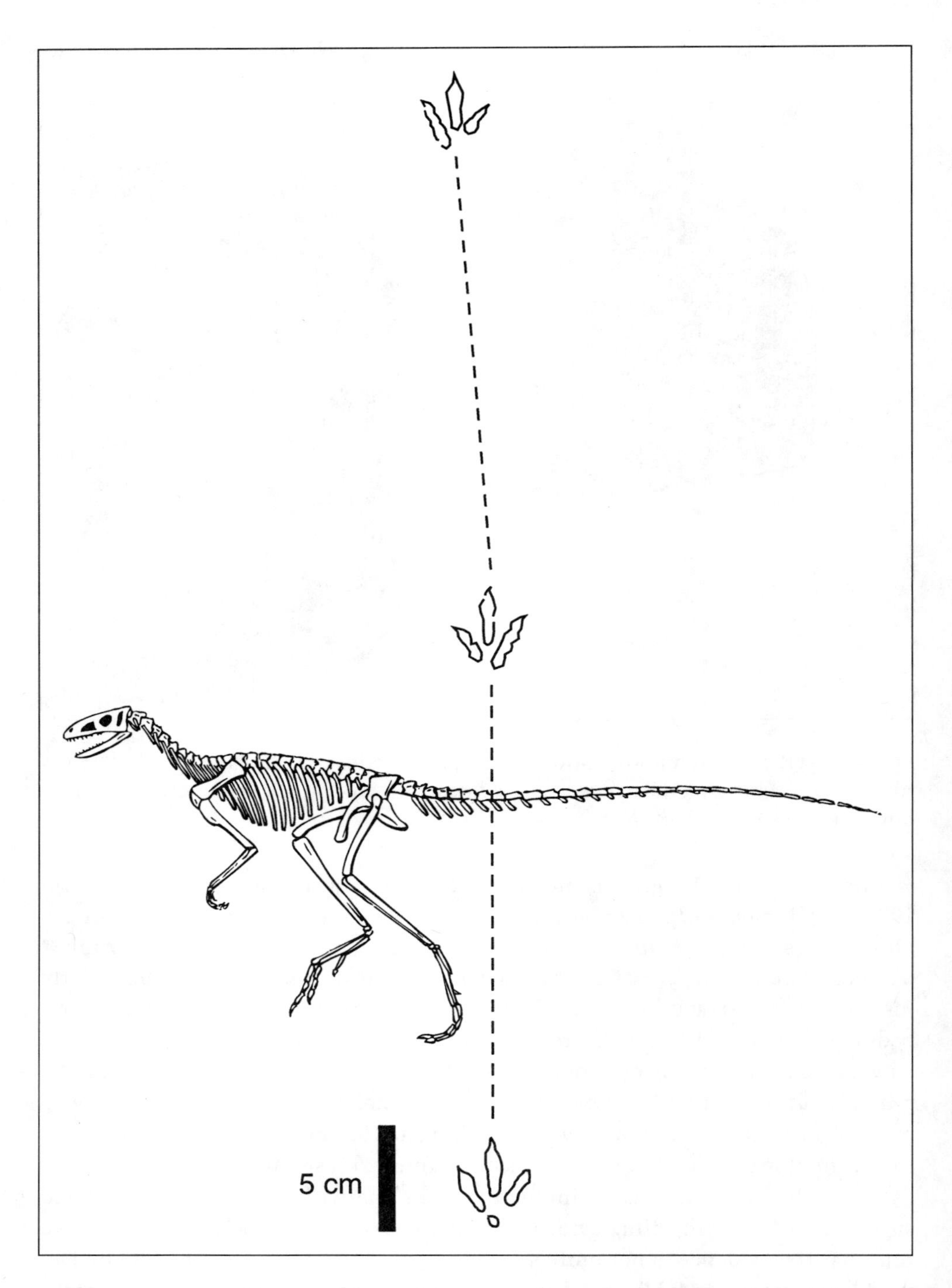

FIGURE 3.13. Dinosaur-like or "dinosauroid" tracks from the Middle Triassic of France may have been made by a dinosaur ancestor such as *Lagosuchus.*

been assigned to *Anchisauripus,* an ichnogenus implying dinosaurian affinity. At first, Demathieu appeared cautious when using the term "dinosauroid tracks" rather than "dinosaur tracks." But, following the argument that if it looks like a dinosaur track, then there is a good possibility that it is a dinosaur track, Demathieu subsequently stated that "All . . . considerations led [him] to infer that the dinosauroid tracks of these areas were made by dinosaurs." Although most vertebrate paleontologists probably prefer to adhere to the cautious bone-oriented hypothesis regarding dinosaur origins (in the Late Triassic), the track-oriented hypothesis of a somewhat earlier origin cannot be ruled out. Even so, we are inclined to point out that hypothesis 3 is very probable; that is, the tracks may have been made by nondinosaurian archosaurs with very dinosaur-like feet. Animals like *Lagosuchus* or *Lagerpeton* known from the Middle Triassic of South America make very suitable candidates (figure 3.13).

FUTURE DIRECTIONS

As indicated in previous chapters, European researchers have suggested that Late Paleozoic (Permian) and Early Mesozoic (Triassic to Early Jurassic) tracks are useful for correlation. Not everyone agrees with this view, and some subscribe to the rather obvious and extreme view that tracks are really useful only where other fossils are absent. We take the view that they always provide some information worth considering.

We have not made an attempt in this chapter to discuss the age of different track assemblages from England, Germany, and elsewhere. We also have not made a detailed analysis of the sedimentary facies in which these tracks occur. However, such analyses can be undertaken successfully at least in local areas. For example, Ryszard Fuglewicz and his colleagues[27] reported what they claim to be the oldest Triassic track assemblages known from Europe. These deposits from the Holy Cross Mountains of Poland contain tracks assigned to the ichnogenera *Capitosauroides* (indicating amphibian affinity), *Chirotherium, Isochirotherium, Brachychirotherium, Synaptichnium,* and *Rhynchosauroides.*

In addition to reporting possibly the oldest track assemblage from the Triassic, Fuglewicz and colleagues[27] made efforts to place the tracks in the context of the ancient Early Triassic paleoenvironment. They reported that the tracks occur at six different levels in a sequence representing river channel deposits, crevasse splay flood deposits, and floodplain suspension sediments, which accumulated further from the channels. The tracks occur mainly in the latter deposits where floods have deposited sediments some distance from the main channels. Studies of current directions in the beds overlying tracks were correlated with trackway trends to show that animals were frequently moving along, or parallel to, the axes of channels (figure 3.14). This is perhaps to be expected and has been noted in many studies of tracksites in North America.[28]

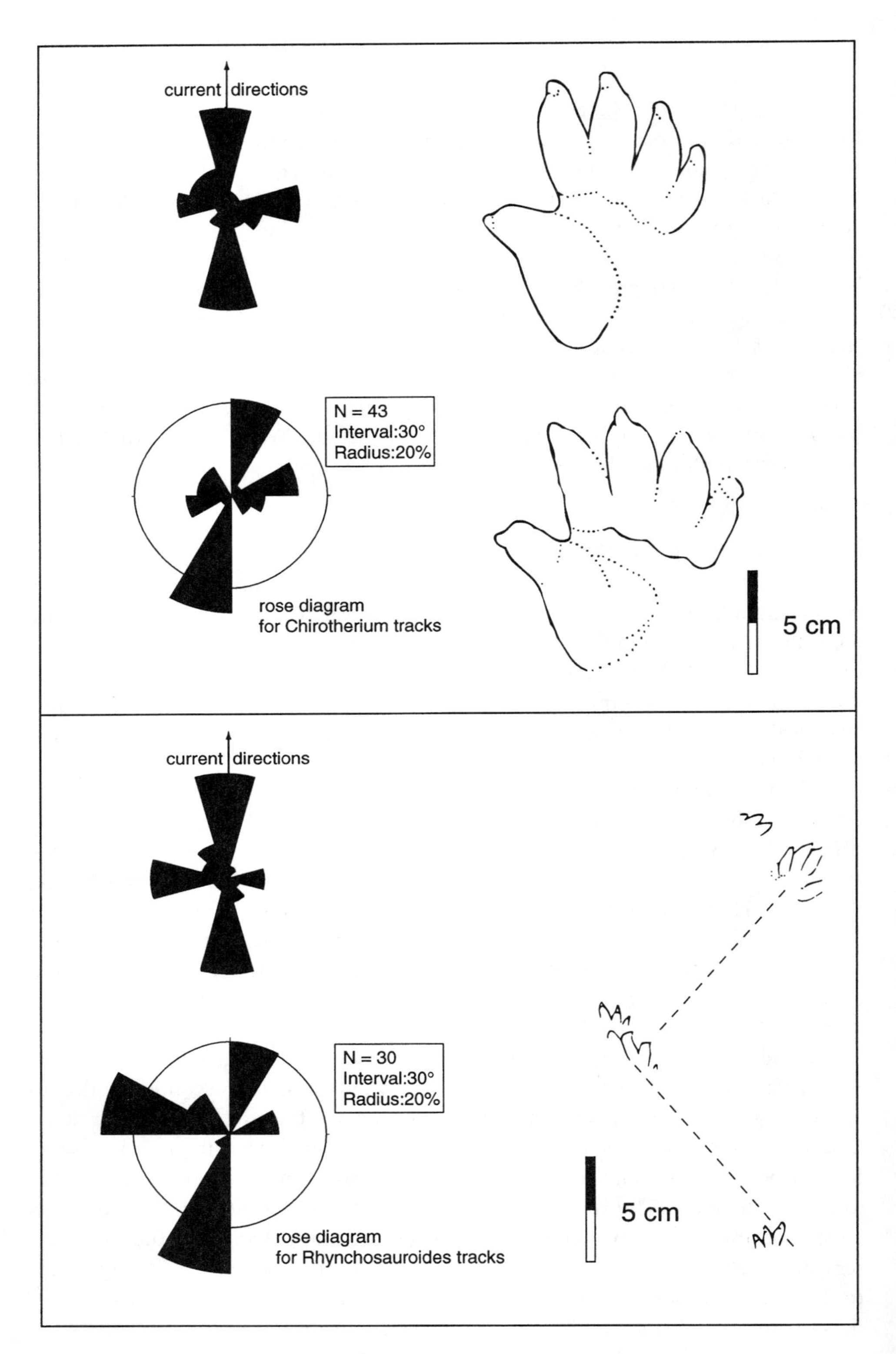

FIGURE 3.14. The orientation of chirothere and *Rhynchosauroides* trackways indicates a preferred direction parallel to river trends (after Fuglewicz et al.[27]). Based on the Triassic of the Holy Cross Mountains, Poland.

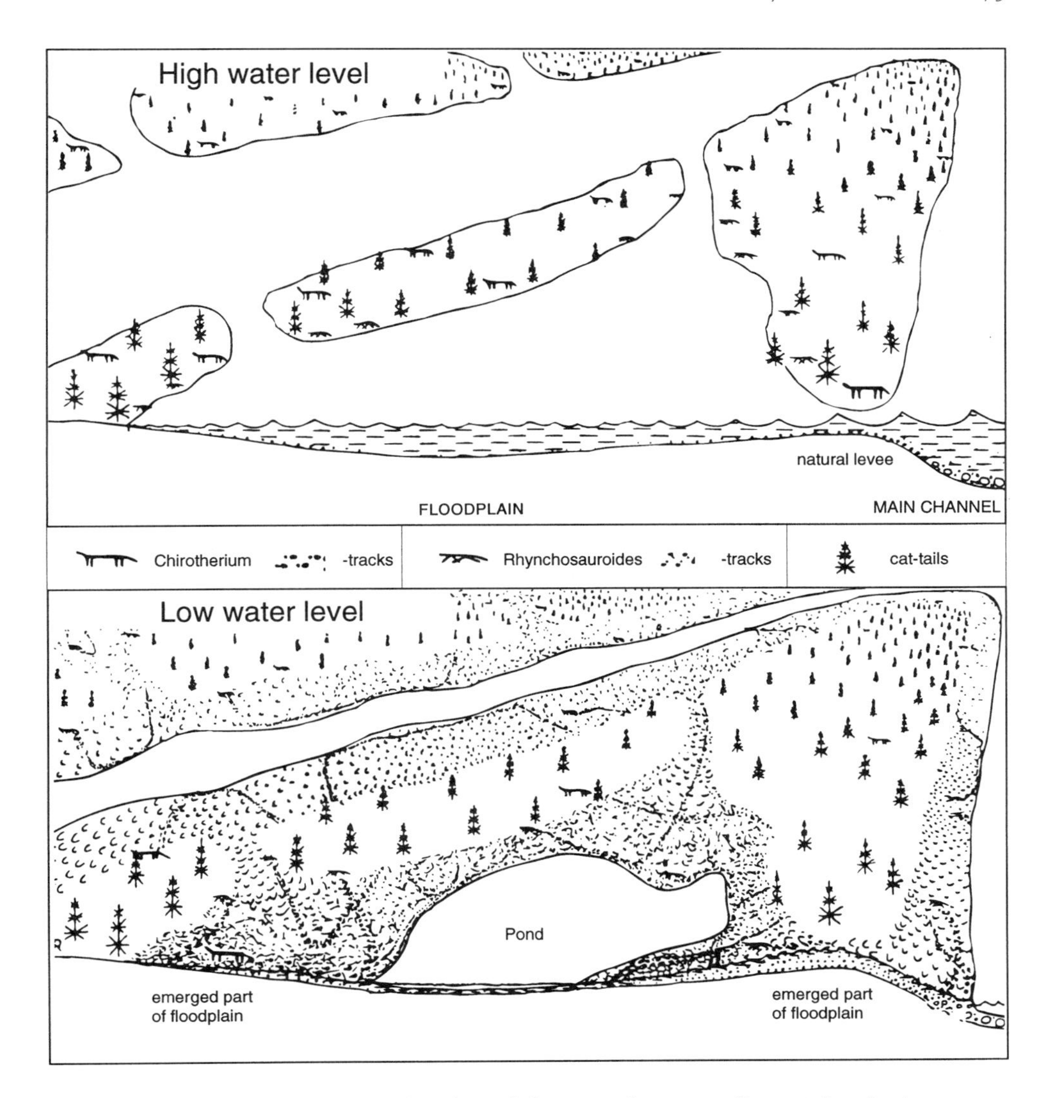

FIGURE 3.15. Careful study of track-bearing deposits allows ichnologists to reconstruct ancient environmental conditions such as low and high water at the time of trackmaking activity. Here the zones or areas where chirothere and *Rhynchosauroides* tracks were made is shown. *After Fuglewicz et al.*[27]

Finally, Fuglewicz and colleagues[27] attempted to reconstruct the ancient environment at times of both low water level and high water level (figure 3.15). Their findings indicated that most footprints were registered at times of low water level, after flooding, as might be expected. Such an integrated study is highly commendable and shows the need for a holistic approach to both the trackmaker and the environment in which it was active. In geologic terms, this means studying the footprints in their sedimentological context.[28]

It is only when several such studies have been conducted in track-bearing deposits of the same age that we will be able to determine the extent to which track types are associated with particular ancient environments. As stated in chapter 1, the distribution of tracks in space and time depends on at least two factors: (1) the age of the deposit, and hence the evolutionary stage of the fauna that existed at that time, and (2) the type of environment, which controls the types of animals which chose particular habitats. Both of these factors can be studied by carefully recording the distribution of bones and tracks in space and time. Both vertebrate ichnology and mainstream vertebrate paleontology are moving in the direction of trying to establish biostratigraphic zones based on recognizable assemblages of tracks and skeletal remains. We can confidently predict that future studies will allow for greater refinement in the zonation of Triassic-age continental deposits, as well as those from other time periods. Discussion regarding whether track zones or zones based on skeletal remains (sometimes called "land vertebrate ages") are most accurate depends on the deposits under consideration to some extent. Ultimately, all information is useful, so both tracks and bones contribute to a better understanding of the ancient vertebrate world.

BIBLIOGRAPHIC NOTES

1. Sarjeant, W. A. S. 1974.
2. Tresise, G. and Sarjeant, W. A. S. 1997.
3. King, M. 1998.
4. Kaup, J. J. 1835.
5. Hamilton, A. 1952.
6. Wendt, H. 1968.
7. Buckland, W. 1858.
8. Von Nopsca, F. 1923.
9. Soergel, W. 1925.
10. Tresise, G. 1996.
11. Demathieu, G. 1970.
12. Demathieu, G. and H. Haubold. 1978.
13. Peabody, F. 1948.
14. Demathieu, G. and H. Haubold. 1982.
15. Parrish, M. J. 1989.
16. Demathieu, G. and H. W. Oosterink. 1983.
17. Haubold, H. 1984.
18. Ellenberger, P. 1976.
19. Demathieu, G. and M. Weidmann. 1982.
20. Ellenberger, P. 1972.
21. Wills, L. J. and W. A. S. Sarjeant. 1970.
22. Thulborn, R. A. 1990.

23. King, M. and M. T. Benton. 1996.
24. Sarjeant, W. A. S. 1996.
25. Demathieu, G. 1989.
26. Courel, L. and G. Demathieu. 1995.
27. Fuglewicz, R. et al. 1990.
28. Lockley, M. G. 1991.

Small *Coelophysis*-like theropod dinosaurs and a large prosauropod make tracks on alluvial sediments that once existed in the area around what is now Cardiff, South Wales. *Courtesy of John Sibbick.*

CHAPTER 4

THE FIRST DINOSAURS: THE LATE TRIASSIC EPOCH

WELSH DINOSAURS AT THE JOLLY SAILOR PUB

Wales is not known for its dinosaurs or indeed for much in the way of vertebrate remains. In fact, as the type locality for all three Lower Paleozoic systems (Cambrian, Ordovician, and Silurian), it is paleontologically most famous for marine invertebrate fossils, including trilobites, brachiopods, and graptolites. Nevertheless, as early as the 1870s a well-preserved set of dinosaur tracks was set in front of the steps of the Jolly Sailor Pub at Newton Nottage, near Porthcawl in South Wales. These tracks were subsequently removed to a location in front of the church (perhaps they were a hazard to those frequenting the pub!), and later, in 1879, were moved to the Cardiff Museum, now the National Museum of Wales.

This trackway from the Jolly Sailor is Late Triassic in age and represents a theropod dinosaur, provisionally labeled "ichnogenus *Anchisauripus*" (figure 4.1). It was first reported in separate papers by T. H. Thomas[1] and W. J. Sollas,[2] and in Thomas's words were "the first of the kind observed on this side of the Atlantic . . . [showing a] . . . relation to the 'Ornithichnites' of the Connecticut new red sandstone." Records from this time indicate that a four-toed track, which we now assign to the ichnogenus *Otozoum* (meaning "giant animal") (figures 4.2 and 4.3), was also recovered from this same area. This is also a track type known from the Connecticut Valley, although from the Lower Jurassic, not the Late Triassic. Although not studied further at the time, the specimens are of considerable significance. *Otozoum* and *Otozoum*-like tracks, for example, are generally attributed to prosauropods.

FIGURE 4.1. The trackway of a theropod dinosaur (compare with *Anchisauripus*) once graced the steps of the Jolly Sailor Pub, at Newton Nottage in South Wales.

In 1977, almost a century after the initial discoveries, several hundred footprints from the Late Triassic were reported at a site near Bendricks Rock in the Cardiff area. These footprints consist predominantly of small dinosaur tracks, about 10 cm long, that we assign to the ichnogenus *Grallator* (figure 4.4). They were found in a large concentration in a relatively small area, and owing to the danger of being lost to theft and erosion, the majority were excavated and taken to the National Museum of Wales, where they are on display. Isolated tracks and trackways have been found in other areas, and in one case a long stride indicates an animal running at an estimated speed of more than 20 km per hour. Although we do not know which dinosaur species made this type of *Grallator* track, the foot of

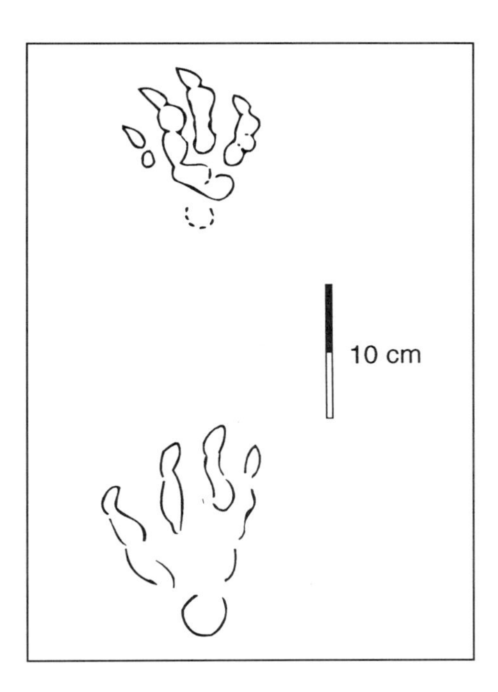

FIGURE 4.2. The distinctive track type *Otozoum* (top) from Wales, compared with Ellenberger's ichnogenus *Pseudotetrasauropus* (below); also compare with figure 4.3.

Coelophysis, a well-known contemporary theropod from North America, fits the footprint well.

The footprints were first described by M. E. Tucker and T. P. Burchette[3] from what has traditionally been described as the Keuper Marls portion of the New Red Sandstone but is now formally classified as the Mercia mudstone group. Tucker and Burchette placed these tracks in their paleoenvironmental context by suggesting that they were made on sheet flood deposits associated with the margins of alluvial fans that accumulated at the base of local uplands (see reconstruction). Further work by the senior author and his colleagues[4] in the 1990s has identified the presence of other archosaur track types assigned the provisional labels *Tetrasauropus* and *Pseudotetrasauropus,* both of which might be of prosauropod origin.

This relatively diverse track assemblage is particularly significant for several reasons. First, it reveals that small dinosaurs were abundant by Late Triassic times, a conclusion that is in line with skeletal evidence (see chapter 3). This phenomenon of small *Grallator* abundance was first noted by Hartmut Haubold in the 1980s and appears to be part of a global pattern. A similar proliferation of small *Grallator* tracks has been reported from rocks of exactly the same age in western and eastern North America, as well as from many sites in mainland Europe, particularly in France, Germany,[5] and Italy (see following sections). Given the uncertainty that attends the identity of the aforementioned Middle Triassic dinosauroid tracks, this Late Triassic small *Grallator* acme zone is probably the best footprint evidence we have that small bipedal theropods were well established over much of

FIGURE 4.3. *Otozoum* from the Porthcawl area of South Wales. *Courtesy of the National Museum of Wales.*

the Northern Hemisphere by this time. Some might argue that certain of these tracks were made by small ornithopods, but majority opinion assigns *Grallator* to a theropod trackmaker.

It is also of considerable ichnologic significance that other elements of the fauna, notably *Tetrasauropus* and *Pseudotetrasauropus*, are also reported from rocks of the same age in mainland Europe, North America, and elsewhere. This means that our track correlations are based on comparisons of track assemblages (or ichnocoenoses), not just individual ichnospecies or ichnogenera. Both *Tetra-*

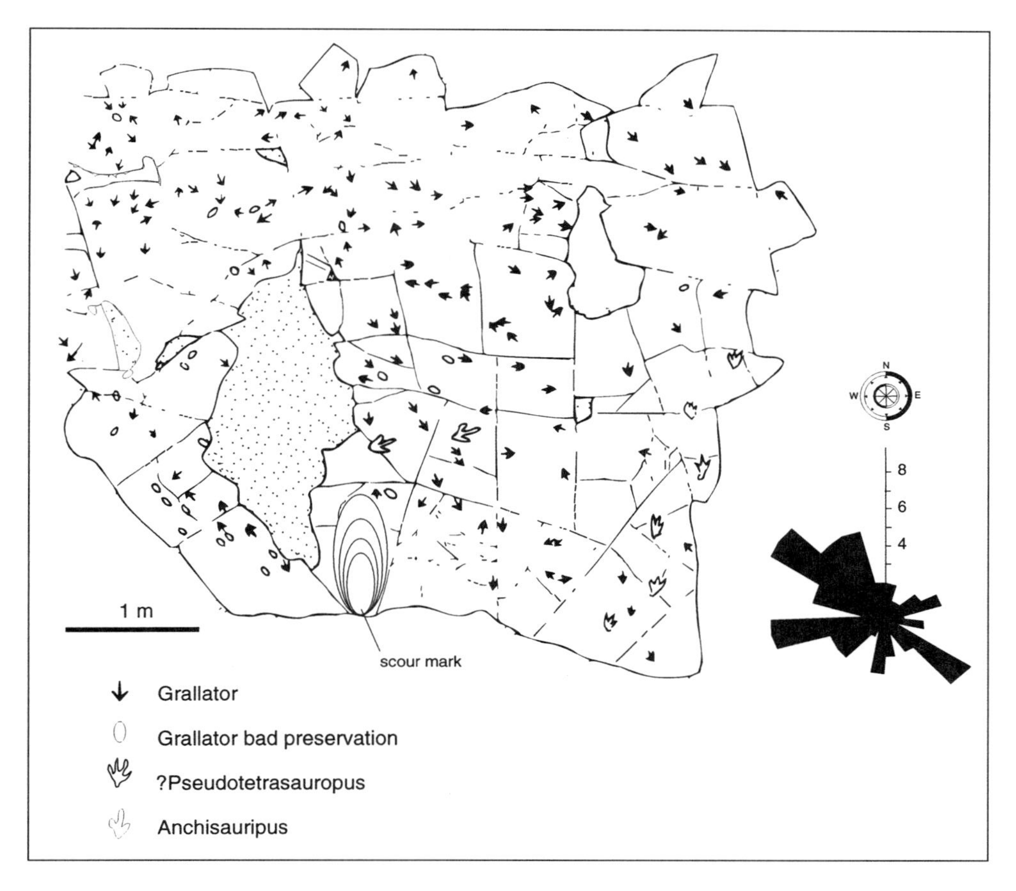

FIGURE 4.4. Large concentrations of small *Grallator* footprints are typical of many Late Triassic track assemblages in Europe and elsewhere. One trackway (left) from a nearby site has a long stride, indicative of running at an estimated speed in excess of 20 km per hour. (See the text for details.)

sauropus and *Pseudotetrasauropus* were originally named in the 1970s by French researcher Paul Ellenberger during the course of his studies of Late Triassic and Early Jurassic footprints from southern Africa.[6,7] His concept of *Tetrasauropus* and *Pseudotetrasauropus* was broad, covering a wide range of tracks of different sizes. (He ultimately named two ichnospecies assigned to *Tetrasauropus* and eight assigned to *Pseudotetrasauropus*!) Although these tracks require further study and revision, it is clear that he regarded many of them as similar to *Otozoum* and interpreted them as the spoor of prosauropods.

Regardless of how confident we are of assigning track types to particular track-making groups, it is clear that the Late Triassic is characterized by a relatively high diversity of footprint types assigned mainly to archosaurs. We interpret the Welsh track assemblages as evidence of a fauna in which small theropod dinosaurs were relatively common. Other elements of the fauna appear to have rarer medium-sized theropods, large and perhaps also small prosauropods, and other archosaurs of unknown affinity. Preliminary efforts to reconstruct the ecology of the fauna suggest that despite the numerical abundance of *Grallator* trackways, representing small carnivores, these animals probably did not make up a large part of the biomass. In some areas tracks are so abundant that surfaces are best described as trampled. Hence, we infer wet conditions underfoot and considerable activity by a diverse fauna.

THE MARCH OF THE PROSAUROPODS

One of the best-known of all Late Triassic dinosaurs is *Plateosaurus,* originally discovered in large numbers at a famous quarry at Trossingen in Germany.[8] It is also acknowledged that the foot of *Plateosaurus* matches well with the footprint *Otozoum* (figures 4.2 and 4.3). The Welsh *Otozoum* is very important in this regard because it is the same age as the German plateosaurs. Thus far the story is simple. Even though *Otozoum* was originally discovered in Lower Jurassic deposits in New England, this is not problematic because prosauropods continued to thrive at that time.

Where the story gets complicated is in the interpretations put on *Otozoum* by some ichnologists. For example, Polish ichnologist Gerard Gierlinski (chapter 5) has suggested that *Otozoum* was made by the armored dinosaur *Scelidosaurus.*[9] We dispute this interpretation because *Otozoum* and related forms are so abundant in the Late Triassic and Early Jurassic at precisely the time the prosauropods radiated so widely. Besides, Gierlinski has changed his interpretation, in a paper in press, and reverted to the traditional prosauropod interpretation. The size range also fits prosauropods, and the occurrence of Late Triassic examples, such as in Wales, does not correlate with the Lower Jurassic *Scelidosaurus.* Further consideration of various other tracks, some resembling *Otozoum* but others quite different, that have been attributed to prosauropods further complicates the picture. Particularly vexing in this regard are the ten or more aforementioned *Tetrasauropus* and *Pseudotetrasauropus* ichnospecies from southern Africa, attributed to prosauropods by Paul Ellenberger. For the purposes of our discussion, our main focus is the extent to which any of these track types have been identified in Europe.

Before proceeding any further we need to introduce the various track types named *Tetrasauropus* and *Pseudotetrasauropus* by Ellenberger.[6,7] As stated in the previous section, he assigned two species to the ichnogenus *Tetrasauropus* (figures 4.5 and 4.6) and eight to *Pseudotetrasauropus* (figure 4.7). He also named *Sauropodopus* (figure 4.7), which literally means "sauropod track," and acknowledged that he also had tracks of the *Otozoum* type. The naming of so many tracks has raised some eyebrows and suggests a case of oversplitting. The remedy suggested by Paul Olsen and Peter Galton was to lump these tracks together again. In an article written for the journal *Paleontographica Africana,*[10] they synonomized or

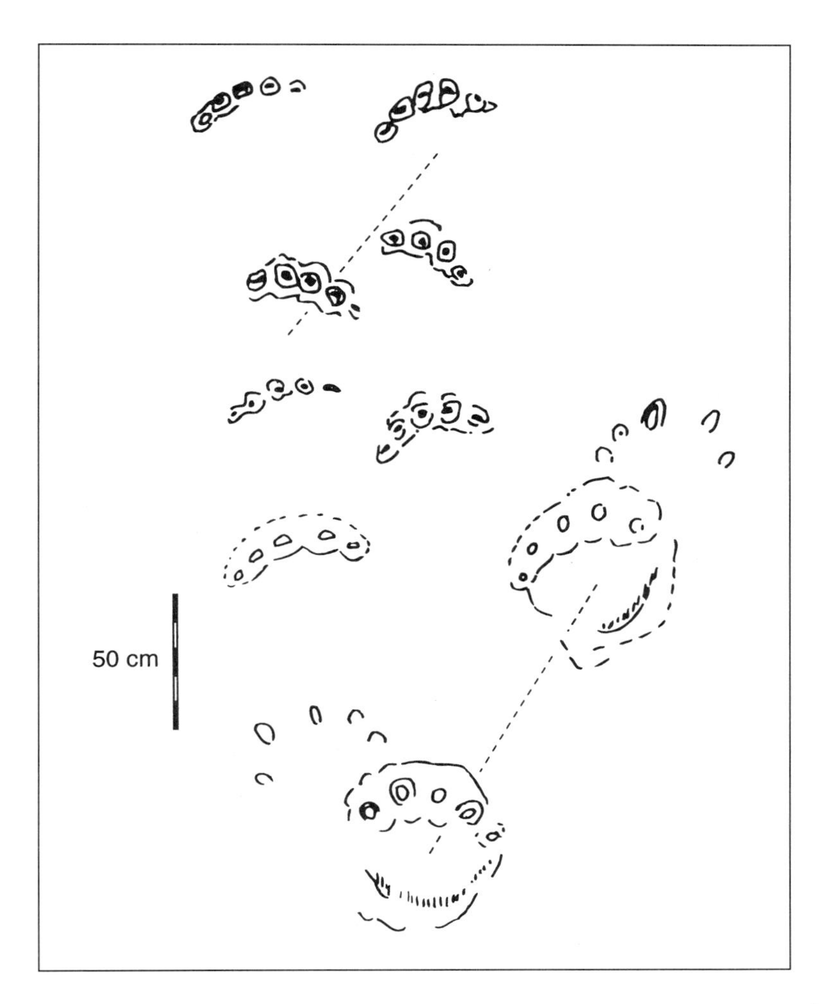

FIGURE 4.5. A comparison of two manus-pes sets of *Tetrasauropus gigas* (below) with a seven-track sequence (above) of *Pentasauropus incrediblis* (or *P. motlejoi*) suggests that they are the same track type. They probably represent a large mammal-like reptile (synapsid) of the dicynodont tribe. Lines connecting tracks indicate pace angulations. *Redrawn after Ellenberger.*[6]

lumped together a large number of Ellenberger's Late Triassic and Early Jurassic species with preexisting names that apply to species from eastern North America, where Edward Hitchcock conducted his classic studies from the 1830s through the 1860s. The problem with this wholesale approach is that they did not go to southern Africa to examine the original localities and specimens, nor did they even go to France to examine Ellenberger's collections. The result is that although some of their lumping may be valid, some is clearly not. In the discussion that follows we shall attempt to sort out some of these problems. We have examined the collections in collaboration with Ellenberger at the University of Montpellier, and we reached substantial agreement on the interpretation of tracks and correlations that

FIGURE 4.6. A large *Tetrasauropus*-like track from the Late Triassic of Wales (Bendricks rock site).

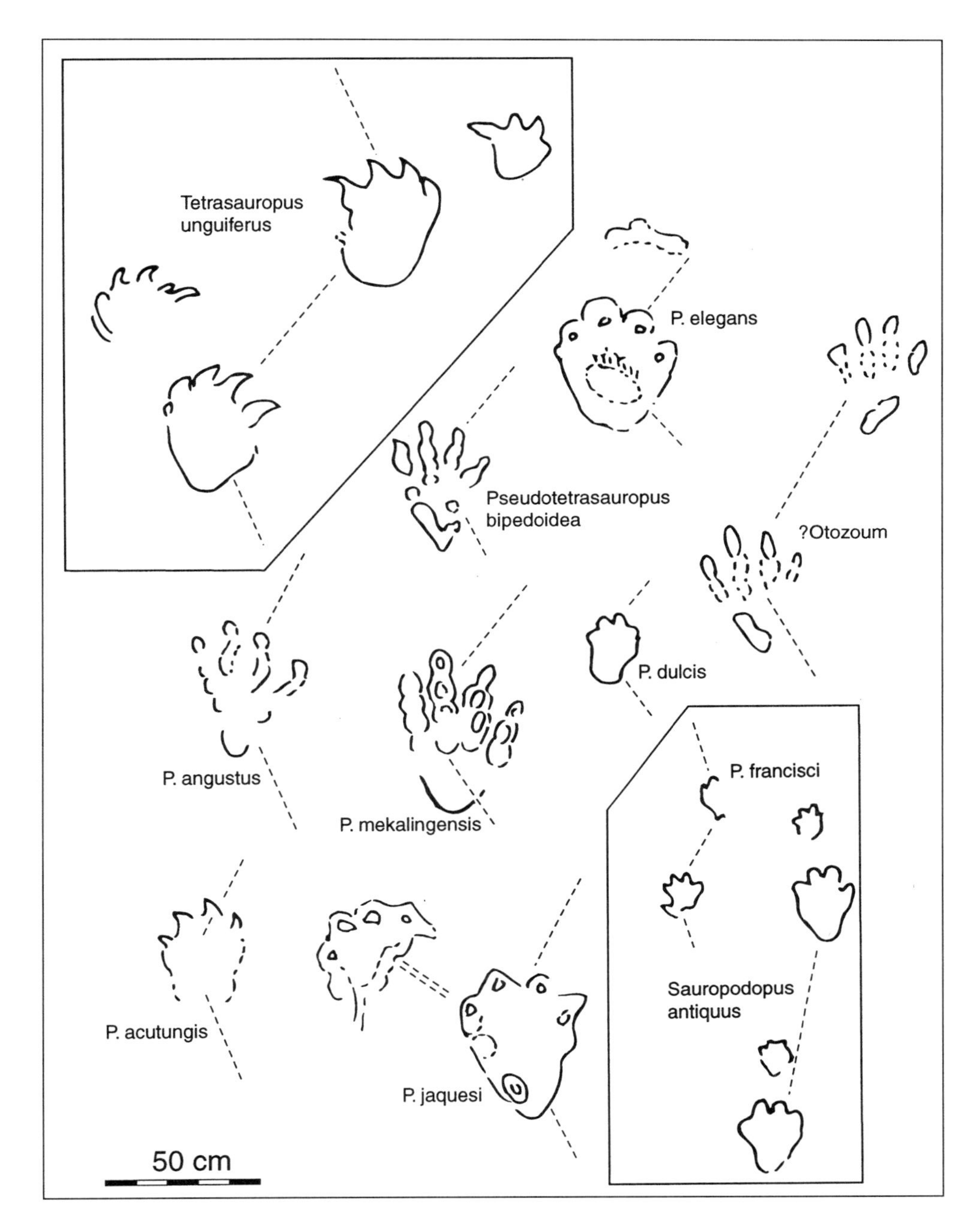

FIGURE 4.7. *Tetrasauropus, Pseudotetrasauropus, Otozoum,* and *Sauropodopus* were all attributed to prosauropods (or sauropodomorphs) by Ellenberger. This interpretation is probably correct in general except that *Sauropodopus* appears to be a chirothere track. It is also probable that *Otozoum* and *Pseudotetrasauropus* are the same track type. Lines connecting tracks indicate pace angulations. *Redrawn after Ellenberger.*[6]

can be made with footprint assemblages in Europe and elsewhere. As is the case for correlations in older Permo-Triassic deposits, Late Triassic correlations continue to reflect the presence of wide-ranging, cosmopolitan faunas.

LEARNING TRACKING FROM THE BUSHMEN: "C'EST L'AFRIQUE"

It is worth noting that Paul Ellenberger was brought up in Bushman country in southern Africa. His father, Victor, was born in a Bushman cave and wrote a book on the tragic end of Bushman culture.[11] Paul himself spent his childhood traveling in Bushman country recording its cave art. He knew the Bushmen as master trackers who could purportedly tell apart any of 40 species of gazelle by their footprints. So he brought to tracking the spirit of the Bushmen, enthusiastically naming each subtly different track type. Ellenberger was the first Frenchman to undertake a serious study of a series of tracksites from a particular region and publish extensively on the subject. Thus, he was largely responsible for establishing a "liberal" French ichnologic tradition that has been followed by such researchers as George Demathieu and Georges Gand (see chapters 2 and 3). One can say that the influence of the Bushmen resulted in a French tradition that favors splitting rather than lumping or, one might say, leads to complication and confusion, rather than simple clarity. As the French might say, "C'est l'Afrique." Although we agree with the perspective of Olsen and Galton[10] that there has been too much splitting of track names, the Bushman perspective is also helpful, especially when we consider that they have often been characterized as the world's best trackers, capable of seeing much that we would miss in the modern world. In short, they remind us that careful observation allows us to see things that might be overlooked, and that tracking is a skill that is acquired and mastered only with practice. In fairness to Ellenberger, it is possible that future studies will validate some of the fine distinctions he drew between tracks that appear morphologically similar at first glance.

Let us then look at Ellenberger's various prosauropod tracks. *Tetrasauropus unguiferus* (figure 4.7) represents a large quadruped with elephant-sized feet. As implied by the term "unguiferous" (meaning "having claws or nails"), this track type has distinctive acute or tapering digit impressions, four of which are seen clearly on the hind footprint. The second and only other species named by Ellenberger is *Tetrasauropus gigas* (figure 4.5), which we regard as different from *T. unguiferus* and more appropriately assigned to *Pentasauropus,* which has five toe impressions and probably represents a large mammal-like reptile, most probably a dicynodont. Ellenberger agreed with this interpretation when we proposed it to him (personal communication 1997).

When we turn our attention to *Pseudotetrasauropus,* we find eight species, of which six apparently represent bipeds (figure 4.7). The two that represent quadrupeds, *P. elegans* and *P. jaquesi,* are similar in size and general appearance to

Tetrasauropus unguiferus and may represent the same species of trackmaker. *P. francisci* is much smaller than the other varieties and has a posterior fifth toe impression that is reminiscent of chirotherid footprints (see chapter 3). In this regard, *Sauropodopus* could be a chirotheroid track owing to its general size, shape, and narrow trackway. It is possible that *P. dulcis* also fits this category. Smallish tracks that resemble *Chirotherium,* but with out-turned pes digits, like *Tetrasauropus,* occur in the Late Triassic of Wales[4] (figure 4.8). This leaves *Pseudotetrasauropus bipedoidea, P. angustus, P. acutunguis,* and *P. mekalingensis,* all of which are approximately the same size, with elongate digit impressions characterized by discrete pads, as seen in *Otozoum.* All apparently represent bipeds, which is also the case for *Otozoum,* except in rare, controversial examples. The general conclusion derived from this analysis is that *Tetrasauropus* represents a quadrupedal trackmaker with robust, fleshy feet that are not separated into elongate digits, and that *Pseudotetrasauropus* is essentially identical to *Otozoum,* whereas *Sauropodus* is of chirotheroid affinity. Thus, there are probably only two varieties of large prosauropod tracks in the Late Triassic footprint assemblages described by Ellenberger from southern Africa that deserve ichnogenus status (i.e., *Tetrasauropus* and *Otozoum* (*Pseudotetrasauropus*). Understanding these tracks, however, may be the key to identifying similar footprints in Europe.

Based on the aforementioned studies in Wales and comparisons made with other regions such as the United States, it appears that there may have been differences between the southern African track assemblages and those found in the Northern Hemisphere. The main difference appears to be that large prosauropod tracks measuring 50 cm or more in length were abundant in southern Africa, but they are rare in Europe. For example, there appears to be only one example of the large footprints of *Tetrasauropus* so far reported from a locality in the Swiss Alps,[12] as described below. Similarly, we know of only one example of large *Otozoum* tracks (compare with the *Pseudotetrasauropus* of Ellenberger), as described in the previous section on Welsh tracks (figure 4.3). No such large tracks are known in North America, although smaller footprints of both types, up to about 30 cm, are known from both North America and Europe.[4] We have provisionally compared the rounded tracks of quadrupeds with *Tetrapodosauropus* and the tracks of bipeds that display discrete, separated digit impressions with *Pseudotetrasauropus.*

Given these interpretations—that large *Tetrasauropus* (more than 50 cm) of the type found in Africa is represented at only one locality in Europe—it is interesting to consider the interpretation put on *Tetrasauropus* by Olsen and Galton.[10] They compared it with a Lower Jurassic track named *Navahopus* from the Navajo Formation of Arizona, which was attributed to a prosauropod by Donald Baird in a paper published in 1980. So confident were they in this comparison that they created the new ichnofamily *Navahopidae,* in which they included both *Navahopus* and *Tetrasauropus.* The problem with this interpretation is that *Navahopus* is a very small track (about 10 cm in diameter) that also has been interpreted as the trackway of a small mammal-like reptile named *Brasilichnium.*[4] This reinterpretation is supported by

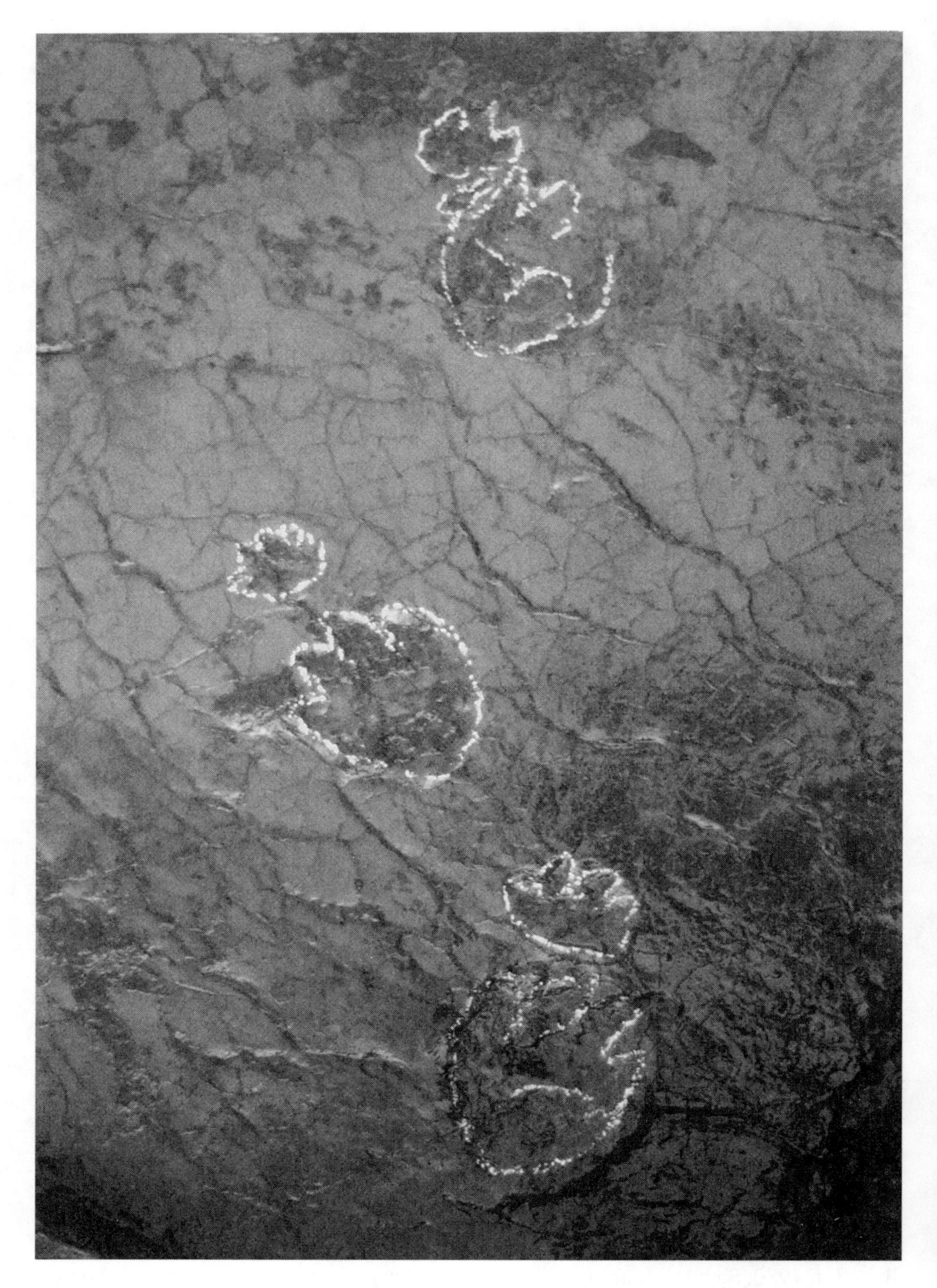

FIGURE 4.8. A small *Tetrasauropus*-like track from the Late Triassic of Wales. After Lockley et al.[4]

two fundamental observations. First, the morphology of the pes of *Navahopus* is indistinguishable from the mammal-like reptile track *Brasilichnium*, which occurs in abundance at the *Navahopus* type locality. The manus track morphology is obscure. Second, despite abundant tracks in the Navajo Formation, no four-toed tracks, other than *Brasilichnium* and one *Navahopus* trackway, are known from the sand dune facies. If *Navahopus* really was a small representative of the types of prosauropods that made giant tracks during the Late Triassic, it was the only one of its kind to leave footprints among swarms of tracks made by mammal-like reptiles in the sand dunes of North America. *Navahopus* is also much younger than *Tetrasauropus*, possibly by as much as 20 million years. Thus, the conclusions of Olsen and Galton are highly questionable, and the giant prosauropod track *Tetrasauropus* should not be considered a synonym of a small, probably mammal-like reptile track from a different epoch, a different continent, and a different habitat.

A BEAUTIFUL BUT ELUSIVE TRACK

As explained in the previous section, Ellenberger's Late Triassic and Early Jurassic footprint collections from southern Africa provide several important keys to understanding tracks of this age in other regions. The track that he named *Kalosauropus* (meaning "beautiful reptile footprint") is an excellent example. When first discovered in southern Africa, it was unknown from other regions. It was subsequently discovered in New Mexico,[13,14] Utah, and Colorado, where it was referred to as *Pseudotetrasauropus*, which it resembles in most respects, although it is much smaller. At that time, however, it was not recognized as an example of *Kalosauropus* because no one other than Ellenberger had ever examined this track type in detail. When we visited Ellenberger's collection at the University of Montpellier, France, we established beyond reasonable doubt that *Kalosauropus* from southern Africa is the same track type that was discovered in North America in the early 1990s and in Europe in 1995, and we here label these tracks correctly for the first time (figure 4.9).

Let us then look closely at *Kalosauropus* in all its beautiful detail and see what can be learned. As shown in figure 4.9, tracks from southern Africa, North America, and Europe are all similar. The distinctive features of *Kalosauropus* are that the digits reveal a large number of well-defined pad impressions. These impressions are most obvious on digit III (the longest) and digit IV, whereas they are often faint on digits II and I. Indeed the impressions of digit II and digit I (the shortest) are often faint and incomplete. In many cases digit I leaves no trace at all, and so can be confused with a typical three-toed track such as *Grallator* if one does not observe that digit III is much shorter, relative to digits II and IV, than in theropod tracks such as *Grallator.*

The tracks of *Kalosauropus* are generally small to medium-sized, ranging from as little as 8 to 11 cm to as much as 23 to 24 cm. The clearer and deeper im-

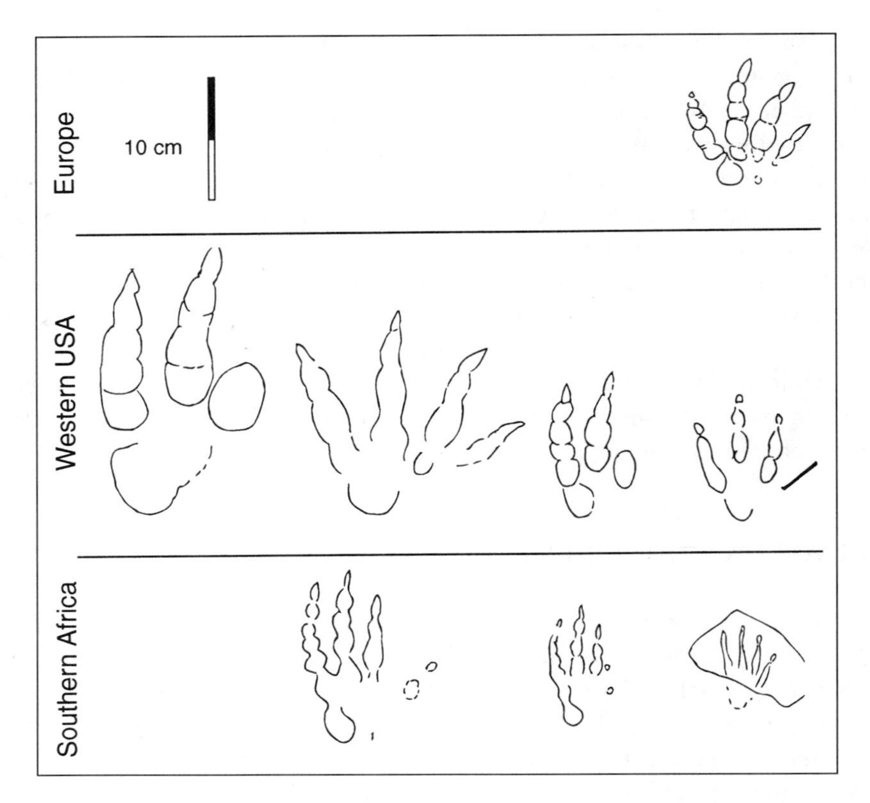

FIGURE 4.9. *Kalosauropus* (the beautiful reptile track) was first discovered in southern Africa by French ichnologist Paul Ellenberger before being recognized in North America and Europe. It could be described as being similar to a miniature version of *Otozoum* or *Pseudotetrasauropus,* and so may also represent small prosauropod species.

pressions left by digits III and IV indicate an animal that carried most of its weight on the outside of its feet. This characteristic is apparently consistent with the fact that *Kalosauropus* also has quite a wide trackway, so it evidently walked with something of a straddle rather than putting one foot directly in front of another. When we add to these observations the fact that the *Kalosauropus* trackmaker always appears to have been bipedal, we have a more complete picture of a somewhat ungainly biped with four toes that favored the outside of its feet.

What type of animal was responsible for making *Kalosauropus* tracks? As already indicated, these tracks are similar to *Pseudotetrasauropus* and *Otozoum,* which might be regarded as spoor of larger relatives. Given that *Pseudotetrasauropus* and *Otozoum* are regarded by most researchers as prosauropod tracks, it follows that *Kalosauropus* might also be attributed to this group. One might argue that *Kalosauropus* represents juveniles, whereas *Pseudotetrasauropus* and *Otozoum* represent adults. Such an argument, however, is undermined by the fact that the two size groups are not found together at the same sites or in the same stratigraphic levels. Moreover, prosauropod species come in a range of sizes that fit the whole range of footprint sizes from the smallest *Kalosauropus* to the largest *Pseudotetrasauropus.* Thus, the beautiful *Kalosauropus* first described by Ellenberger appears to be quite distinctive and deserving of its name, even though its identity and origin have proved elusive.

HIGH-ALTITUDE TRACKS IN THE SWISS ALPS

Our discussion of Ellenberger's large, purported prosauropod tracks from southern Africa provides a pertinent introduction to one of Europe's most spectacularly located high-altitude tracksites. The site at Piz dal Diavel (meaning "devil's peak") in the Swiss Alps was discovered by geologists of the Federal Polytechnic Institute in the 1960s, but it took 20 years until serious attempts were made to take a close look at this remote alpine tracksite situated at 2450 m (8000 feet) above sea level. A team from the University of Zurich mapped the surface (figure 4.10), which measures 30 by 60 m, and made casts and molds of some of the tracks.[12]

The tracks include a single large trackway of the *Tetrasauropus* type, consisting of 25 consecutive footprints measuring about 60 cm long. The step is short, about 1 m, and the corresponding stride is about 2 m, and the trackmaker has been interpreted as a prosauropod, probably a plateosaurid. This is the first example of a large *Tetrasauropus* trackway from Europe, although a smaller example (about 30 cm long; figure 4.6) has been reported from Wales.[4] It appears, therefore, that large prosauropod tracks are rare in the Late Triassic of Europe, even though plateosaurid remains are abundant at Trossingen in Germany.

Thirteen other trackways have been attributed to theropod dinosaurs with feet in the size range of 25 to 30 cm. Such tracks are large in comparison with other Late Triassic theropod tracks, which are often diminutive. The tracks are about the same size as the Welsh tracks that we labeled *Anchisauripus* (figure 4.1). Although the Swiss examples appear to have a somewhat longer step and stride, both the Swiss and Welsh examples appear to have a trackway that is noticeably wider than most theropod trackways. It is tempting, therefore, to suggest that they are both of the same type in terms of size and trackway width (or pace angulation). Such a wide trackway also hints at a posture that is not as erect as that achieved by most theropods.

Studies of the track-bearing surface reveal that it consists of a dolomitic limestone that contains fish remains, tiny shells of crustaceans (ostracods), and plant debris, suggesting limey mud flats in a coastal region. The surface also reveals mud cracks. Such a setting is somewhat different from the ancient environment recon-

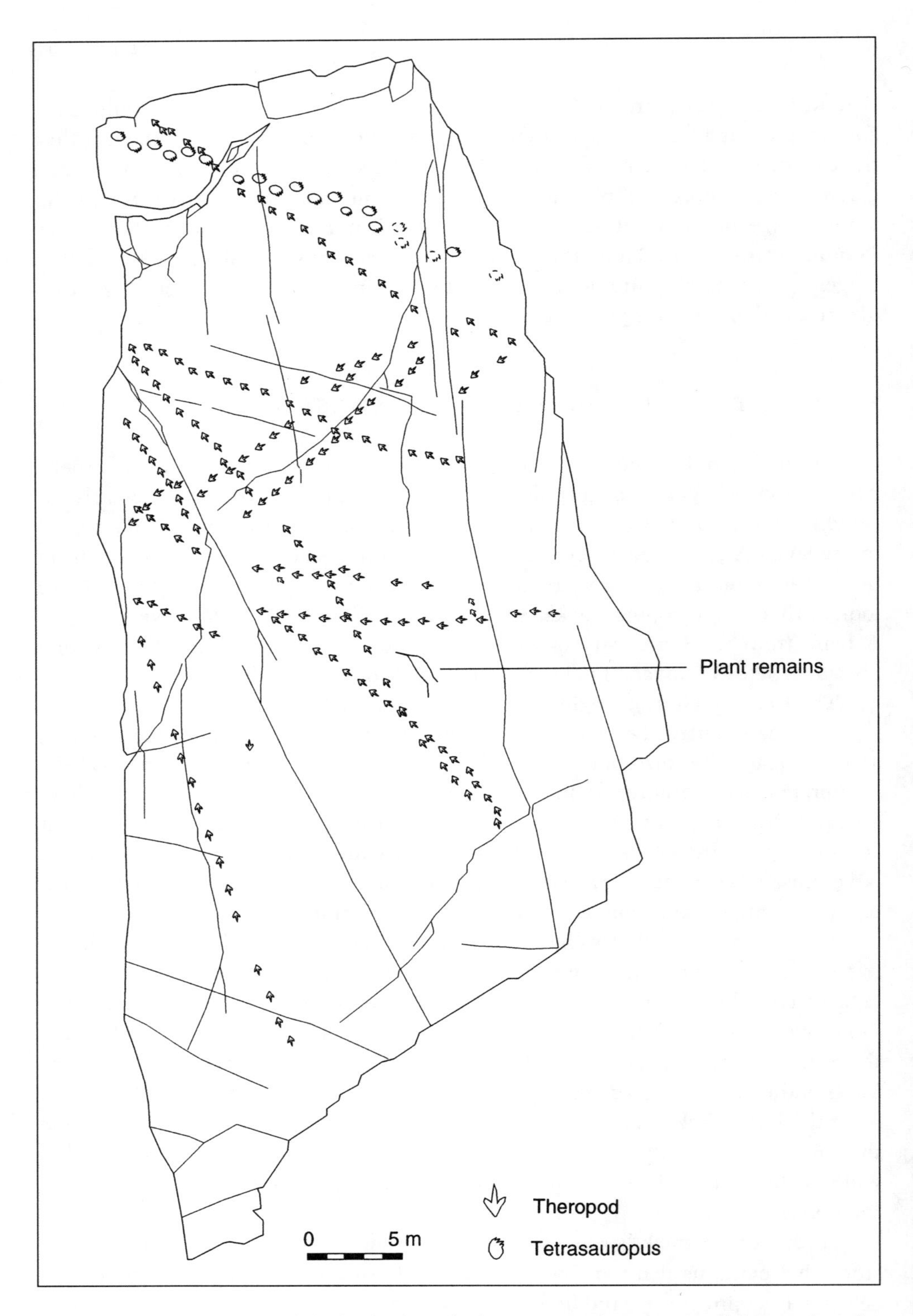

FIGURE 4.10. Map of the high-altitude dinosaur track site at Piz dal Diavel in the Swiss Alps, revealing the largest prosauropod trackway in Europe. (*Redrawn from Furrer 1993.*)

structed for the Welsh trackway assemblages, which appear to have been formed beside braided stream deposits that formed as a result of periodic flooding from alluvial fans accumulating at the margins of nearby uplands. Both settings, however, appear to represent relatively arid or semi-arid environments and provide some support for the hypothesis suggesting that this was the preferred habitat of prosauropods.

VON HUENE'S LITTLE DINOSAUR TRACK: *COELUROSAURICHNUS*

We can begin this story with the punch line. *Coelurosaurichnus* is not a valid name and so should be abandoned. To understand this problem, however, we need to review the history, beginning with the discovery and naming of the track from a site in northern Italy. The track (figures 4.11 and 4.12) was named in 1941 by the German paleontologist Friedrich von Huene,[15] whom we have already discussed in connection with *Prestosuchus* and *Plateosaurus*.

There are two reasons why the name *Coelurosaurichnus* is inappropriate, one based on the rules of nomenclature and the second based on our understanding of the concept of coelurosaurs. From an ichnologic viewpoint the first rule is more important. It is clear that the original specimen of *Coelurosaurichnus* is not well preserved and that it is simply a small three-toed dinosaur track. It was found in strata that, all over Europe, are full of small tracks, usually assigned to *Grallator*.[4,5,16,17] Thus, it became clear that this track is just a poorly preserved *Grallator*

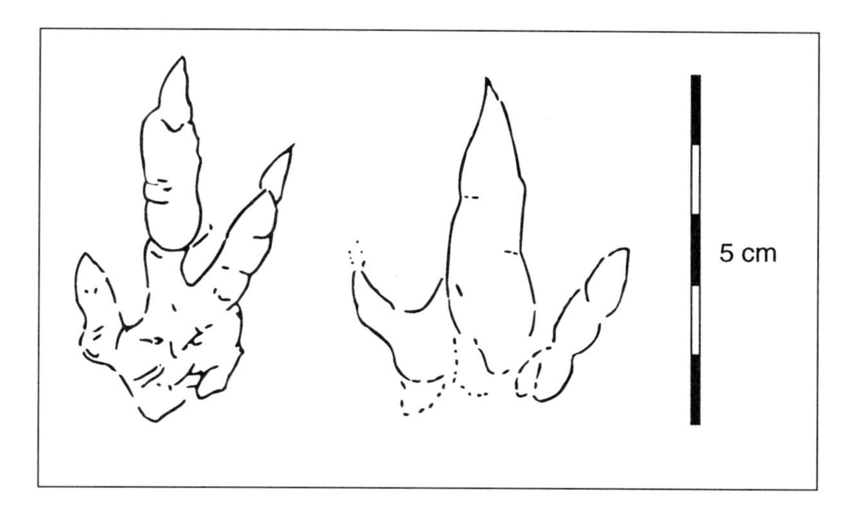

FIGURE 4.11. Von Huene's little track (left), *Coelurosaurichnus,* was not made by a coelurosaur in the strict sense and is in fact a poorly preserved example of the track type more commonly known as *Grallator* (accurately redrawn on right).

FIGURE 4.12. Photograph of *Coelurosaurichnus. Courtesy of Giuseppe Leonardi.*

track and does not deserve a separate name.[18] In short, von Huene should have compared the track with *Grallator,* which was well known at the time,[19] and realized that a new name was not necessary.

The inappropriateness of the name *Coelurosaurichnus,* therefore, is already established on the grounds that it is a synonym (technically a junior synonym) of *Grallator.* Since the name *Coelurosaurichnus* was first coined, it became popular, at least among some European workers, because there was a long tradition of differentiating small, lightly built, or gracile theropods from large, robust forms by dividing them into the two categories of coelurosaur and carnosaur. Modern theropod family trees present a threefold classification, with primitive ceratosaurs branching off first, leaving a bigger group, known as "tetanurines," consisting of coelurosaurs and carnosaurs. The catch is that what we now call "coelurosaurs" are not known until the middle of the Mesozoic (Late Jurassic), so the concept of a Late Triassic coelurosaur is no longer tenable. Note, however, that this is not the technical reason for abandoning the name. *Coelurosaurichnus* will always remain associated with von Huene's little dinosaur footprint. But it cannot now be used to describe anything else without implying that we mean *Grallator.*

Before the proposal to abandon the name *Coelurosaurichnus* was published,[18] a number of tracks had also been labeled as *Coelurosaurichnus*, even though many of them were different from von Huene's specimen. We are currently working with Giuseppe Leonardi, who first spotted von Huene's error, to reexamine this material and try to establish appropriate names. To give an idea of the extent and intricacy of the problem, we note the following: since the publication of von Huene's 1941 paper[15] establishing the ichnogenus *Coelurosaurichnus*, more than a dozen addition species have been erected.[20] Species such as *C. palissyi* and *C. sabinesis*[21,22] and *C. largenierensis* and *C. perrauxi*,[23] all from France, are reported from rocks of Middle Triassic age. After the English debate (see chapter 3), this again opens up the question "What types of dinosaur-like trackmakers are making so many apparently distinctive footprints so early in the Triassic?" We predict that eventually the English debate is likely to be replayed to some degree, with a serious reevaluation of what these tracks are and what they really represent being necessary for an improved understanding of the complex *Coelurosaurichnus* problem.

FUTURE DIRECTIONS: DIGGING DEEPER IN THE LATE TRIASSIC

It would be inappropriate to leave the reader with the impression that all Late Triassic tracks can be lumped together. Most of those we have examined are from late in the Late Triassic (Norian-Rhaetian stages) rather than from early in the Late Triassic (Carnian stage). The reason for this is that these younger sedimentary sequences are apparently much richer in tracks than the older stratigraphic units. This is particularly true in North America, where dozens of younger Late Triassic (Norian-Rhaetian) tracksites have been described,[13] whereas older, Carnian sites are virtually unknown. Although it is wonderful to have so many late Late Triassic sites available for study, it is the early Late Triassic sites that would be the most interesting in terms of understanding the faunas that existed at the time the dinosaurs were emerging. We have already seen (see chapter 3) that there is debate about the timing of the origin of dinosaurs, but in terms of the track record, there is somewhat of a gap at this crucial time. This is perhaps not so much a real gap as a gap in our understanding, for many tracksites exist, such as those yielding purported *Coelurosaurichnus* tracks of various types. It is just that the literature is somewhat obscure and localized. In this age of synthesis and global communication, we hope to see such problems resolved before long.

It is appropriate, therefore, that we should end this chapter with a note of optimism by noting that the gap is starting to be filled. Future studies will dig deeper into the Late Triassic, as they have already begun to do, to provide evidence of the

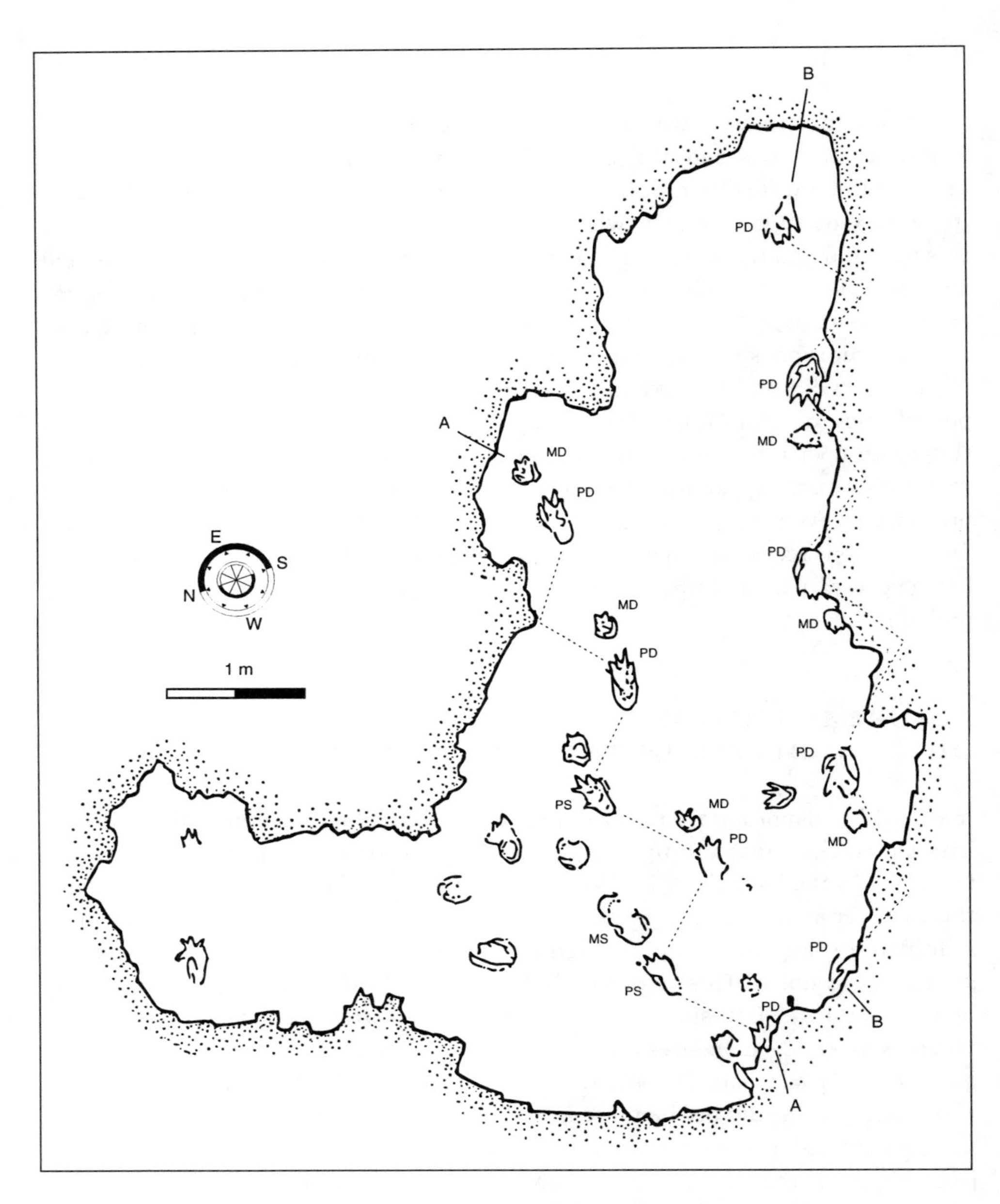

FIGURE 4.13. New tracks, as yet unnamed, from the Dolomite region, promise to reveal more about the nature of early Late Triassic (Carnian) track assemblages. *After Dalla Vecchia.*[24]

earliest faunas of this epoch. It is appropriate also to end in northern Italy, where the *Coelurosaurichnus* problem originated. Fabio Dalla Vecchia[24] has shown that a number of interesting tracksites (figure 4.13) have been identified from the Carnian part of the Late Triassic in the Dogna Valley region of the Dolomites, classic geologic terrain in northeastern Italy near the Austrian and Slovenian frontiers. To date, Dalla Vecchia has not assigned new names to these track types in his preliminary publications, but it appears probable that in the near future we will have a better understanding of the composition of track assemblages from the Middle Triassic, early Late Triassic (Carnian), and late Late Triassic (Norian-Rhaetian), not just in Italy but in other parts of Europe.

At present, we can at least conclude as follows. There seems little doubt that dinosaur footprints were abundant in many regions by the Late Triassic epoch, although they are found sometimes with tracks of other nondinosaurian archosaurs and other reptiles. The fauna appears to have been dominated by saurischian (lizard-hipped) dinosaurs, that is, theropods and prosauropods. The ornithischians (bird-hipped dinosaurs) have not yet been identified by their tracks in the Late Triassic, although, as we shall see, we recognize such footprints in the Early Jurassic (see chapter 5). This probably reflects the much greater abundance and earlier proliferation of saurischians at the beginning of the age of dinosaurs. The ornithischians come into their own later in the Mesozoic.

A ONCE GREEN AND PLEASANT LAND

Although no longer technically part of Europe, for many years Greenland was politically linked to Denmark. During the Late Triassic, it was also closely connected to Europe and situated, moreover, in a subtropical latitude. Recent work on Late Triassic vertebrate assemblages from the Fleming Fjord Formation in Jameson Land in east Greenland has revealed the presence of tracks.[25] Most tracks can be assigned to *Grallator* and range in size from 4 to 28 cm. Figure 4.14 shows several dozen trackways. Many show elongate metatarsus impressions, which indicate that the animals were sinking deeply into the soft substrate. Steve Gatesy has studied these tracks to determine the pattern of foot emplacement and extraction as the animals traversed this substrate.[26] In some places, tail drags also have been recorded.

There are several trackways of large animals with "approximately circular prints (36–53 cm in diameter)" which are attributed to large quadrupedal archosaurs.[25] These tracks appear similar to those reported from Wales (figure 4.6) and Switzerland (figure 4.10) and may be attributable to prosauropods and possibly to the ichnogenus *Tetrasauropus*. The genus *Plateosaurus* has been recovered from these deposits.

Clearly, the logistic problems of field work in Greenland have precluded extensive track discoveries in the past. However, the finds in Jameson Land clearly point

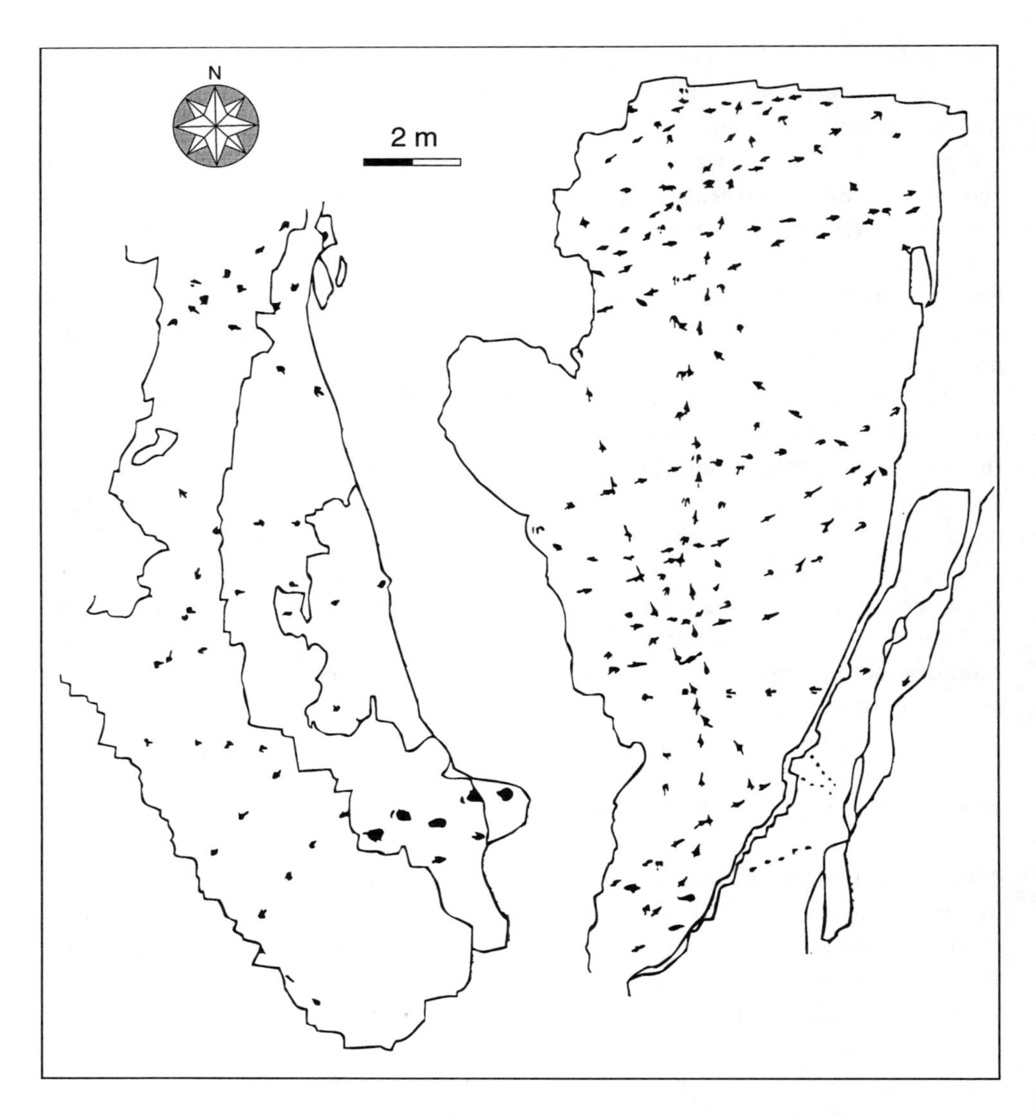

FIGURE 4.14. Dinosaur tracks from Jameson Land, east Greenland. *After Jenkins et al.*[25]

to another part of ancient Europe that has considerable potential for track discovery. (The reader is referred to chapter 8 for more arctic dinosaur track discoveries.)

BIBLIOGRAPHIC NOTES

1. Thomas, T. H. 1879.
2. Sollas, W. J. 1879.

3. Tucker, M. E. and T. P. Burchette. 1977.
4. Lockley, M. G., M. J. King, S. Howe, and T. Sharp. 1996.
5. Haubold, H. 1986.
6. Ellenberger, P. 1972.
7. Ellenberger, P. 1974.
8. Weishampel, D. B. 1984.
9. Gierlinski, G., 1995.
10. Olsen, P. and P. Galton, 1984.
11. Ellenberger, V. 1953.
12. Furrer, H. 1993.
13. Lockley, M. G. and A. P. Hunt. 1995.
14. Lockley, M. G., V. F. dos Santos, and A. P. Hunt. 1993.
15. Von Huene, F. 1941.
16. Haderer, F. O. 1988.
17. Haderer, F. O. 1992.
18. Leonardi, G. and M. G. Lockley. 1995
19. Hitchcock, E. 1858.
20. Chure, D. and J. McIntosh. 1989.
21. Gand, G. 1976a.
22. Gand, G. 1976b.
23. Haubold, H. 1984.
24. Dalla Vecchia, F. 1996.
25. Jenkins, F. et al. 1994.
26. Gatesy, S. and K. M. Middleton. 1998.

A theropod dinosaur drinks from a sauropod footprint. An imaginative scenario based on an Early Jurassic tracksite in northern Italy. Courtesy of Kaspar Graf.

CHAPTER 5

EARLY JURASSIC

NOAH'S RAVEN VISITS EUROPE

For historical reasons, Early Jurassic dinosaur tracks are among the best known anywhere in the world. They were first made famous by Edward Hitchcock, who in the 1830s through 1860s made an intensive study of tracks in the Connecticut Valley region of the eastern United States.[1] At that time Early Jurassic dinosaurs were not known, nor was the true antiquity of the Early Jurassic (back to 208 million years) appreciated. Hitchcock therefore concluded that the tracks were those of giant prehistoric birds, possibly related to, or resembling, the moa of New Zealand or the elephant bird of Madagascar. In a lighthearted and symbolic way they were referred to as antediluvian tracks of Noah's raven. It was not until after his death that Early Jurassic dinosaurs were discovered, and the footprints were reinterpreted as those of dinosaurs by Yale paleontologist Richard Swan Lull.[2] Although in one sense Hitchcock had technically been wrong, his work is considered monumental, and he is regarded as the founder of the science of vertebrate ichnology. In another sense, Hitchcock was right all along because the tracks are mainly those of theropods, which, among dinosaurs, are now regarded as the closest relatives of birds.

Among the many names that Hitchcock and Lull gave to the tracks of the Connecticut Valley region, the two ichnogenera *Grallator* and *Eubrontes* are probably the best known (figure 5.1). According to their interpretations, each ichnogenus contains several ichnospecies. According to recent reevaluation of the material, by Paul Olsen of Columbia University, *Grallator,* which is generally represented by small foot-

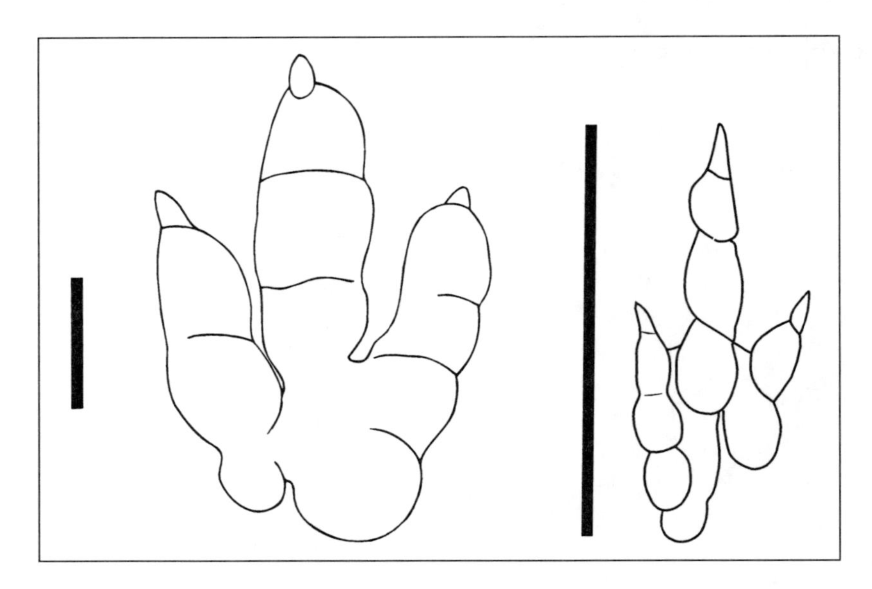

FIGURE 5.1. *Eubrontes* (left) and *Grallator* (right). *After Hitchcock*[1] *and Lull.*[2] Scale bars 10 cm.

prints, may be a juvenile of the larger *Eubrontes* morphotype and must be considered in conjunction with a third ichnogenus, *Anchisauripus,* in any thorough study.[3] Olsen's interpretations are noteworthy because of the excessive splitting of genera by researchers in the past (see chapters 2 and 4). But the problem of discriminating these different track types is a difficult one, and so far ichnologists have been reluctant to completely revise the large collections of material amassed by Hitchcock.

Partly as a result of this lack of comprehensive study in recent generations, there has been a tendency, among casual ichnologists, to assign a large number of tracks to such categories as *Grallator* and *Eubrontes,* whether or not they actually originated from the Lower Jurassic, and regardless of whether they have been subjected to careful scrutiny. As a result, the names *Grallator* and *Eubrontes* have been applied to tracks from all over the world, in some cases including footprints that are both older and younger than the Lower Jurassic. In the case of Late Triassic tracks, the name *Grallator* may be correct in some cases (see chapter 4).

It also appears that both *Grallator* and *Eubrontes* have been correctly identified in the Lower Jurassic track record in Europe by such serious researchers as the French ichnologists Albert de Lapparent and George Demathieu, as well as the Polish ichnologist Gerard Gierlinski, whose contributions are discussed below.

FRANCE: THE LE VEILLON SITES

The first thorough study of Lower Jurassic footprints in Europe was undertaken by Albert de Lapparent and Christian Montenat in the 1960s when abundant tracks were discovered on the west coast of France, about 40 km northwest of LaRochelle.[4] This locality, on the northern shores of the Bay of Biscay, is known as Le Veillon in the Vendée region (la côte Vendéenne). Although the area had been studied by geologists since 1912, the tracks were not reported until 1963. When they were investigated by de Lapparent and Montenat, however, they were found to be very abundant and exposed at least seven different stratigraphic levels. These researchers referred to the track-bearing beds as belonging to the Infralias. In some terminology this is equivalent to the Rhaetian and so is part of the very latest Triassic, strictly speaking. It is evident, however, that the tracks from Le Veillon span an interval of time (seven track layers) of some duration. Moreover, the lower beds represent terrestrial or at least estuarine deposits, whereas the upper beds, which are calcareous (limestones), show the influence of the widespread marine transgression of the Early Jurassic or true Lias. Consequently, de Lapparent and Montenat asserted that the track beds of Le Veillon span the Triassic-Jurassic boundary. We include them here in our chapter on the Early Jurassic while acknowledging that the oldest track layers may be latest Triassic (Rhaetian) in age.

The result of the Le Veillon study was a substantial memoir published by the Geological Society of France, in which nine ichnospecies were named. Seven of these, including four ichnospecies attributed to *Grallator* and *Eubrontes* and three assigned to the new inchnogenera *Saltopoides, Anatopus,* and *Talmontopus,* were attributed to tridactyl dinosaurs. Two others, *Batrachopus* and *Datutherium,* were attributed to quadrupedal pseudosuchians, and two others of uncertain affinity were illustrated but not named.

In light of our present knowledge of latest Triassic and Early Jurassic tracks, *Grallator, Eubrontes,* and *Batrachopus* are widely recognized as characteristic ichnogenera. To the best of our knowledge the other forms—*Saltopoides, Anatopus,* and *Talmontopus,* which were named solely on the basis of the Le Veillon discoveries, and *Datutherium,* which is a name proposed for a Middle Triassic footprint[5]—have not been reported from any other sites. This suggests that the recognition of nine, or perhaps as many as eleven, distinct track types (ichnospecies) at this site may be an overestimate.

Before dismissing this high-diversity estimate as a case of oversplitting, we should give credit to de Lapparent and Montenat for a thorough study in which they measured hundreds of tracks and made sensible efforts to discriminate the different track types. The tridactyl tracks have a considerable size range. Their smallest ichnospecies, *Grallator oloneusis,* which has an average foot length of only 4.3 cm (minimum 3.8 cm), was described on the basis of 160 tracks as being very similar to *G. tenuis* and *G. gracilis* from Connecticut. Similarly, an intermediate-sized and, as the name suggests, variable form, *Grallator variabilis,* with an average foot length of about 12 cm, was named on the basis of at least 60 measured tracks. *G. variabilis* is described as being comparable or "homologous" with *G. cuneatus* from Connecticut. The largest form, *Grallator maximus,* based on "several trackways," has an average footprint length of about 28 cm and is compared casually with *Anchisauripus* from Connecticut. The largest tridactyl footprints, assigned to *Eubrontes veillonensis,* measure up to 46 cm in length (average 34 cm) based on a sample of about 15 trackways (figure 5.2). This form was described as being "most close" to *E. giganteus* from North America.[4] Although we agree with de Lapparent and Montenat that these four types are distinctive, we suspect that most if not all could be assigned names based on the Connecticut Valley ichnofaunas, as their comparisons indicate. As we shall see in the sections that follow, Demathieu has attempted to do this for other Lower Jurassic faunas that are similar in composition.

When we examine the newly named ichnogenera, we find that *Saltopoides,* with a foot length of 15.5, and *Anatopus,* with a length of 8 to 9 cm, fall within the size range of *Grallator variabilis. Saltopoides,* however, which is based on only two trackways, reveals one example with an exceptionally long step (172 cm for a foot length of 15.5 cm), which seems to indicate a running animal. Based on the speed estimation formula of British Zoologist R. McNeill Alexander,[6] we estimate a running speed of about 11 km per hour.

Little can be said about *Anatopus palmatus,* except that it appears to be based on footprints that are incompletely preserved. Specifically, the lateral digits (II and IV) appear short, but this is evidently because only the tips (claws and distal ends) are preserved. De Lapparent and Montenat also inferred the presence of a trace of interdigital webbing in this track type. We have reservations about this interpretation because such apparent indications of webbing are often the result of weathering of curved or "bulged" surfaces that develop between digit impressions as a result of the deformation of the substrate by the foot emplacement. In other words, most purported web traces are pseudo–web traces that are the result of preservational phenomena. In fact, there are no convincing examples of dinosaur tracks that show compelling evidence of webbing. We think that *Anatopus* is probably an incomplete footprint, possibly allied to *Grallator variabilis,* and probably does not warrant being placed in a separate ichnogenus.

Talmontopus also falls in the size range of *Grallator* (*G. maximus*) and small *Eubrontes,* and it is longer (27 cm) than wide (23 cm), which is typical of these and other presumed theropod tracks. As we shall also see, it is similar to the theropod track named *Grallator* (*Eubrontes*) *soltykovensis* from rocks of the same age in southern Sweden. The interpretation of de Lapparent and Montenat that this may be the track of an ornithopod is therefore open to question. The differentiation of theropod and ornithopod tracks in the Lower Jurassic, as well as in other epochs, is difficult. Traditionally, the criteria used are morphologic similarity to *Grallator* (implying theropod affinity) and similarity to *Anomoepus* (implying ornithopod affinity).

The recognition of *Batrachopus,* which was first named by Hitchcock[1] from Connecticut, indicates that the fauna from Le Veillon was not composed only of bipedal dinosaurs. *Batrachopus* represents some form of small quadrupedal archosaur, possibly a pseudosuchian, according to de Lapparent and Montenat. If incompletely preserved, it is hard to distinguish from the small chirothere track *Chirotherium lulli* known from rocks of approximately the same age in eastern North America.

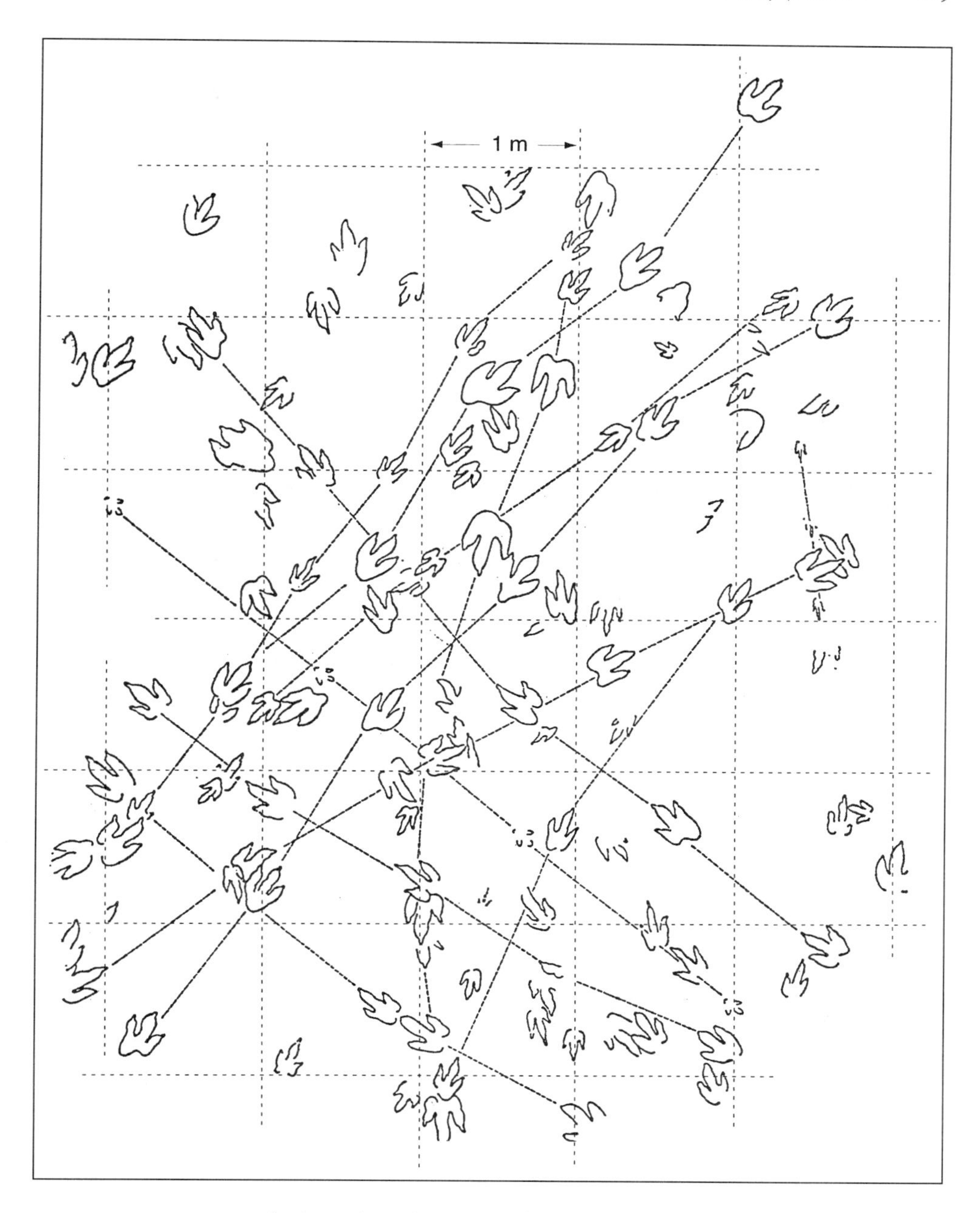

FIGURE 5.2. Tracks from the Infralias du Veillon. *After de Lapparent and Montenat.*[4]

FIGURE 5.3. A slab with *Eubrontes* and *Grallator* from the Infralias du Veillon can be seen at the Dinosaur Museum, Aathal, Switzerland. *Courtesy of Köbi Siber.*

In their synthesis of the Le Veillon ichnofauna, de Lapparent and Montenat observed that "the three most abundant genera—*Batrachopus, Grallator,* and *Eubrontes*—are equally well known in Connecticut" [our translation] (1967, p. 34).[4] It appears therefore that the fauna that is represented in the Lower Jurassic of France is quite similar, if not almost identical, to that found in eastern North America. The differences implied by the new names introduced for the French tracks are therefore somewhat misleading examples of provincial taxonomy.

Given the relative proximity of France and New England during the Early Jurassic and the cosmopolitan nature of the fauna, it is not surprising that the track assemblages are similar. However, the environments in which the tracks were made were not identical. In North America the tracks were preserved mainly on the muddy shores of rift valley lakes, whereas in France the tracks are found in limestones and other sediments, which represent a marginal marine environment. The presence of mud cracks, raindrop impressions, and the casts of salt crystals suggests an arid setting, possibly an estuary, that was subject to fluctuating water levels that caused the sediments to be exposed periodically, creating conditions suitable for trackmaking. The sediments at some levels are rich in plant fragments, suggesting a well-vegetated hinterland. In such a rich setting, inferred de Lapparent and Montenat, "the animals were not simply passing Le Veillon by accident, but were in their natural habitat" [our translation] (1967, p. 37).[4] The impression of abundance among these dinosaurs is given not just by the fact that tracks occur at several levels, but by the high concentrations observed on certain surfaces. For example, at one site at least 15 tracks are recorded crisscrossing a small area in many different directions (figure 5.2).

As a footnote to the Le Veillon story, we can note the reminiscences of Albert de Lapparent as told to Phillipe Taquet and summarized in his book, *Dinosaur Impressions.*[7] Many of the tracks were hard to study owing to their location in the intertidal zone, so the spring equinoxes in 1965 and 1966 were chosen as the best low-tide times to study the footprints. However, the team successfully extracted a slab 6 m long that was reconstructed in the museum of Sables-d'Olonne. At the time of this research, the site was characterized as the "best in France." Unfortunately, the announcement on television of this important discovery attracted too many curious visitors and inexperienced rock hounds, who added to tidal erosion. Today most traces have been removed by the seas and by vandalism.[7] Nevertheless, a fine slab (figure 5.3) found its way to the Dinosaur Museum in Aathal, Switzerland.

FRANCE: THE CAUSSES REGION

Our summary of the Le Veillon track assemblages provides a useful introduction to another French region that is rich in dinosaur tracksites, namely, the Causses region comprising the southern part of the Massif Central in the area north of Montpellier. Here Georges Demathieu and his colleague Jacques Sciau described various footprints from three localities: Saint-Lèons, Saint-Laurent de Tréves, and Sauclières.[8–10] In his first paper,[8] discussing the two former localities, Demathieu used statistics to help determine whether tracks were distinct from those found in the Connecticut region. His conclusion was that a new species, *Grallator lescurei,* could be identified on the basis of material from Saint-Lèons, whereas the material from Saint-Laurent was best assigned to the known North American species *Grallator minisculus.* In a subsequent paper discussing the sites at Sauclières,[10] Demathieu and Sciau identified four species of dinosaur tracks: *Grallator variabilis,* the new ichnospecies *Grallator sauclierensis* (figure 5.4), *Grallator minisculus* and *Dilophosauripus williamsi,* and the nondinosaurian ichnospecies *Batrachopus deweyi,* a Connecticut Valley ichnospecies. Clearly, the presence of *G. variabilis, G. minisculus,* and *Batrachopus deweyi* (figure 5.5) indicates that the Sauclières ichnofauna has elements in common with the other sites from the Massif Central, Le Veillon, and the Connecticut Valley. The presence of *Dilophosauripus williamsi,* an ichnospecies from Arizona that is close to *Eubrontes,* extends the thread of common ichnospecies to western North America, where *Grallator* and *Eubrontes* also occur.

It is probably fair to conclude that although French ichnologists have made efforts to be objective and quantitative in their descriptions of track assemblages from Le Veillon and the Massif Central region, their decisions in naming new ichnospecies have been somewhat subjective. For example, all the common Le Veillon ichnospecies are regarded as similar to, or in some cases very close to, Connecticut Valley species, whereas the rarer forms are placed in new genera. The reasons for naming new species and genera, therefore, are not altogether clear.

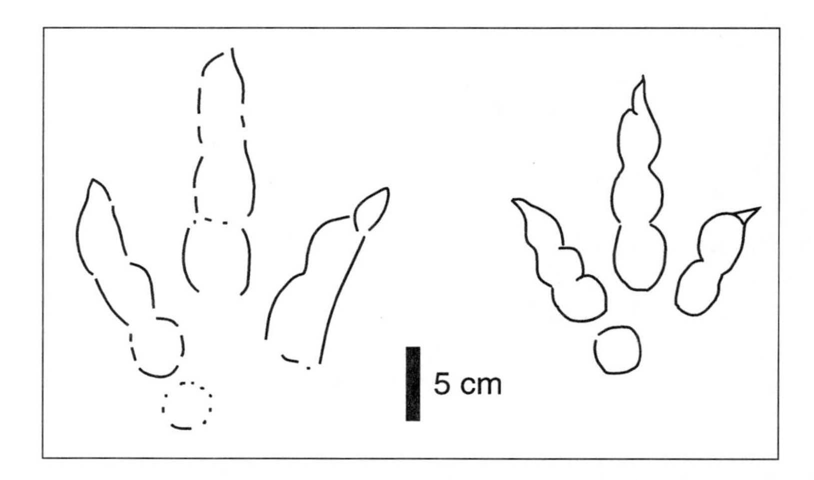

FIGURE 5.4. Tracks from the Lias (Hettangian) of the Causses. *After Demathieu.*[9]

There is certainly little discussion of distinctive features that warrant the establishment of new ichnospecies. In short, the ichnospecies labels are something of a compromise. By this we mean that the use of familiar ichnogenera (*Grallator, Eubrontes,* and *Batrachopus*) acknowledges similarity, whereas the use of new ichnospecies names conveys the suggestion that the track assemblages are distinct in some way. But only a moment's consideration will reveal that no reliable criteria have been presented to demonstrate that the samples are distinct at the species level. We know only two pertinent facts with certainty: (1) the samples (from France and America) are different, and (2) they have been given different names at the species level.

All other features such as age and ichnogenus labels, and possibly even the detailed morphology, are essentially the same. Thus, the conclusion can be drawn that the ichnospecies names are essentially geographic labels that identify the sites from which the tracks in question originate.

This conclusion only applies to the Le Veillon assemblage because a case can be made that the ichnospecies *G. lescurei* from Saint-Lèons was distinguished from ichnospecies from other areas as the result of a careful and objective study.[8] Similarly *G. sauclierensis* from Sauclières was described and discussed in relation to other species, and criteria for distinguishing it were presented.

It should be stressed that our characterization of certain ichnospecies as mere geographic labels that separate one assemblage from another without actually demonstrating morphologic differences in no way implies a criticism of the authors involved, nor does it imply that the ichnospecies are not valid, although we have questioned certain names (as above). The naming of ichnospecies must be understood in a historical context. As outlined at the beginning of this chapter, in the early nineteenth century Hitchcock[1] originally named many ichnospecies, using what most late-twentieth-century ichnologists would probably call mainly subjective criteria. Most of these names were preserved in the revisions published by

FIGURE 5.5. *Batrachopus* tracks from the Lias (Hettangian) of the Causses. *After Demathieu and Sciau.*[10] These tracks also resemble *Antipus.*

Lull.[2] The result is that we have the impression that there are probably too many names and that the ichnotaxonomy is duplicative to some extent.

How do we get this impression? The answer is mainly by intuition and word of mouth. Since Lull's day, modern paleobiology has tended to look at individual morphologic features as part of a spectrum of morphologic variation. The result of this sample-based or statistical population perspective is a tendency toward lumping rather than splitting, and the intuition that Hitchcock and Lull were probably splitters. But no one has comprehensively reanalyzed the Hitchcock collection by statistical or other means to establish specific, quantitative (or qualitative) morphologic criteria that can help test the idea that the ichnofauna is oversplit.[11] Until this is done to the satisfaction of the ichnologic community, we also cannot begin to determine which names are valid and which are not. Thus, French track researchers, like everyone else, have no choice but to compare their local track assemblages with the classic Connecticut Valley material in the way that they have done. They can only say that it looks "similar to" the American material or that it does not. Short of revising American track ichnotaxonomy, prior to or during their research into their own material, they can only describe their own assemblages and make educated, subjective decisions as to whether to introduce new names (which could later prove to be synonyms of the American names) or whether to use the existing American names, which might not ultimately prove valid, either in relation to other American names or in relation to the characteristics of the sample being described. Such a convoluted situation can be resolved only by returning to the source of the problem: the adequate description and untangling of the ichnotaxonomic complexities, both historical and morphological, that surround the literature on the original Connecticut material. Again, the French research is adequate under these circumstances and a reflection of ichnotaxonomic history.

Sticklers may argue that there is more to a full understanding of Lower Jurassic tracks than a statistical or morphometric analysis of the shapes of the classic types such as *Grallator* and *Eubrontes* from Connecticut. Ideally, careful study of the context and preservation of tracks should be undertaken prior to morphological analysis. This way, we would hope to unequivocally establish which tracks were true tracks and which owed part of their form to extramorphologic (i.e., preservational) influences.

FRANCE: SANARY SUR MER

Another French tracksite worthy of attention is at Sanary sur Mer on the Mediterranean coast about 40 km east of Marseille. One of the most distinctive footprints from this site is a slender-toed tridactyl footprint resembling the track of a large bird (figure 5.6). Such footprints are not common in the Lower Jurassic, where the footprints of the *Grallator* and *Eubrontes* type are so abundant, but it does appear that such slender-toed tracks are geographically widespread. Footprints of approximately the same size, shape, and age are known from Hungary, where they have been named *Komlosaurus* after the town of Komlo,[12] from North Africa,[13] from South Africa,[14] and from the western United States.[15] At this stage in current ichnologic research it is not clear whether these footprints are all of the same type or varieties of a more general type, as is the case with *Grallator* footprints. What is clear, as this and another example (below) show, is that distinctive new track types that have yet to be fully described are still being discovered in the Early Jurassic of Europe and elsewhere.

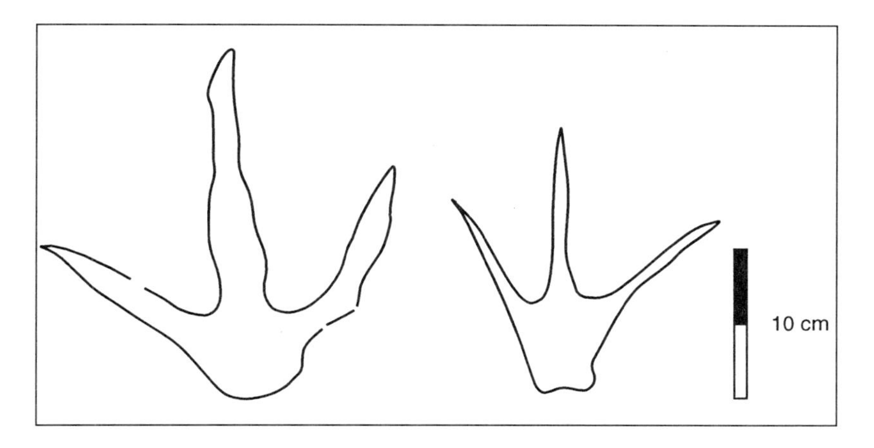

FIGURE 5.6. Slender-toed track from Sanary sur Mer (left), courtesy of Paul Ellenberger, compared with *Komlosaurus* from Hungary (right). See Kordos.[12]

TRACKS FROM SWEDISH COAL MINES AND RAILROAD TUNNELS

From the breezy shores of the Bay of Biscay and the pastoral settings of the Massif Central, we can move to a rather different setting: that of a coal mine in the region of Scania in southern Sweden. The discovery of fossil footprints preserved in the roofs of coal mines is somewhat unusual in Europe, although famous examples of abundant tracks are known from the Cretaceous of the western United States.[15] The first report of tracks underground was published in 1991 by Swedish geologists Anders Ahlberg and Mikael Siverson when they reported a dinosaur footprint attributed to *Grallator* (*Eubrontes*) from the Höganäs Formation in the Helsingborg Railway tunnel in Scania.[16] To the best of our knowledge, this is the first report of a dinosaur track in a railway tunnel. For security reasons it was not possible to remove the tracks from the tunnel for fear of endangering the integrity of the engineering, so a replica of the track was made and reproduced in bronze and exhibited at the railway station. Subsequent reports of underground tracks from this region were published by Polish ichnologist Gerard Gierlinski and Anders Ahlberg in 1994,[17] based on discoveries from coal mines in the Höganäs-Bjuv area that date back to the 1950s.

In the latest Triassic and Early Jurassic, the region that is now southern Sweden was undergoing a change from an arid to a humid setting, marked by the deposition of the Höganäs Formation, which consists of sandstones, kaolin-rich mudstones, and coals, representing swampy or wetland settings in a coastal plain environment.[17] Footprints of dinosaurs occur in both the latest Triassic (Rhaetian) and Earliest Jurassic (Hettangian) portions of this formation. As we did in our review of the Le Veillon tracks, it is convenient to treat these footprints that straddle the boundary in this chapter.

Having firmly established that various species of *Grallator, Eubrontes,* and *Batrachopus,* typical of the classic Lower Jurassic deposits or "classic Liassic"[1] of the Connecticut Valley region, are also typical of deposits of this age in France, we can consider the labels applied to tracks of the same age in southern Sweden. The footprints in this region have been attributed to *Grallator* (*Eubrontes*) *cf. giganteus* and *Grallator* (*Eubrontes*) *soltykovensis* by Gierlinski and Ahlberg (figure 5.7). The former species is reported from the older, Triassic part of the Höganäs Formation, the latter from the younger, Jurassic part. The stereotyped view of paleontologic vocabulary is that it is wordy and difficult because of so many long, double-barreled names, and we have tried in many cases to use only single generic (i.e., genus) names. But what is the significance of these triple-barreled names? Again, as is often the case in the study of Lower Jurassic tracks, the answer goes back to the footprint assemblages of the Connecticut Valley. In an attempt to understand the relationship between the many species of *Grallator, Anchisauripus,* and *Eubrontes* named by Hitchcock[1] and Lull,[2] Columbia University paleontologist Paul Olsen plotted all type specimens on a single graph[3] and showed that they appear to represent a continuum from small (*Grallator*) to large (*Eubrontes*). He therefore proposed that they really all belong to the same ichnogenus, *Grallator.* His study, however, was preliminary and based on fairly simple measurements of track size and shape, so he did not propose abolishing the names *Anchisauripus* and *Eubrontes* altogether. Instead, he proposed the naming of what we might call three subgenera or, strictly speaking, three subichnogenera, namely, *Grallator* (*Grallator*), *Grallator* (*Anchisauripus*), and *Grallator* (*Eubrontes*). The use of these names by Gierlinski and Ahlberg represents another example of introducing Connecticut Valley track terminology into the European ichnologic literature.

We are in general agreement with Gierlinski and Ahlberg that the Swedish tracks are similar to those from the Connecticut Valley. Therefore, the use of the label *Grallator* (*Eubrontes*) *giganteus* focuses on similarities and has the advantage of not introducing any new terminology. *Grallator* (*Eubrontes*) *soltykovensis,* as the name applies, is based on a track originally named by Gierlinski in 1991[18] from the locality of Soltykow in Poland (see following section), and subsequently recognized in 1994 in Sweden. In our opinion this track is similar to *Talmontopus tersi* from Le Veillon and *Grallator sauclierensis* from the Massif Central, as well as tracks from the western United States[15] (figure 5.7) and tracks from China.[19] The question that remains unresolved is which of the original Hitchcock collection tracks, if any, it is most similar to. If it can be shown that a valid name exists among the Connecticut Valley ichnospecies, then all of the various ichnospecies assigned to this morphotype might prove to be incorrect (i.e., redundant). Pending a revision or study of the appropriate material from the Connecticut Valley region, the most constructive contribution we can make is to point out some of these similarities so that they are taken into consideration in future studies.

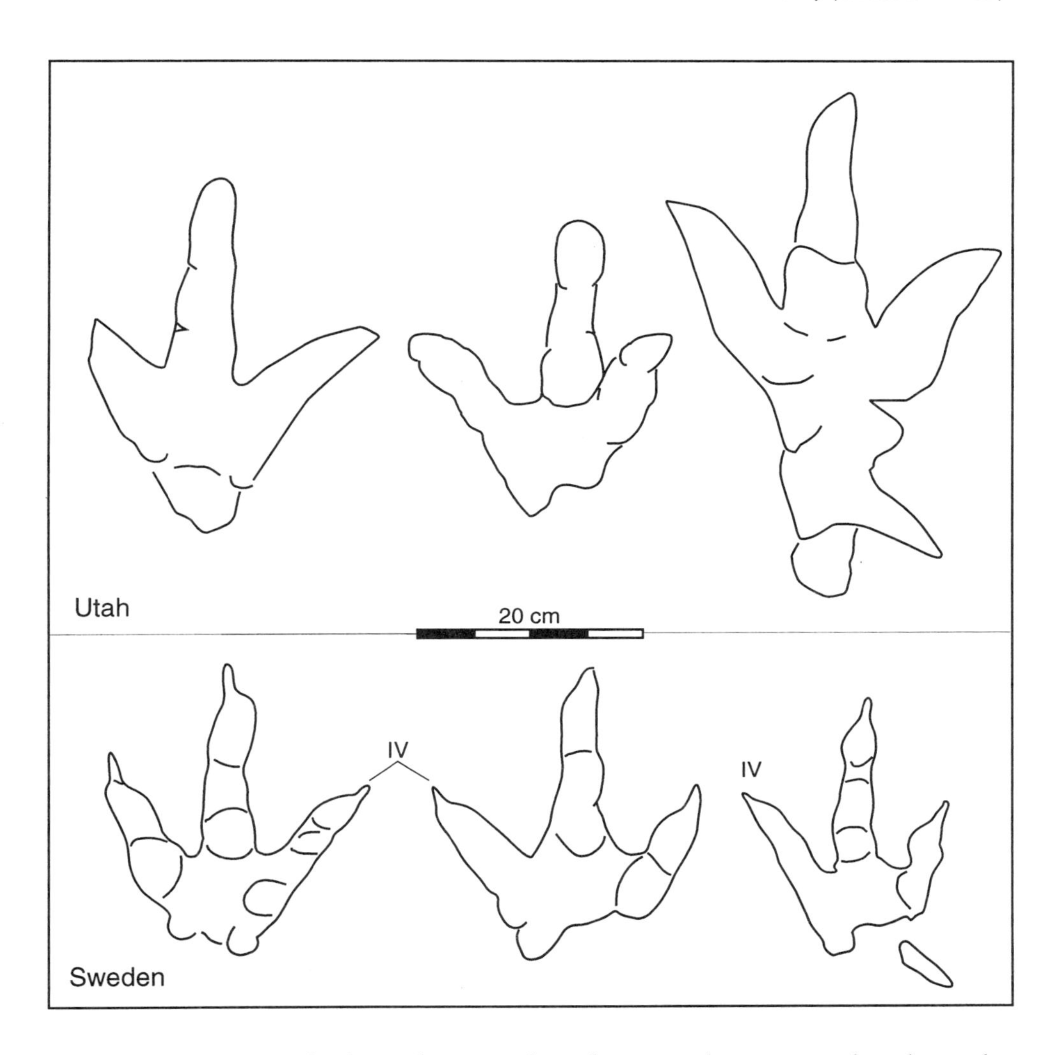

FIGURE 5.7. Tracks from the Lias of southern Sweden compared with similar tracks from Utah.

TRACKING IN THE HOLY CROSS MOUNTAINS, POLAND

As noted in the previous section, Gierlinski described the theropod footprint *Grallator (Eubrontes) soltykovensis* from Early Hettangian deposits near Soltykow in the Holy Cross Mountains of Poland, in a paper published in 1991.[18] He also described another new species, *Grallator (Grallator) zvierzi,* named in honor of Jadwiga Zwierz, head of the museum of the Geological Institute in Warsaw, and reported that it originated from somewhat younger, Late Hettangian deposits. We have already commented that *Grallator (Eubrontes) soltykovensis* is similar to tracks of the same age in France. *Grallator (Grallator) zvierzi,* according to Gier-

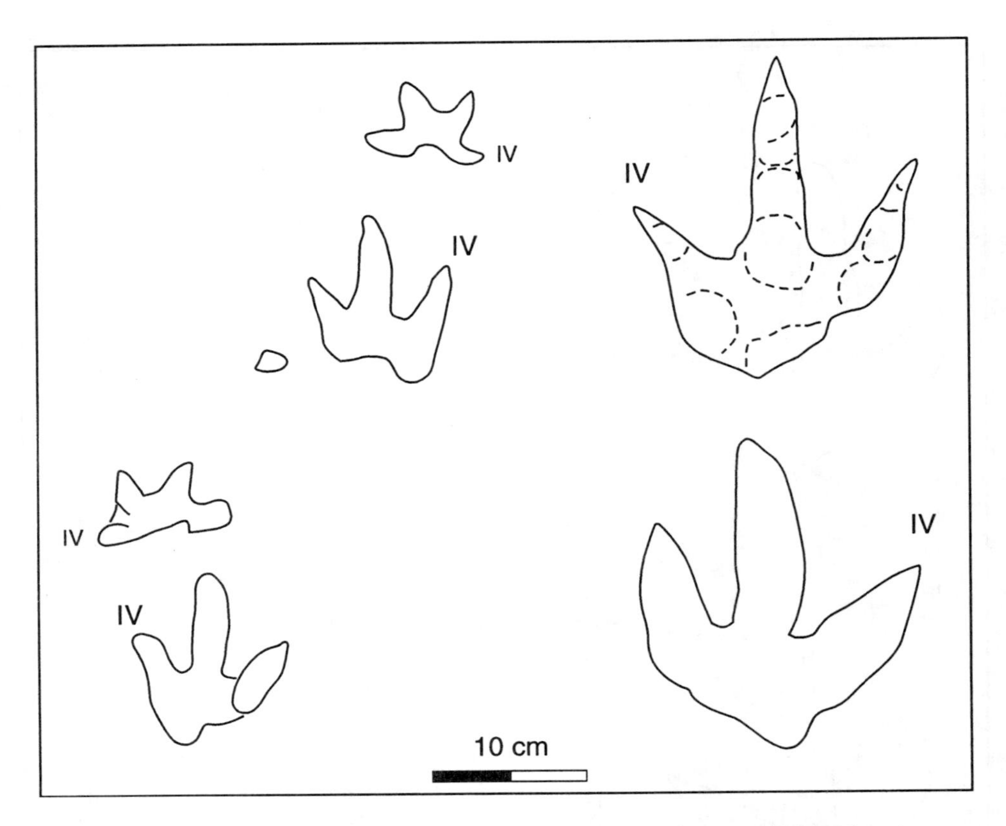

FIGURE 5.8. Tracks from the Lias of Poland. *Anomoepus* (left), showing two manus-pes sets, and (right) *Grallator* (*Eubrontes*). *After Gierlinski.*[18]

linski, is similar to *G. cuneatus* and *G. tenuis* from Connecticut and *G. variabilis* from France, as well as various other *Grallator* tracks from around the world. What is most interesting is that the Holy Cross footprint fauna includes not only the various theropod tracks assigned to *Grallator* and *Eubrontes* but also footprints of entirely different types. These have been assigned to the ichnogenera *Anomoepus* and *Moyenosauripus.* The former track type (*Anomoepus*) was named in 1848 by Hitchcock[1,20] and is now widely accepted as an example of an ornithopod track type. This interpretation is based on the presence, in some trackways, of distinctive four-toed or five-toed front footprints, which are consistent with ornithopod morphology but cannot possibly be reconciled with the anatomy of the theropod hand. The name *Anomoepus* means "anomalous feet," highlighting the disparity between front and back footprints (figure 5.8). We agree with Gierlinski that certain tracks from the Holy Cross Mountains (figure 5.8) can be assigned to *Anomoepus;* in this case he named the new ichnospecies *Anomoepus pienkovskii* in honor of the Polish geologist Gregorz Pienkovskii. Gierlinski contends that the Polish ichnospecies differs from the classic Connecticut forms in having a larger manus footprint.

FIGURE 5.9. Typical configuration of many classic *Anomoepus* tracks from North America and *Moyenosauripus* tracks from southern Africa. *Redrawn from Lull[2] and Ellenberger.[14]*

The new ichnospecies *Moyenosauripus karaszevskii,* named by Gierlinski in the same study[18] in honor of Wladyslaw Karaszewski, is also reported from the same Late Hettangian deposits as *Anomoepus.* Before discussing our interpretation of this trackway, it is necessary to review what is known of *Moyenosauripus.* The ichnogenus was originally named by Ellenberger[21] on the basis of very well preserved tracks from southern Africa, which he recognized as being similar to *Anomoepus* from Connecticut. It is now generally accepted that the track type is similar to *Anomoepus,* as has been reiterated by several researchers.[22,23] Among the distinctive features of *Moyenosauripus* (named after a site called Ile de Moyeni) are the disparity in size and shape between hind and front feet, as in *Anomoepus,* and a distinctive elongate heel impression in many examples (figure 5.9), with a short impression of digit I (the hallux) preserved on the inside of the trackway in

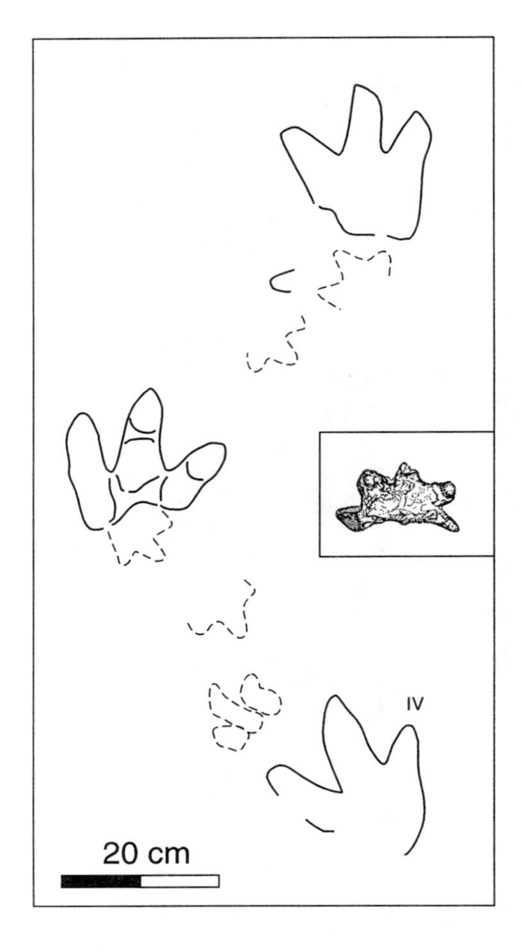

FIGURE 5.10. Purported *Moyenosauripus* from the Lias of Poland with a manus (inset) from a different trackway is not the same as *Moyenosauripus* from Africa. This manus track resembles an ankylosaur footprint. Compare with figures 5.9 and 5.10. *Redrawn from Gierlinski.*[18]

some cases. All these features are undoubtedly characteristic of *Anomoepus,* supporting the case that the two ichnogenera are similar, if not essentially impossible to differentiate at the level of the ichnogenus and ichnospecies.

Gierlinski evidently agreed with this viewpoint to the extent that he proposed that five of the seven ichnospecies of *Moyenosauripus* named by Ellenberger, from Africa, should be transferred to *Anomoepus.* However, he also suggested that the remaining two ichnospecies (*M. natator* and *M. levicaudata*) should remain in *Moyenosauripus.*[18] Gierlinski asserted that these two ichnospecies from Africa, as well as *M. karaszevskii* from Poland (figure 5.10), have a different morphology because they display only "two phalangeal pads on digit III"[18] in contrast to the three found in *Anomoepus.* Having examined *Moyenosauripus* in Ellenberger's collection, we agree that it is similar to *Anomoepus,* but we think that Gierlinski's *M. karaszevskii,* with only two pads on digit II, is probably a completely different ichnogenus. Unfortunately, his material does not have well-preserved manus tracks in association with the pes. In fact, only one manus track is reported and attributed to this ichnospecies (figure 5.10).

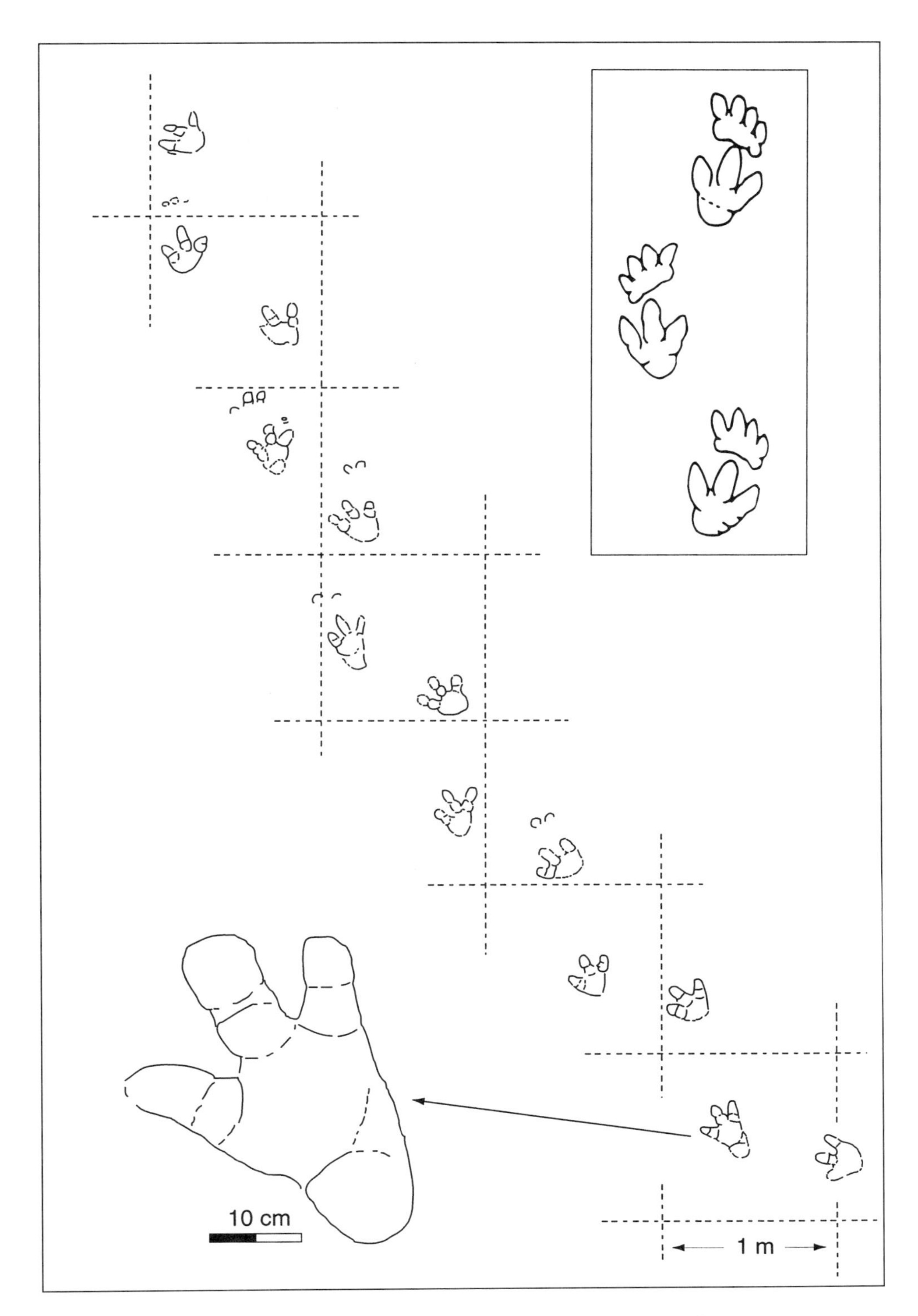

FIGURE 5.11. Trackway of a stegosaur-like species from the lowermost Jurassic (Hettangian) of the Dordogne region, France. Sporadic, faint traces of pes digit I not shown. Predicted stegosaur trackway is shown (box) for comparison. *Redrawn from Le Loeuff et al.,[25] and Thulborn.[23]*

The key to understanding this track (*M. karaszevskii*) may lie in the analysis of an entirely new discovery described in the following section. In short, this new discovery of a well-preserved Hettangian trackway from the Dordogne region of France (figure 5.11) confirms the presence of a trackmaker that is larger and more robust than *Anomoepus* (and *Moyenosauripus*), fundamentally different in morphology, and not of ornithopod affinity. This track type is stegosaur-like and appears to be thyreophoran in origin (see following section). The implications of this conclusion are complex. *Moyenosauripus* is the same as, or very like, *Anomoepus*, and cannot be redefined in the way that Gierlinski proposed. His new species, *karaszevskii*, is perhaps distinct from other tracks in the Polish sample but has orphan status because it lacks a suitable name. Gierlinski recently suggested that this track may also have been made by an early thyreophoran and not an ornithopod, as he previously inferred.[24] Thyreophoran (meaning "shield bearer") refers to the armored and plated dinosaurs (see below).[25]

Gierlinski has also reported the tracks of an animal that he considers to be a sauropod from the Lower Jurassic of the Holy Cross Mountains.[26] The tracks are quite small by sauropod standards, consisting of oval-shaped pes impressions about 30 cm long and 20 cm wide, with a narrow gauge. The tracks also show a manus that conforms to the typical sauropod pattern. The only other sauropod tracks known from Europe are those reported from northern Italy (see section below).

We cannot end this section without a word of praise for Gierlinski. The tracks he has described are not found on large, conveniently exposed surfaces. Rather, slabs have to be excavated and put back together piece by piece. Gierlinski has done a great service by identifying the first good examples of *Anomoepus* tracks from Europe. It seems he may also have discovered another distinctive large form of possible stegosaurian affinity. He has also opened up the question of how we differentiate the tracks of early ornithopods from those of other ornithischians (stegosaurs and ankylosaurs or thyreophorans). We shall examine this question further in the following section.

TRACKWAY EVIDENCE FOR THE EARLY ORIGIN OF STEGOSAURS

For thirty years, since their marriage, Jean and Colette Picaud have lived down on the farm on the edge of Corgnac, a small village in the northeast of the Dordogne. Mme Picaud (*née* Petit) and her brother Jean-Pierre had played in the courtyard when the farm belonged to their grandmother years before that.

During that time all three were unaware of the importance of the slab of rock in front of the farmhouse. The shape in it that they had always thought resembled an enormous footprint was just that—only it was the print of a dinosaur that lived at the beginning of the Jurassic period over two hundred million years ago.

This story, taken from a popular magazine on France,[27] tells a familiar story. Dinosaur tracks are often right under our noses, literally in our front or back yards. Yet we often do not notice them unless we know what we are looking for. The story goes on to tell how, on a visit back to the farm, Jean-Pierre's childhood memories were revived, and he took a closer look at the rock pavement that makes up the courtyard. As he explored the surface he concluded that he was looking at footprints similar to some he had seen on a recent trip to Esperaza in the foothills of the Pyrenees. He decided to consult the experts and enlisted the help of Jean Le Loeuff, director of the Dinosaur Museum at Esperaza. Le Loeuff soon confirmed that the tracks are indeed those of dinosaurs.

According to the magazine article, the tracks have been billed as the "best in France" and have been identified as being of several different types, attributed to a large ornithopod (ornithischian) and two theropods, one of which was assigned to *Grallator.* The site has yet to be studied in detail, although replicas of the main tracks have been made. In our collaborations with Le Loeuff, we have concluded that the ornithischian trackway (figure 5.11) is probably attributable to a stegosaur, one of the well-known variety of plated dinosaurs that became abundant in the Late Jurassic. As we shall see, the correspondence between the tracks and stegosaur feet is almost perfect, except for one small but significant detail.

So what are the features that so strongly suggest stegosaurian affinity? First, the hind footprint is quite large (about 40 cm long, including the heel, and 28 to 30 cm wide; figure 5.11). The toes are short and blunt, with only two distinct pads on each toe. Moreover, these pads, especially the proximal ones, which are taken to represent first phalanges that attach to the metatarsus (ankle), are very short and wide. This morphology is entirely consistent with the known morphology of stegosaur hind feet. Other characteristics of the trackway are a stride that averages about 1.23 m and slight inward rotation of the axis of the hind foot. None of these features is really characteristic of the best-known ornithopod tracks from the Lower Jurassic, namely, *Anomoepus* (and *Moyenosauripus*) described in the previous section (figure 5.10). Apart from exhibiting the impression of digit I and a phalangeal formula of 2–3–4, corresponding to digits II, III, and IV, *Anomoepus* and *Moyenosauripus* generally have a distinctive elongate heel, when it is preserved, and a generally gracile appearance and small size (foot length and width are typically in the range of 12 to 18 cm), therefore averaging about half the size of the stegosaurian prints. Some examples of *Moyenosauripus,* however, are larger. It is clear that the same disparity in size and shape affects the manus tracks, which in the stegosaurian trackway are large and robust, but in *Anomoepus* are small and slender by comparison. The weight of evidence favors a stegosaurian trackmaker interpretation for the French trackway, but the jury is still out, and the term "basal, thyreophoran" is less specific and less controversial.[25]

There is a technical ambiguity regarding the affinity of these tracks. Although the hind foot is clearly three-toed, at least functionally, some researchers[25] claim to have seen faint traces of digit I (the hallux) in some, but not all, tracks. Because loss of the hallux is one of the features that distinguishes stegosaurs from other thyreophorans (ankylosaurs), we have a completely stegosaurian footprint throughout much of the trackway, with a hint of a nonstegosaurian hallux in a few of the prints. If three toes were stegosaurian and one was not, was it truly a stegosaurian foot? What we can say is that the reduction of phalanges from 3–4–5 to the stegosaurian condition (2–2–2) for digits II, III, and IV had already taken place before the hallux (digit I) was lost. That such rapid changes in foot evolution occurred so early in the Early Jurassic could hardly have been predicted without the discovery of such clear footprint evidence.

Another reason the French discovery is so important, and possibly deserving of the title of the "best" tracks in France, is that a complete stegosaur trackway has never previously been reported, even though we now think the Polish trackway described in the previous section (figure 5.10) may be of this type. What is even more exciting is that the trackway dates from the very earliest stage in the Jurassic (the Hettangian), about 204 to 208 million years old, yet the oldest known skeletal remains of stegosaurs, from China and England, are no more than 165 to 175 million years old. Thus, the trackway evidence appears to push back the origin of a stegosaur-like foot by as much as 30 to 40 million years. Prior to this discovery there had only been two reports of isolated tracks of stegosaurs from 150 million year ago, Late Jurassic deposits in North America. One of these reports[28] features only hind footprint casts, and the other,[15] a single front footprint cast that has been given the name *Stegopodus*. Not everyone may agree that these American Upper Jurassic tracks are stegosaurian in origin, but at least we can all agree that stegosaurs were abundant at that time. But what of the French tracks appearing so early in the Jurassic? Surely these are likely to prove even more controversial.

In such cases, where the evidence is at variance with conventional wisdom, it is important to give all the reasons why the trackway is assigned to a group not yet thought to exist on the basis of skeletal remains. The reasons in this case are fairly straightforward. The hind footprints provide an almost perfect match for stegosaurian feet. In short, among the large quadrupedal ornithischian dinosaurs known to have existed at any time during the Jurassic, there are three main groups: the armored dinosaurs or ankylosaurs, the plated dinosaurs or stegosaurs, and the ornithopods, which developed into the large so-called duck-billed varieties such as *Iguanodon* by the early Cretaceous (see chapter 8). A fourth ancestral basal thyreophoran group includes forms such as *Scelidosaurus* found in the Lower Jurassic.

The ankylosaurs had four toes on their hind feet. By contrast, the hind feet of stegosaurs had only three blunt toes, each with only two phalanges. The proximal phalanges, attached to the ankle bones (metatarsals) were short, and the distal phalanges or unguals, which would have been sheathed by "toe nails," were slightly longer but still quite short and blunt. This is exactly the pattern we see in the trackway from the Dordogne region. Moreover, the front footprints, which are not so completely preserved because the animal overstepped them with its hind footprints, also show three toe impressions similar in size to the pes. This fits the stegosaurian pattern very well because stegosaurs had three large digits on the manus (digits I, II, and III) and two smaller digits (IV and V), as the North American example shows. The isolated manus track reported from Poland (figure 5.10) is similar to the French examples but looks ankylosaur-like.

Because of the lack of known stegosaur trackways prior to this discovery, track expert Tony Thulborn reconstructed a stegosaur trackway (figure 5.9).[23] His reconstruction (really a prediction) is very close to the French trackway, a result that reflects well on Thulborn's ichnologic predictive skills. Of particular note is that Thulborn has inferred a fleshy heel pad on the hind foot, as seen in slightly more elongate form in the real trackway. Thulborn made his predicted stegosaur slightly flat-footed (i.e., with outward rotation of the foot axis), whereas the real trackway suggests a slightly pigeon-toed or "duck-footed" stegosaur with noticeable inward rotation. Otherwise there are no significant differences in morphology. The greater overlap of hind on front feet in the French trackway is a function of variable gait.

Just as we dismissed the possibility of an ankylosaurian interpretation, so too the possibility of an ornithopod interpretation appears to be out of the question, despite what was written in the popular article.[27] First, there is no evidence of such large ornithopods in the Early Jurassic. In fact, tracks of comparable size that have been attributed to ornithopods do not appear until the Cretaceous. Second, ornithopods had many more phalanges in their hind toes, and so their digit impressions were longer. In the case of large Cretaceous ornithopods, which had very well padded feet, their tracks reveal a single large coalesced, lozenge-shaped pad, never two phalangeal pads. The front footprints of large ornithopods, if they are preserved at all, for many ornithopods were bipedal, are much smaller than the pes tracks, and consist of small hoof-shaped tracks that clearly indicate that the digits of the manus were tightly bound by skin or integument. Thus, by a process of elimination, neither large ornithopods nor ankylosaurs appear to have been responsible for the tracks from Corgnac. This leaves only very stegosaur-like animals as possible candidates, or some hitherto unknown relatives.

Although it is surprising to find such ancient stegosaur-like trackways, we should remember that there are other examples of trackways of particular groups of animals that are much older than their bones (see chapter 2). Put another way, there is no reason why the oldest bones known also represent the time at which the group originated. In fact, only through pure serendipity could we expect to find the remains of the very earliest representatives of a particular group. This is particularly true when it comes to creatures such as dinosaurs that are sometimes known only from a handful of skeletons. Clearly the trackway from Corgnac is of usual significance in this regard. It provides a key to our understanding of the early evolution of stegosaurs. Stegosaurs belong to the ornithischian group of dinosaurs that includes both ornithopods and ankylosaurs, as noted above. Both of these groups have Early Jurassic representatives that share certain characteristics such as armor plates. Thus, the discovery of trackway evidence for Early Jurassic stegosaur relatives is not so surprising. If we believed that the oldest bones were in fact these known from the Middle Jurassic, we would have to explain the late appearance of stegosaurs relative to other ornithischians. The occurrence of stegosaur-like tracks in Poland and France suggests that proto-stegosaurs were widespread in Europe early in the Jurassic, thus underscoring the importance of the track record.

A final and obvious, but important, conclusion derived from the French trackways is that stegosaurs were quadrupedal. This is perhaps intuitively obvious to most researchers, and it is certainly the way they appear in most reconstructions. Certain researchers, including Robert Bakker, author of the controversial book *Dinosaur Heresies,*[28] have advocated that they may have been bipedal or facultative bipeds. The track evidence lends no support to this hypothesis.[29]

THE FIRST SAUROPODS? EVIDENCE FROM ITALY

On the margin of the Dolomite region, near the towns of Trento and Rovereto in northern Italy, in the year 883, a massive earthquake-induced landslide carried millions of tons of Jurassic bedrock into the valley below. The scar or "ravine" left on the slopes above has become known as the Lavini di Marco. This local landmark was also the site of bloody battles in World War I, and shell impact craters can still be identified and are sometimes confused with other geologic features such as dinosaur tracks. In recent years the site has become famous for its dinosaur tracks. Unlike the tracks discussed in previous sections of this chapter, the Lavini de Marco footprints are associated with carbonate (limestone) rather than clastic (sandstone) environments, giving us an insight into the dinosaurs that frequented such habitats.

The tracks from this site were discovered only a few years ago and first described in detail by Giuseppe Leonardi, Marco Avanzini, and their colleagues,[30–32] who concluded that the tracks were those of sauropods (figure 5.12), ornithopods, and theropods. We visited the site and determined that the main track types are those of theropods and sauropods only (figure 5.13), a conclusion supported by subsequent work in the region.[33] The presence of sauropod and theropod tracks in limestone substrates appears to be a recurrent theme throughout the Jurassic and Cretaceous, and has given rise to the suggestion that there was a brontosaur tracks facies or "*Brontopodus* ichnofacies."[34] Sauropods, therefore, had a preference for such habitats, which represented semi-arid to arid climatic regimes in subtropical to tropical low latitudes.[35] This is a pattern we shall see repeated in the Middle and Upper Jurassic and in the Cretaceous (see chapters 6 through 8).

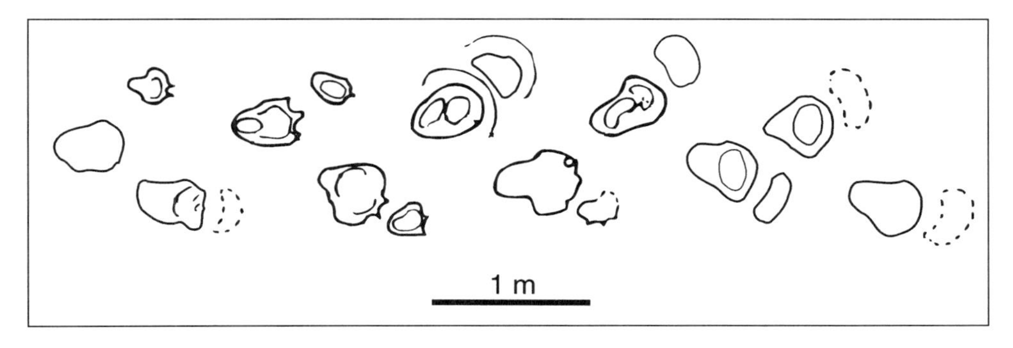

FIGURE 5.12. Sauropod trackway from the Lavini de Marco site. *Redrawn from Lanzinger and Leonardi*[30]*; Leonardi and Avanzini.*[31]

Detailed studies of the tracks in this region are ongoing, so few track names have been assigned. Some of the larger theropod tracks have been labeled *Eubrontes*[36] and others have been referred to as *Grallator.* The sauropod tracks are relatively small to medium-sized (45 to 50 cm in length) and indicate animals with a narrow gauge (*Parabrontopodus* type[36,37]). Preliminary reports[30–32] that suggest the presence of large ornithopods are probably incorrect. It has been fairly well established that large ornithopod tracks of the type often attributed to *Iguanodon*-sized species are not known from the Jurassic with any certainty and appear abundantly only in the Cretaceous (see chapter 8) some 50 million years later than those discussed in this chapter. Thus, a large ornithopod with footprints 35 to 40 cm in diameter, as proposed in some of the initial studies, would be highly anomalous for this epoch.

One of the most interesting of the aforementioned studies of the Lavini di Marco site was the attempted reconstruction of the ancient environment by Avanzini and colleagues,[33] in which they showed that the dinosaurs were crossing a tidal flat at times when it was temporarily exposed. Traditionally, geologists have regarded the types of limestone sequence found at Lavini di Marco as what are known as "platform carbonates"—that is, shallow marine deposits that contain only marine fossils. The discovery of tracks in several layers forces geologists to rethink this interpretation and to conclude that the platform was periodically emergent above sea level. Such periods of emergence have significant implications for geology. Not only must geologists accept that these areas that were once considered marine were part of the coastal continental landscape, at least periodically, but they must also look at the influences of fresh water and vegetation growth on the flora and fauna of these coastal regions. In geology, the way in which fresh water, rather than sea water, influences limestone is a major field of study that has implications for how quickly the sediment is dissolved or cemented. Such factors, in turn, may influence the potential of such sediments for producing oil and gas.

Avanzini and colleagues found that despite the draining of the platform to leave the tidal flats periodically exposed, the main influence on the sediments was

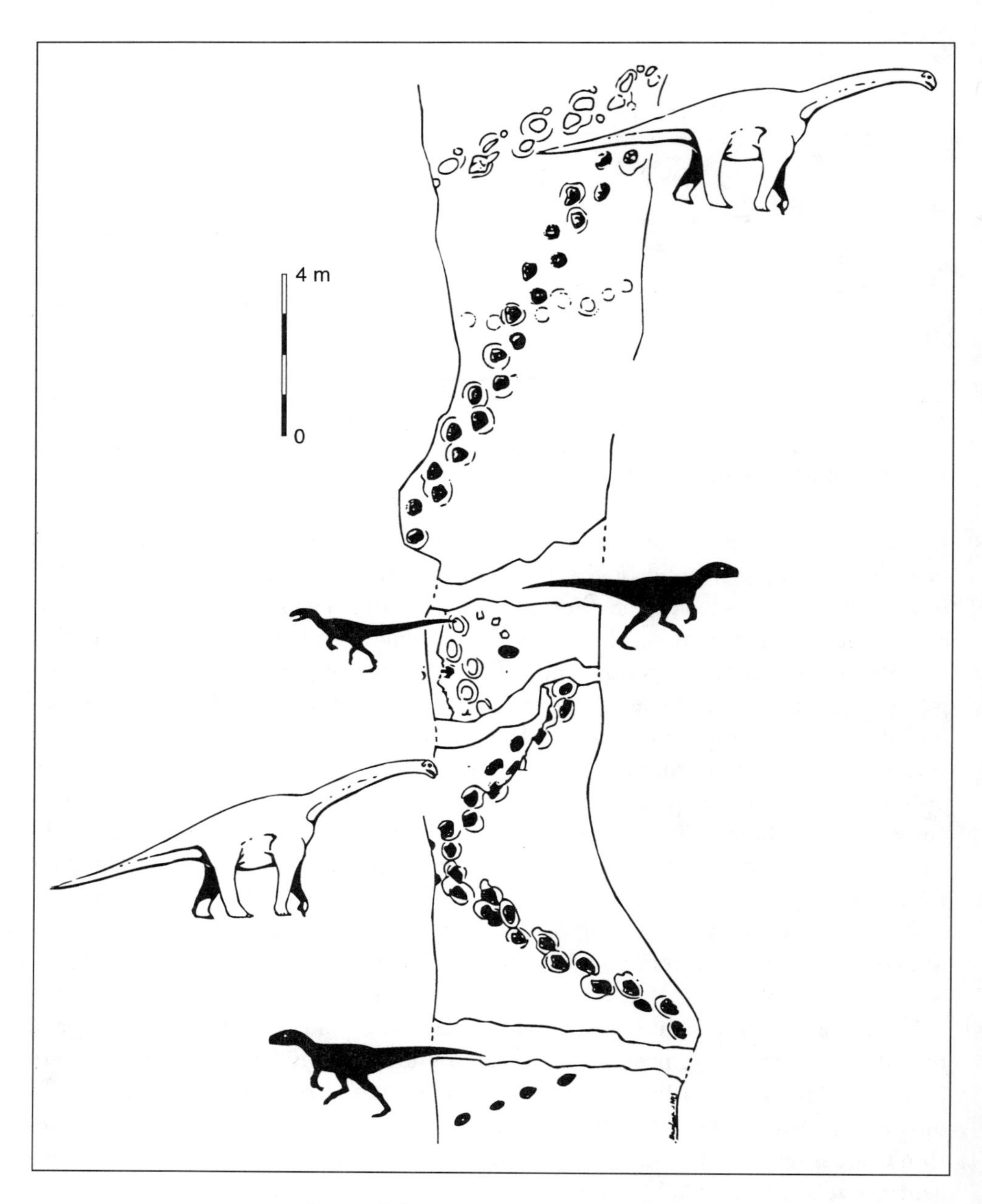

FIGURE 5.13. Map of part of the Lavini de Marco site suggests a sauropod-theropod track assemblage.

by sea water, not fresh water. Fresh water runoff from rain and rivers may have stimulated the growth of continental vegetation in localized areas, providing food for dinosaurs. More significantly, we learn that owing to the arid to semi-arid setting, evaporation was intense, leading to rapid cementation of the limey mud flats into solid limestone and dolomite. Here again we learn a very interesting lesson that helps answer an often asked question: "How were the tracks preserved?" Quite rightly, the nongeologist finds it hard to imagine that tracks made while the tide was out are preserved after it floods back in. Avanzini's study suggests that the lime mud hardened quickly under the influence of cementation, caused by evaporation, a chemical process, rather than mere drying, or baking, which can be undone by redissolving the dry sediment. This is the same principle used in quick-drying cement. The study concluded that "early cementation in semi arid and arid settings could play a major role in the fossilization of dinosaur footprints on carbonate tidal flats."[33] Given that dinosaur tracks are common on such carbonate flats throughout the Jurassic and Cretaceous, such conclusions have far-reaching implications that may help explain preservation at many sites discussed in subsequent chapters.

BIBLIOGRAPHIC NOTES

1. Hitchcock, E. 1858.
2. Lull, R. S. 1953.
3. Olsen, P. E. 1980.
4. Lapparent, A. F. de and G. Montenat. 1967.
5. Montenat, C. 1967.
6. Alexander, R. McN. 1976.
7. Taquet, P. 1998.
8. Demathieu, G. 1990.
9. Demathieu, G. 1993.
10. Demathieu, G. and J. Sciau. 1995.
11. Lockley, M. G. In press.
12. Kordos, L. 1983.
13. Ishigaki, S. 1986.
14. Ellenberger, P. 1972.
15. Lockley, M. G. and A. Hunt. 1995.
16. Ahlberg, A. and M. Siverson. 1991.
17. Gierlinski, G. and A. Ahlberg. 1994.
18. Gierlinski, G. 1991.
19. Zhen, S., R. Li, C. Rao, N. Mateer, and M. G. Lockley. 1989.
20. Hitchcock, E. 1848.
21. Ellenberger, P. 1974.
22. Olsen, P. E. and P. Galton. 1984.

23. Thulborn, R. A. 1984.
24. Gierlinski, G. In press.
25. Le Loeuff, J. et al. 1998.
26. Gierlinski, G. 1997.
27. Anderton, S. 1997.
28. Bakker, R. T. 1996.
29. Lockley, M. G. and A. P. Hunt. 1998.
30. Lanzinger, M. and G. Leonardi. 1991.
31. Leonardi, G. and M. Avanzini. 1994.
32. Avanzini, M. and G. Leonardi. 1993.
33. Avanzini, M., J. van den Dreissche, and E. Keppens. 1997.
34. Lockley, M. G., A. P. Hunt, and C. Meyer. 1994.
35. Lockley, M. G., C. Meyer, A. P. Hunt, and S. G. Lucas. 1994.
36. Avanzini, M. 1995.
37. Lockley, M. G., J. O. Farlow, and C. A. Meyer. 1994.

A large Middle Jurassic Sauropod makes tracks in the area that is now near Fatima in Central Portugal. *Courtesy of Mauricio Anton.*

CHAPTER 6

THE DARK AGES: MIDDLE JURASSIC

DINOSAURS IN THE GREAT DELTAS OF YORKSHIRE

The good news about dinosaur tracks from the Yorkshire region of England is that they represent an epoch, the Middle Jurassic, that is not well known for producing skeletal remains. Hence the information they provide is potentially useful in reconstructing the faunas that existed in an otherwise dark age in paleontologic history. Examples include the recently described possible sauropod track *Deltapodus* and the enigmatic trace *Ravatichnus,* both described below. The bad news regarding the Yorkshire tracks is that many are isolated examples of three-toed traces that are not easy to identify or distinguish, at first glance, from tracks found in other older, or younger, Mesozoic deposits.

The Middle Jurassic deposits of the Yorkshire coast have traditionally been regarded as estuarine and deltaic deposits and were formerly named the "Great Estuarine Series." They consist of various fluvio-deltaic deposits exposed in the sea cliffs around Whitby. These cliffs are actively eroded by the North Sea, and most of the tracks that come to light do so as the result of blocks falling out of the cliffs onto the rocky foreshore. Here they are often within the intertidal zone and so are accessible for study and replication only between tides and in suitable weather conditions.

Tracks were first discovered and reported from the Yorkshire coast in the 1890s and early 1900s when Harold Broderick[1] described various three-toed tracks from a site near Whitby (figure 6.1). Since that time there have been many other sporadic

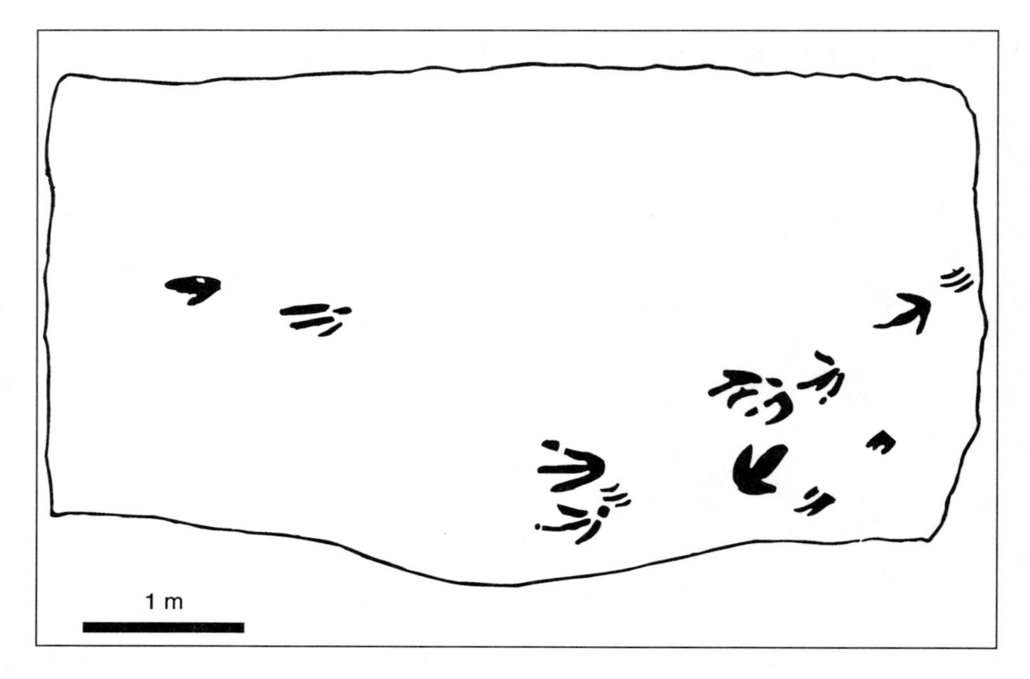

FIGURE 6.1. Middle Jurassic tracks from Yorkshire. *After Broderick.*[1]

discoveries of small three-toed tracks, as summarized by Cyril Ivens and Geoffrey Watson in their booklet, *Records of Dinosaur Footprints on the North Coast of Yorkshire.*[2] Such summaries are useful for telling the modern ichnologist how many tracks have been found and the locations from which they originated. Such local publications are also classic examples of the tradition of detailed historical documentation prevalent in so many natural history disciplines in Europe. But such documentation does not tell us whether the tracks are similar to footprints found in other regions and whether such comparisons have been attempted. In this case, the only person who had attempted a comparison was Bill Sarjeant (see chapters 2 and 3), who compared one of the tracks described by Broderick with *Satapliasauras* from the Cretaceous of Russia.[3] Although we applaud Sarjeant's efforts to compare the Yorkshire tracks with other material,[4] we have two reservations about this particular comparison. First, *Satapliasaurus* from Russia is not well known and has not been described from a complete trackway. Second, work on both Triassic and Jurassic track zones strongly suggests that correlations are convincingly demonstrated only between rocks of the same age. Given that Sarjeant did not intend to make a categorical statement about the similarity of the Yorkshire tracks to those from Russia, we advocate caution in attaching much significance to this comparison.

It is entirely probable that our inability to easily match the Yorkshire tridactyl tracks with footprints from elsewhere is simply the result of such obvious factors

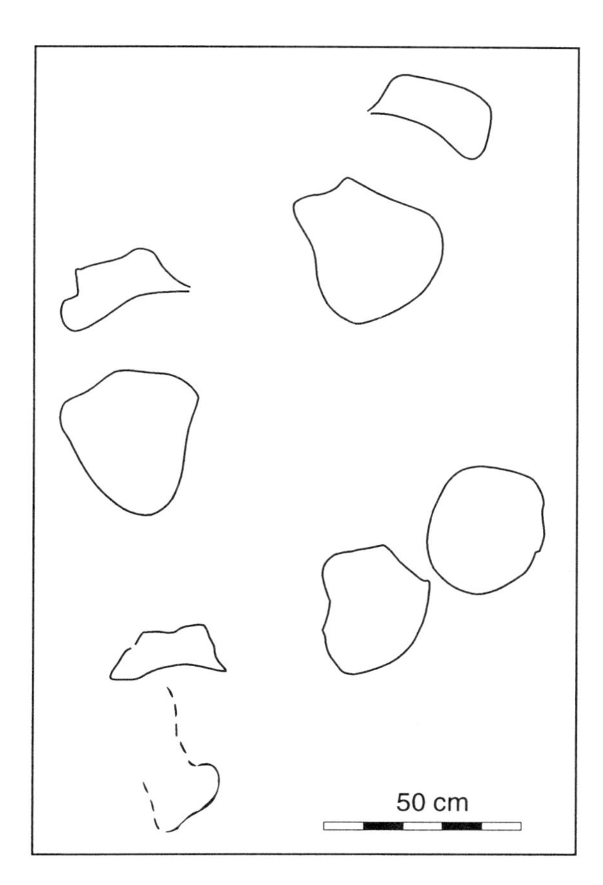

FIGURE 6.2. *Deltapodus brodericki* from the Middle Jurassic of Yorkshire. *After Whyte and Romano.*[6,7]

as lack of study and lack of other documented examples of the right age (i.e., Aalenian to early Bajocian). We know of no well-documented assemblages of tracks of this age. A recently documented track assemblage from the later Middle Jurassic (Bathonian) of Utah contains distinctive footprints that appear identical to tracks of the same age in Buckinghamshire, England.[5] Such track correlations, which are the first ever proposed for Middle Jurassic footprints, suggest the potential for correlation of the Yorkshire footprints once further work has been completed.

Research on such correlations is being undertaken by Mike Romano and Martin Whyte from the University of Sheffield. Although they have not yet completed an up-to-date description of the three-toed footprints, they have already discovered and named the trackway of a large quadrupedal animal (figure 6.2). This distinctive set of tracks in which the hind footprint is roughly triangular, giving the name *Deltapodus,* and the fore footprint is crescent-shaped is so far unique in the track record. At this time the only possible trackmaker candidates appear to be sauropods, stegosaurs, or ankylosaurs. Whyte and Romano suggest that the trackmaker was most probably a sauropod.[6,7] We have expressed doubts about this interpretation[8] owing to the three-toed configuration of the hind footprint and the

FIGURE 6.3. *Ravatichnus,* an enigmatic five-toed trace, was first reported from the Middle Jurassic of central Asia (left), but appears also to have been found in the Jurassic of Yorkshire (right). *Redrawn from Lockley et al.*[9]

low frequency of reports of sauropod tracks in such deltaic deposits. However, we can offer no particularly compelling alternative theory except to say that, in our opinion, the tracks might be of thyreophoran origin (i.e., ankylosaurid or stegosaurid). The hind footprints match the predicted pattern for stegosaur tracks and remind us that thyreophoran footprints like the recently discovered French trackway (see chapter 5) date back to the Early Jurassic.

Finally, we should note that the Sheffield group has uncovered evidence of a very distinctive trace consisting of five rounded impressions arranged in a semicircle. Such a trace, named *Ravatichnus* (figure 6.3), is known from Middle Jurassic fluvio-deltaic deposits in Central Asia.[9] Hence, a correlation may be made using tracks, despite a complete lack of knowledge of the origin of the trace. This correlation is only the second made on the basis of Middle Jurassic tracks.

GLIMPSE OF A DINOSAUR FROM THE DARK AGES

In the same way that we can tell the story of giant Middle Jurassic sauropods from Portugal and describe many of their characteristics without having any skeletal remains of the actual animal in hand, so we can also tell the story of a small Middle Jurassic carnivore from England and Utah. Although the tracks (figure 6.4) were first discovered in England, they were intensively studied in North America, and so the story is most easily told from that side of the Atlantic. In North America the lack of Middle Jurassic dinosaur fossils is even more striking than in Europe. There are absolutely no known remains from the entire continent, so the discovery of abundant tracks is obviously significant. The sites in question are found in the Carmel Formation on the borders of Dinosaur National Monument in eastern Utah,[5] where the formation is thought to be Bathonian in age. The tracks named *Carmelopodus* (meaning "track from the Carmel Formation") are all quite small, ranging from pigeon-sized footprints (about 4 cm long) up to the size of the tracks of a small emu (15 to 20 cm).

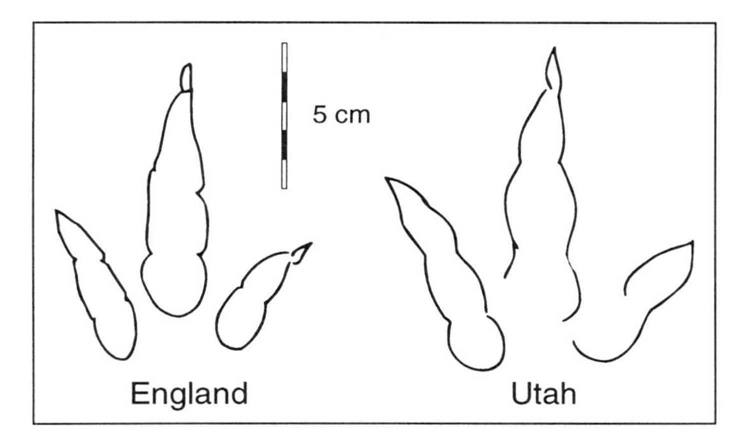

FIGURE 6.4. *Carmelopodus,* a small theropod track from the Middle Jurassic of Buckinghamshire, is apparently identical to tracks from the Carmel Formation of eastern Utah. The tracks indicate an animal with a short metatarsal IV. *Redrawn from Lockley et al.*[5]

Judging from the abundance of Carmel footprints, comprising at least 60 trackways from a scattering of localities, the trackmaker was quite common. The track anatomy is also distinctive and apparently unique to this species. Whereas typical footprints of carnivorous dinosaurs (theropods) reveal a foot pad formula of 2–3–4 corresponding to digits II, III, and IV, this track type indicates the formula 2–3–3. From this we can conclude that the trackmaker had a shortened ankle bone on digit IV (or short metatarsal IV), and we can predict that the skeletons of individuals of this species of trackmaker will show this clearly, if we ever find any. By happy coincidence, *Carmelopodus* has been recorded in rocks of the same age (Bathonian) in the Forest Marble in Buckinghamshire in England.[10]

To our knowledge this is the first example of a reliable correlation of Middle Jurassic tracks between North America and Europe, with the possible exception of the aforementioned *Ravatichnus.* Apart from demonstrating that the trackmaker of *Carmelopodus* footprints was widespread, it is evidence that the widening Atlantic had not entirely separated North America from Europe. Although this was a dark epoch in the age of dinosaurs, tracks such as *Carmelopodus* shed light on hitherto unsuspected details of theropod foot anatomy and may foreshadow future discoveries of foot skeletons with this distinctive anatomy.

THE FIRST IBERIAN SAUROPODS

In 1990, when Vanda Faria dos Santos first took an interest in the few poorly documented tracksites in Portugal, she could not have anticipated the remarkable accelerated pace of discovery and documentation that would pursue her studies for the years that followed. Her initial aim was simply to document known sites as a small research project related to her geology degree and discuss their significance in the context of emerging new perspectives on dinosaur ichnology. Since that time, the Portuguese track record has emerged as one of the most important in Europe, giving us new insight into the value of tracks for interpreting morphology and behavior, especially among sauropods. Portugal also has provided fascinating case histories in environmental sciences, as large tracksites have been discovered and set aside for posterity. In this short time span dos Santos, in conjunction with colleagues such as ourselves, has helped coauthor dozens of scientific papers, including a volume entitled *Aspects of Sauropod Paleobiology.*[11] As we shall see, the Portuguese track story extends into subsequent chapters of this book (see chapters 7 through 9) and is by no means complete.

From a geologic perspective, the Portuguese dinosaur footprints story begins in the Middle Jurassic at a site near Fatima, today a destination for religious pilgrims. Between May and October of 1917, hundreds claimed to have witnessed an illuminated vision of the Virgin Mary in a field near Fatima. According to many observers, this was not an isolated vision, for numerous pilgrims testified that they witnessed supernatural illuminations at the same site for several months.

There is no telling how the legend of Fatima might have evolved had giant brontosaur tracks been known in 1917, but this was not the case. They were not discovered until 1994. The present tracksite is essentially the floor of a large quarry, excavated in predominantly marine limestones of Bajocian-Bathonian age. (In recognition of the owner, the quarry is also known as "Pedreira do Galinha" rather than simply "the Fatima site"). It has come as something of a surprise that rock sequences traditionally regarded by geologists as wholly marine in origin somehow reveal abundant dinosaur tracks (see chapters 5 and 7). The simplest explanation, and the one that also retains the marine interpretation, is that the limestone layers (beds) are essentially marine. That is, they were laid down underwater, albeit in shallow water. But the sea level fluctuated, exposing surfaces periodically. It was on these "emergent surfaces," as geologists call them, that the dinosaurs walked. Simply put, although the limestone layers were deposited underwater, the surfaces that separate them represent hiatuses, or times of nondeposition, evidently when the sea level dropped, leaving the limey substrates exposed (see chapter 5). Footprints were "registered" at these times when sediment was not being laid down.

The quarry floor, which represents one of these surfaces, is approximately 150 m wide and more than 200 m long (30,000 m^2) and conveniently surrounded on one side by a semi-circular platform or cuvette of overlying strata that provides an excellent vantage point to view the site. In fact, the site was discovered by an observant caver walking along this rim. At the time, however, much of the floor of the working quarry was covered in rock dust. It was after only the quarry owners cleaned the surface that the full extent of the trackways became evident (figure 6.5).

FIGURE 6.5. Photograph of the Fatima tracksite, the largest Middle Jurassic site presently known and now a national monument. The brontosaur trackways include two segments that measure 142 and 147 m, the longest on record.

The most spectacular trails are two segments of brontosaur trackway that measure 142 and 147 m. At the time of their discovery in 1994, these were the longest trackways anywhere in the world.[11] The previous record, of 141 m, had been held by another Portuguese trackway at Carenque in the outskirts of Lisbon (see chapter 9). It is not the record length of the trackways that is important, even though they qualify for the *Guinness Book of Records.* What makes them exceptional is that they are the largest sauropod tracks known from the Middle Jurassic and that they have a very distinctive morphology (figures 6.6 to 6.8). The two distinctive features are the large size of the front footprint (manus) relative to the hind footprint (pes) and the presence of a large claw impression on the inside of the manus track.

These features have significant implications for the study of sauropod tracks because until recently it was thought that they were all more or less the same. Had this been the case, there would have been little prospect of ever differentiating the footprints made by different groups (families) within the sauropods. We now know, however, that sauropod trackways vary in width and can be classified as either narrow gauge or wide gauge.[12] Wide-gauge trackways, of which the Fatima tracks are a good example, are found in both the Jurassic and the Creta-

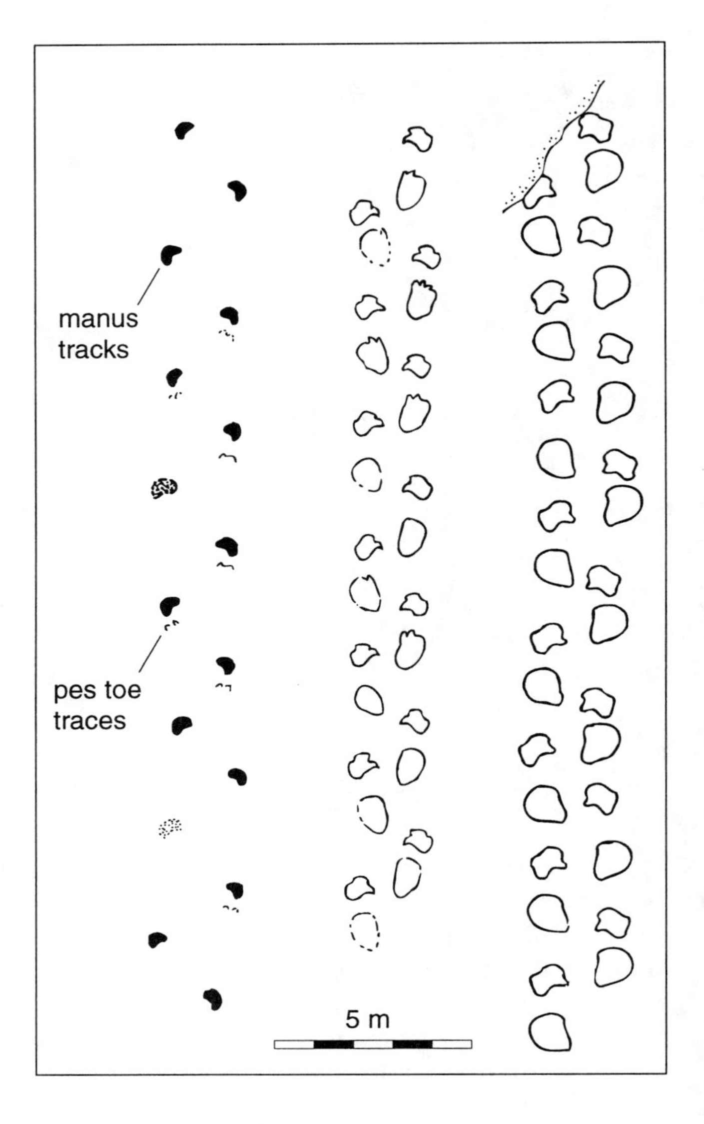

FIGURE 6.6.
Detail of sauropod trackways from the Fatima site. *After dos Santos et al.*[11]

ceous, whereas narrow-gauge tracks appear to be confined almost exclusively to the Jurassic.[8]

When we look at the difference in size, also known as the "degree of heteropody," between front and back footprints in sauropod trackways, we also notice considerable variation. The Fatima tracks are a good example of sauropod footprints in which the manus is exceptionally large. In fact, they are the largest manus tracks ever recorded and the largest relative to the pes. The ratio of the area covered by the manus and pes, respectively, is about 35:65 or 1:2. This compares with ratios of 25:75 (1:3), 20:80 (1:4), or even 17:83 (1:5) seen in other sauropod footprints (figure 6.7). Although the Fatima tracks have not been given a new

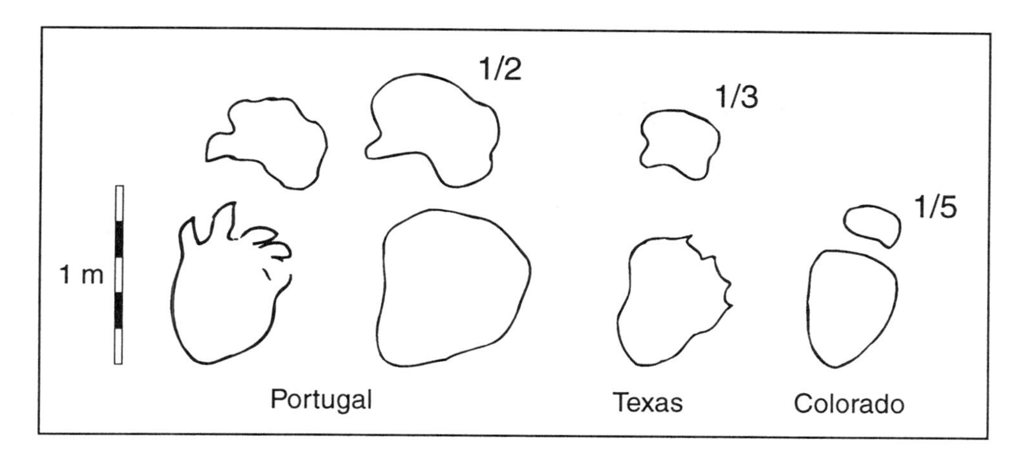

FIGURE 6.7. Details of sauropod tracks from the Fatima site. Note the large size of the manus impression relative to the pes track (ratio 1:2). Also note the impression of the well-developed manus claw. *After dos Santos et al.*[11]

name, a case can be made that they are distinctive enough to be recognized as a new ichnospecies.

It is not just the size of the manus tracks that is noteworthy. They also exhibit the impression of a large claw (digit I) that is situated on the internal or medial side of the track (figure 6.7). It has always been something of a puzzle as to how this claw was carried by sauropods, for in almost all footprints an impression was lacking. This lack of a manus claw trace has led to various hypotheses, including the explanation proposed in 1980 by Gerard de Beaumont and Georges Demathieu[13] that the claw and other finger and wrist bones rotated backward, making the plantar surface on which these animals walked (figure 6.9). We do not agree with this hypothesis because the sauropod metacarpus (wrist) is so obviously designed for weight bearing. The Fatima tracks demonstrate that the claw was in contact with the ground and situated in the position where one would predict that the distal end of erect metacarpal I was incorporated into the plantar surface, or sole, of the foot. In other words, the knuckle-walking hypothesis is not supported by the trackway evidence.

In summary, we can conclude that the Fatima tracksite is one of the most important sauropod footprint sites in the world. Not only is it the largest Middle Jurassic sauropod tracksite, but it is the one at which tracks are best preserved. The tracks are wide gauge and have the largest manus tracks ever reported, both in actual size and in proportion to the pes footprints. In addition, the tracks provide clear evidence of the way in which the manus claw was carried, thus refuting a controversial interpretation. Although no other track types have been documented from the Fatima site, dos Santos (written communication) has reported the discov-

FIGURE 6.8. Close-up of sauropod trackway from the Fatima site.

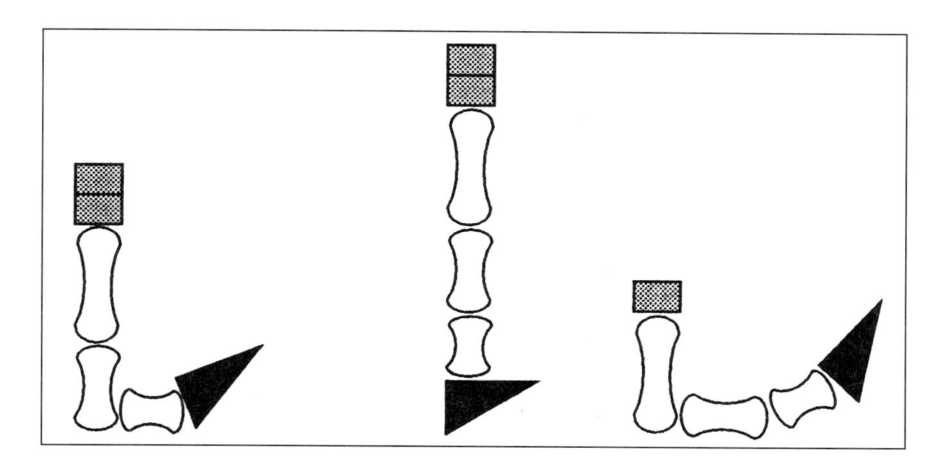

FIGURE 6.9. The controversial interpretation of Beaumont and Demathieu[13] that sauropods rotated their wrists to the rear (posterior) was proposed to explain the lack of manus claw impressions. This interpretation is not widely accepted or supported by the track evidence.

ery of a large theropod track (63 cm in length), which may be the largest on record in the Middle Jurassic.[14] The importance of the Fatima site also has been recognized by the Portuguese government, which has taken steps to ensure its preservation by declaring it a natural monument (see appendix).

MR. POOLEY'S ENIGMATIC TRACK DISCOVERY

In 1975, Bill Sarjeant described an enigmatic small footprint from the Stonesfield Slate near Burford in Oxfordshire. He named it *Pooleyichnus burfordensis* in honor of Mr. C. Pooley, who had discovered the footprint in 1886.[15] The track, although not well preserved, evidently represents a small five-toed animal with a foot only about 5 cm long (figure 6.10). As noted by Sarjeant, prior to 1975 there were virtually no reports of any small, five-toed Middle Jurassic footprints from anywhere in the world. In fact, the only small Jurassic footprints that Sarjeant cited were some Late Jurassic pterosaur footprints from North America[16] and the small three-toed Middle Jurassic theropod tracks discussed elsewhere in this chapter.

Sarjeant suggested that the track might be that of a mammal or a mammal-like reptile, in part because the remains of early mammals are known from the Stonesfield Slate.[15] It should be noted, however, that the Stonesfield Slate, not a true slate, also contains the remains of other vertebrates, including dinosaurs and pterosaurs. In our opinion, the track shows some similarities to the hind footprint of a pterosaur or large lizard, but this is by no means certain. If the track is mammalian in origin, as suggested by Sarjeant, then it would be one of the largest mammal

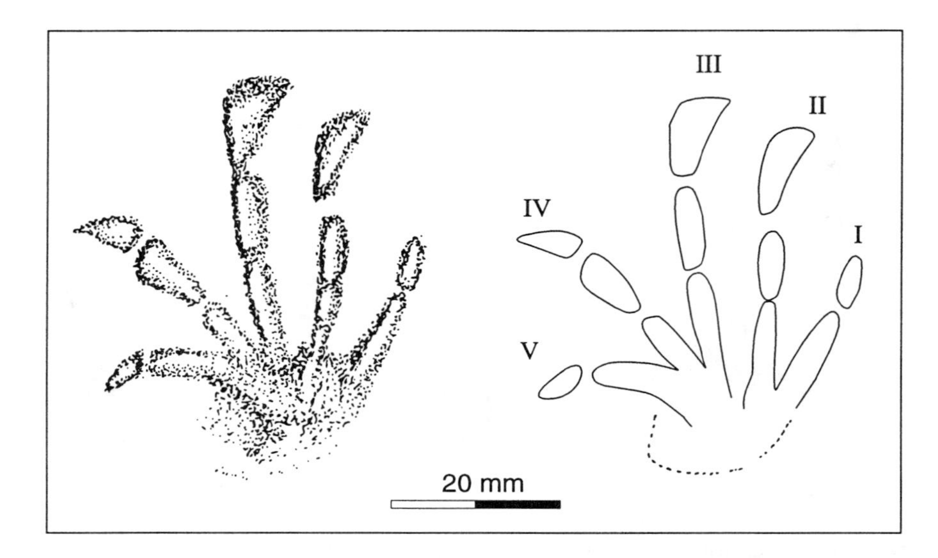

FIGURE 6.10. *Pooleyichnus burfordensis,* an enigmatic track from the Stonesfield Slate, near Burford, Oxfordshire, England. *After Sarjeant.*[15]

tracks recorded from the Mesozoic, for all other reports of mammal tracks are of animals with feet only 1 to 2 cm in length. Mr. Pooley's enigmatic track discovery remains just that—enigmatic—until we find more and better examples.

DINOSAUR TRACKS FROM THE WESTERN ISLES

When we think of the windswept Hebrides and other North Atlantic islands, we do not usually think of dinosaurs. In 1984, however, J. E. Andrews and J. D. Hudson[17] reported the first Jurassic dinosaur footprint ever reported from Scotland. The track, which originates from the Leate Shale Formation within the Great Estuarine Group, is also the first evidence of dinosaurs from Scotland. The footprint is tridactyl and preserved as a rather indistinct cast, probably an underprint, or, more precisely, a transmitted cast. It is reported to be about 45 cm long and 55 cm wide, but details do not allow us to say whether it is the right or left foot that is represented. Such dimensions are not typical of theropod tracks, which are usually wider than long, but given the imperfect preservation, the measurements are only approximations. Moreover, there are no ornithopod tracks or trackmakers from the Middle Jurassic that approach this size, so the interpretation of theropod affinity seems reasonable given the data available.

A decade after the discovery of this dinosaur footprint, dinosaur bones, consisting of the partial remains of a probable sauropod femur, were reported[18] from approximately the same stratigraphic level in the Great Estuarine Group in the formation immediately overlying the Leate Shale. In the same year, 1995, a partial theropod tibia was reported from the Lower Jurassic of Skye.[19] This rather meager record of Scottish dinosaurs is instructive of the realities of field paleontology, and we can draw the conclusion that the solitary track, although providing limited information, is just as informative as the scraps of bone. The Skye footprint provides a typical example of the utility of ichnology in providing evidence of vertebrate groups in geologic locations that had not previously produced bones. Once tracks are discovered, the potential to find bones may soon be demonstrated, as in this case, even though the bones suggest the presence of a different group of animals. Similarly the presence of bones in a formation indicates the potential to discover tracks. So dinosaurs are now known from both the trace fossil and body fossil records of Scotland, and it is the combined record that is most informative.

THE TIP OF THE ICHNOLOGIC ICEBERG

Although the Middle Jurassic track record is meager in comparison with the rich record documented in the Lower and Upper Jurassic (see chapters 5 and 7), it is far better today than it was a little more than a decade ago, when most of this chapter could not have been written. A large new site has been discovered in Oxfordshire. A group from Cambridge University, working on the site in conjunction with a team from Oxford University, made the following report:

> The site is a currently active gravel quarry and landfill a few miles north of Oxford. Unfortunately this means that the exposed trackways are being damaged by the trucks that come in and out of the quarry and that they are destined to disappear under a huge pile of rubbish. The emphasis is therefore to quickly collect and record as much data as possible. This is quite a task because of the size of the site. The total area it covers is approximately 0.5 km^2, although the floor of the quarry where the footprint bed is exposed is only about half that area. The limestones that are being quarried at the site are Bathonian in age, belonging to the White Limestone Formation, and were deposited in a marginal marine environment. The footprint bed itself is a highly bioturbated, peloidal, micritic limestone.
>
> Work done to date at the site has revealed 37 more or less continuous dinosaur trackways, a couple of which would seem to be traceable for a distance of 300 m. Individual footprints are often quite well defined, toes and claws being clearly discernible, and the trackways may be attributed to both sauropods and theropods. The former are more common, accounting for about two thirds of the trackways found so far, and both pes and manus prints are identifiable, the pes prints being typically of the order of 1 m in length. The theropod footprints are typically about 0.6 m in length [figures 6.11 and 6.12] and one of the theropod trackways was clearly made by a running individual. It is also apparent from the surveying done to date that

FIGURE 6.11. Trackway of a large theropod from the Middle Jurassic of Oxfordshire (compare with figure 6.12).

the trackways display some degree of parallelism. The trackways tend to have a general north-south orientation, with the direction of travel being northerly in the majority of cases.

Alex Burton, Cambridge University, Earth Sciences Department
April 1998

Based on this information, it is clear that this new Oxfordshire tracksite adds significantly to the Middle Jurassic track record of England and Europe and that the trackways so far observed are among the longest on record. Already it is interesting

FIGURE 6.12. Photograph of trackway of a large theropod from the Middle Jurassic of Oxfordshire (compare with figure 6.11).

to note the association of large sauropod and large theropod tracks on carbonate substrates. It is also interesting that the theropod tracks are wide gauge. They originate from the same deposits as the famous *Megalosaurus,* the first dinosaur ever described.[20] Ichnologists and dinosaur trackers everywhere await the full results with eager anticipation, and we look forward to comparing the site with the Fatima tracksite in Portugal, where large theropod tracks (63 cm long) have also been reported.

BIBLIOGRAPHIC NOTES

1. Broderick, H. 1909.
2. Ivens, C. and G. Watson. 1995.
3. Sarjeant, W. A. S. 1970.
4. Gabouniya, L. K. 1951.
5. Lockley, M. G., A. P. Hunt, M. Paquette, S-A. Bilbey, and A. Hamblin. 1998.
6. Whyte, M. A. and M. Romano. 1993.
7. Whyte, M. A. and M. Romano. 1994.
8. Lockley, M. G., C. A. Meyer, A. P. Hunt, and S. G. Lucas. 1994.
9. Lockley, M. G., C. A. Meyer, R. Schultz-Pittman, and G. Forney. 1996.
10. Delair, J. B. and W. A. S. Sarjeant. 1985.
11. Santos, V. F. dos, M. G. Lockley, C. A. Meyer, J. Carvalho, A. M. Galopim de Carvalho, and J. J. Moratalla. 1994.
12. Lockley, M. G., J. O. Farlow, and C. A. Meyer. 1994.
13. Beaumont, G. and G. Demathieu. 1980.
14. Santos, V. F. dos. 1998.
15. Sarjeant, W. A. S. 1975.
16. Stokes, W. L. 1957.
17. Andrews, J. E. and J. D. Hudson. 1984.
18. Clark, N. D. L., J. D. Boyd, R. J. Dixon, and D. A. Ross. 1995.
19. Benton, M. J., D. M. Martill, and M. A. Taylor. 1995.
20. Buckland, W. 1824.

Seven small sauropods travel as a group across a Late Jurassic coastal plain that is now part of a limestone sequence at Cabo Espichel in Portugal. *Courtesy of Paul Koroshetz.*

CHAPTER 7

THE AGE OF BRONTOSAURS: LATE JURASSIC

MEGALOSAUR TRACKS

Although known as the age of brontosaurs, the Late Jurassic Epoch in Europe is historically important for the discovery of theropod tracks, and it is here that we begin this chapter. We note also that these tracks may be (at least in some cases) of megalosaurid origin, and may be of the same type or similar to those noted in the final section of chapter 6.

It is well known among vertebrate paleontologists that *Megalosaurus* was the first dinosaur ever named, in the year 1824.[1] The namer of this famous dinosaur was the Reverend William Buckland (see chapter 2), who described this creature on the basis of a piece of lower jaw from the Middle Jurassic Stonesfield Slate of Oxfordshire, England. As noted in chapter 6, tracks are now known from strata of this age in this same area. Despite the fame of *Megalosaurus,* it is not at all well known, and the concept of the megalosaurid family is considered a "taxonomic wastebasket"[2] and its relationship to other theropods is not well understood.

Despite uncertainty about the concept of megalosaurs or megalosaurids, which results from the scarcity of skeletal remains, ichnologists have claimed that megalosaurid tracks are abundant in the early part of the Late Jurassic Epoch and that their tracks indicate very large animals with a somewhat primitive gait that were widely distributed throughout the Northern Hemisphere. This appears to be a classic case of a track record that is more extensive and complete than the corresponding bone record. But before we examine the details of this evidence and the

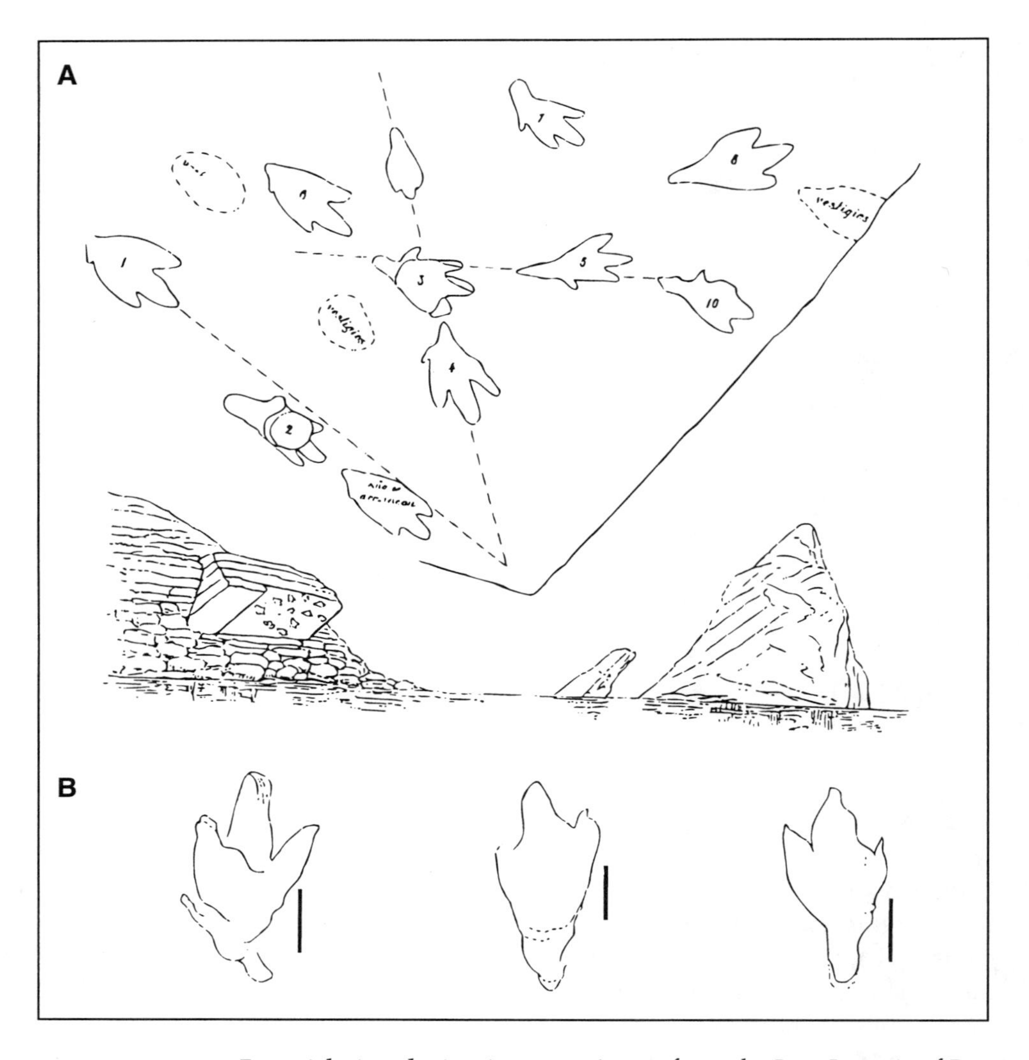

FIGURE 7.1. *Eutynichnium lusitanicum* specimens from the Late Jurassic of Portugal include material from the original type locality at Cabo Mondego, showing an outcrop and map (A) after Gomes and (B) from left to right (MNHN-MG-P261, -264, and -263 are National Museum of Natural History specimens). *After Gomes.*[5]

strength of these conclusions, we need to step back and look at the concept of megalosaur tracks as it has evolved in the ichnologic literature.

The concept of megalosaur tracks appears to originate with observations made by Albert de Lapparent and colleagues[3,4] in the 1950s when they took a second look at large theropod tracks from the Late Jurassic at Cabo Mondego in Portugal that had originally been reported in 1884 and described by Gomes in 1915–16[5] (figure 7.1). These may have been the first Late Jurassic tracks reported from Europe. Following the observations of these researchers, Lessertisseur,[6] another French ich-

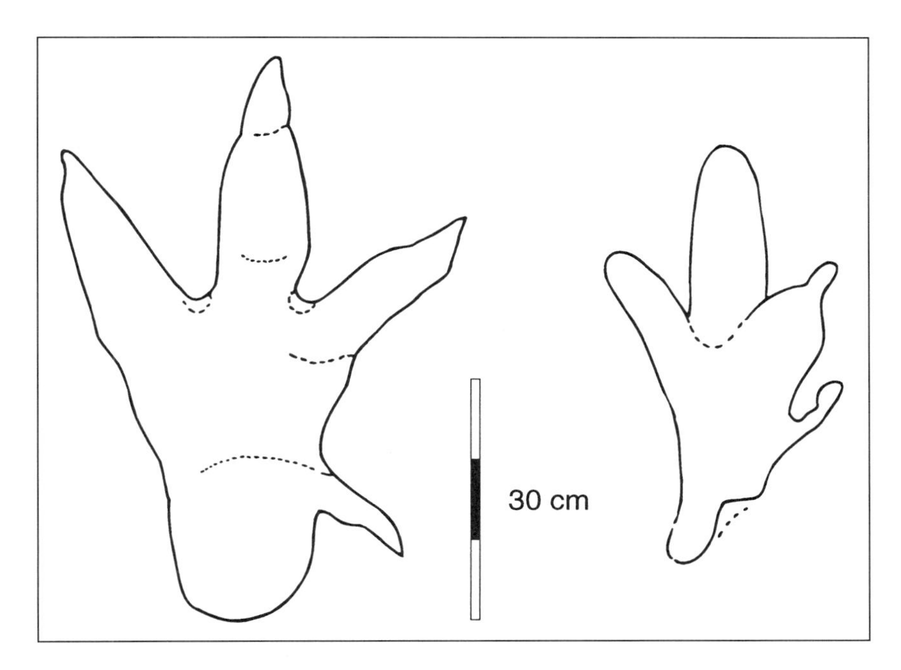

FIGURE 7.2. Line drawings of *Megalosauripus* (right) after Lessertisseur[6] and *Eutynichnium lusitanicum* Nopsca[9] after Haubold.[11] Note that *Megalosauripus* was renamed *Buckeburgichnus maximus* by Kuhn (1958).

nologist, coined the name *Megalosauripus,* spelled with an "i" to illustrate the principle of naming tracks. Although he did not actually describe any tracks under this label, he did refer to the material from Cabo Mondego as being of probable megalosaurid origin. The spelling of this name is mentioned because other researchers later spelled it with an "o" in reference to quite different footprints. Lessertisseur also coined the name *Tyrannosauripus* (also spelled with an "i") without reference to any actual material.[6,7] Evidently, he was an ichnologic theoretician, interested in examining the principle of giving precise, exact, or appropriate names to all tracks. "Il serait à souhaiter qu'on attribue à toutes ces empreintes des noms précis, ne serait qu'en accolant un suffixe au nom de leur auteur présumé (*Megalosauripus, Tyrannosauripus*)."[6]

In the case of *Megalosauripus,* however, Lessertisseur confused the issue by placing a purported megalosaur track from the Lower Cretaceous (Wealden beds of Germany), illustrated by Abel,[8] alongside a Late Jurassic example from Portugal. This mixing of tracks of different ages was done despite the fact that megalosaurs are not known from the Lower Cretaceous. To complicate the matter further, no description was provided and no specific name was assigned. The ichnogenus was therefore a *nomen nudum* (meaning "naked name") on at least two counts: (1) lack of a species name and (2) lack of an explicitly described type specimen. A *nomen nudum* in this case indicates that a label was given to a fossil without an actual

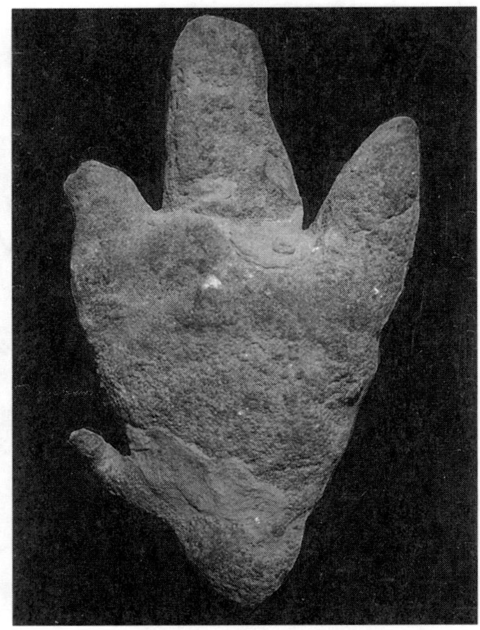

FIGURE 7.3. Photographs of large theropod tracks from Cabo Mondego *in situ* (left) and in the National Museum of Natural History (right).

known specimen ever existing to match the name or concept. If a genus name is given without a species name, as in this case, it is also a *nomen nudum.* The Cretaceous track (see chapter 8) reveals a pronounced hallux impression (figure 7.2) and is at least similar to tracks from the Late Jurassic of Portugal in this respect. It turns out, however, that another ichnospecies name, *Eutynichnium lusitanicum,* had been coined for the Portuguese material in 1923 by the famous Baron Franz von Nopsca.[9] Again, however, the name is problematic owing to the lack of a type specimen and description. It is nevertheless clear that the name *lusitanicum* connects these tracks with Portugal at an early date (Lusitania is an old name for the area that is now Portugal).

In the 1960s, 1970s, and 1980s, several investigators formally applied the label *Megalosauropus* (spelled with an "o") to Jurassic and Cretaceous material from as far afield as Australia,[10] Texas,[11] Croatia,[11] Germany,[12] and Uzbekistan,[13] and other tracks from England,[14] Spain,[15,16] and Brazil were informally referred to as megalosaurid in origin. By the 1990s it became apparent that the concept of megalosaur tracks needed to be revised. As a result of studies of Late Jurassic tracksites in Portugal and studies of similar footprints in North America and Asia, it became apparent that tracks that may be assigned to megalosaurid dinosaurs with a higher degree of confidence occur in rocks of this age on three continents.[17,18]

A reexamination of the material from Portugal attributed to *Megalosaurus* by de Lapparent and colleagues[3,4] led us to tentatively use the name *Megalosauripus*

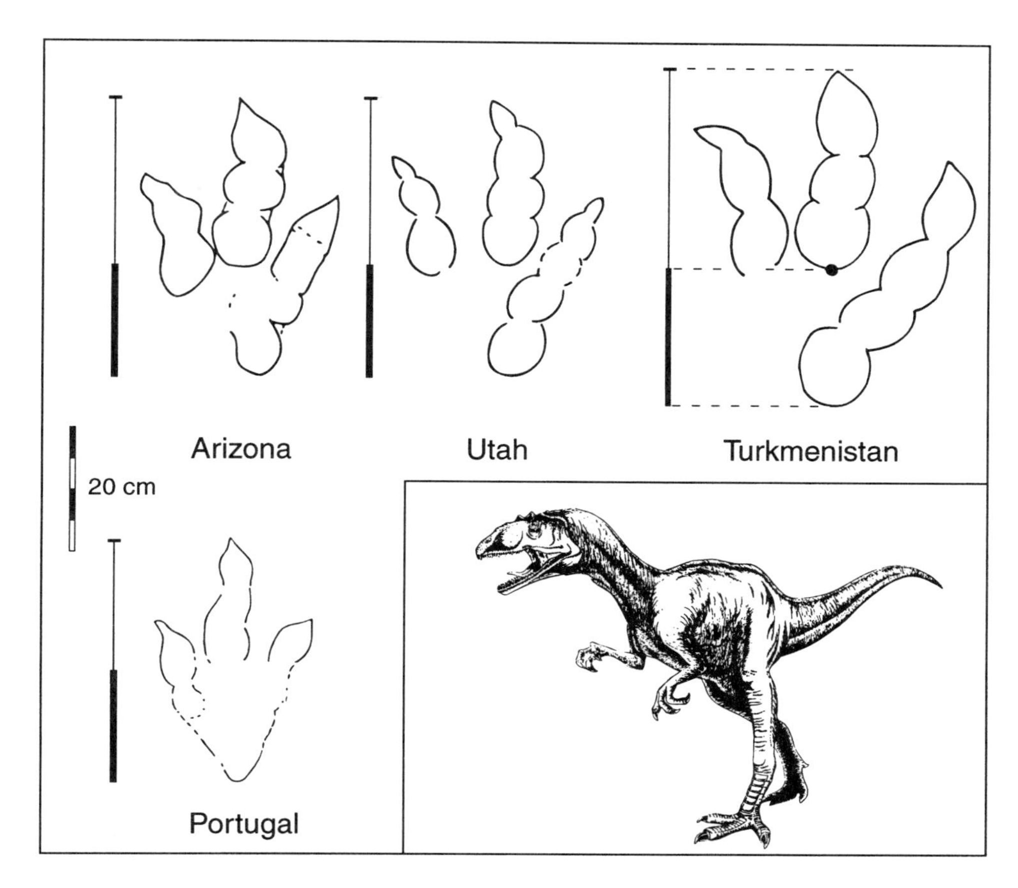

FIGURE 7.4. Comparison of *Megalosauripus* from Portugal (Cabo Mondego), North America (Arizona and Utah), and Asia (Turkmenistan). *Redrawn from Lockley et al.*[17]

(with an "i") for certain large track specimens.[17] For reasons of historical consistency, the name *Eutynichium lusitanicum* is reserved for the specimens from Cabo Mondego described by Gomes[5] and *Megalosauripus* is used for certain large tracks from other localities.[18] The name *Megalosauropus* (spelled with an "o") is now reserved for Cretaceous material from Australia, which is different from the Jurassic material.[10]

Now that we have a label for these purported megalosaurid tracks, we can return to a discussion of what the tracks and trackways actually look like and what they can tell us. Then we can discuss whether the evidence really indicates that the tracks are megalosaurid in origin. The Portuguese sample of *Megalosauripus* is perhaps not the best in the world, but it is historically important in our understanding of the whole discussion. At the Cabo Mondego site, the original sample of *Eutynichnium* tracks, now preserved in the National Museum of Natural History in Lisbon (figure 7.3), mainly represents casts of feet that sank deeply into the substrate, leaving

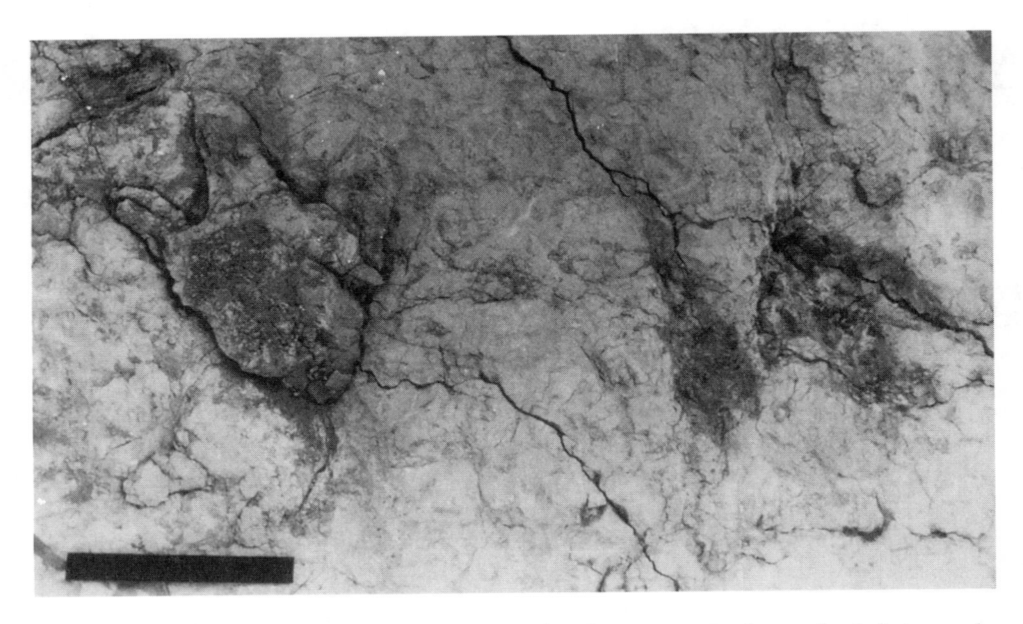

FIGURE 7.5. *Megalosauripus* from Zambujal Quarry. Track on the left is partly unfilled (compare with figure 7.6B); track on the right lacks infill (scale bar 50 cm).

impressions of the metatarsus and hallux (figure 7.1). Most other tracks—from Portugal, Spain, North America, and Asia (figure 7.4)—are shallower, much larger impressions, without traces of the metatarsus and hallux. So the material from the Cabo Mondego track sample may be different from the larger tracks in the region. They are only about 40 cm long by 30 cm wide, excluding the heel and hallux. Their trackways also appear to be narrow.

Other tracks that we attribute to megalosaurids, from Zambujal Quarry[19,20] and Praia de Cavalo,[21] are much bigger, measuring up to 77 cm by 60 cm and 67 cm by 56 cm, respectively (figures 7.5 and 7.6). When taken in conjunction with a megalosaurid track from Asia that measures 72 by 55 cm, these are the three largest theropod footprints known from the Upper Jurassic anywhere in the world. Although size alone is not a foolproof criterion for establishing that the tracks are megalosaurid in origin, it is a factor to be considered, and recent discoveries of Late Jurassic megalosaurid skeletal remains from North America suggest that these animals were very large. Preliminary estimates suggest that some may have reached 45 feet in length.[22] We can also compare these footprint dimensions with those recently reported for large Middle Jurassic theropods from Portugal and England (see chapter 6).

The trackway pattern is also of considerable interest because in many, but not all, cases it is highly irregular. By this we mean that the trackway is sometimes very wide, with great variability in step length. For example, the trackway from Praia de Cavalo has been described as that of a limping animal because the right-

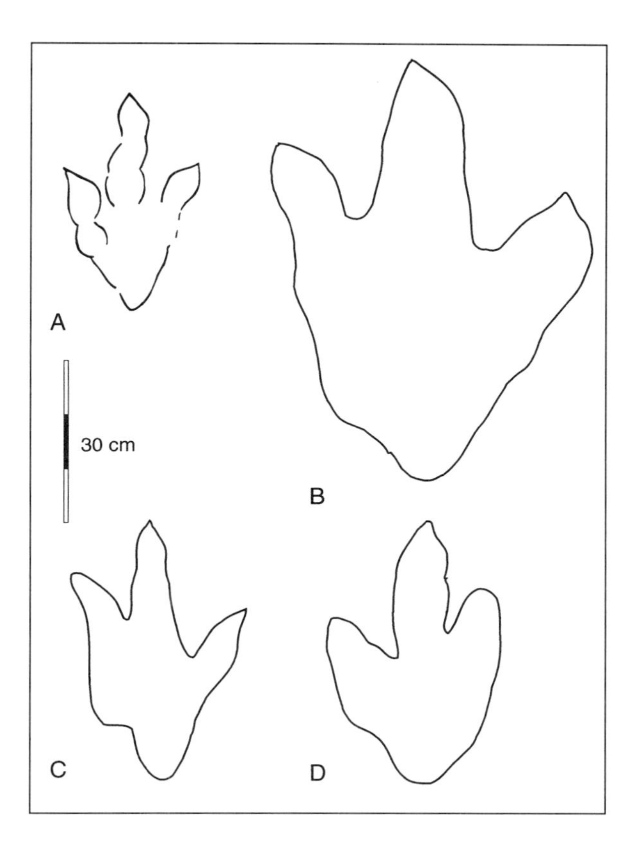

FIGURE 7.6. Line drawings of large, elongate *Megalosauripus*-like tracks from the Upper Jurassic of Cabo Mondego (A) and Zambujal Quarry (B) with similar footprints (C and D) from Asturias, Spain (after Valenzuela et al.[27,28]). C and D may be compared with *Hispanosauropus* (after Mensink and Mertmann.[25]).

to-left step averages 180 cm, whereas the left-to-right step averages 206 cm, a difference of 13%.[21] Trackways from North America and Asia also show this irregularity, sometimes to an even greater degree (figure 7.7). Such pronounced irregularities and wide trackways have never been reported for any other assemblage of theropod tracks. The aforementioned studies of North American Late Jurassic megalosaurids have resulted in the conclusion that these creatures had a "more primitive body form—longer more flexible torso, shorter hind legs, and more flexible ankle bones" than most other theropods.[22] Such observations, arrived at independent of our track studies, appear entirely consistent with the trackway evidence. In short, large, short-legged, long-bodied animals are likely to have left wide, irregular trackways. Based on these observations, the case for a megalosaurid origin seems worthy of consideration, and, moreover, the name *Megalosauripus* appears appropriate.

Finally it is worth mentioning that the appearance of these giant carnivores, the largest yet reported from the Jurassic in terms of their colossal foot sizes, coincides with the heyday of the brontosaurs, the largest dinosaurs known. Therefore, it seems that the track record may be giving us clues to an ecologic relationship between brontosaurs and megalosaurs. There has been much conjecture about

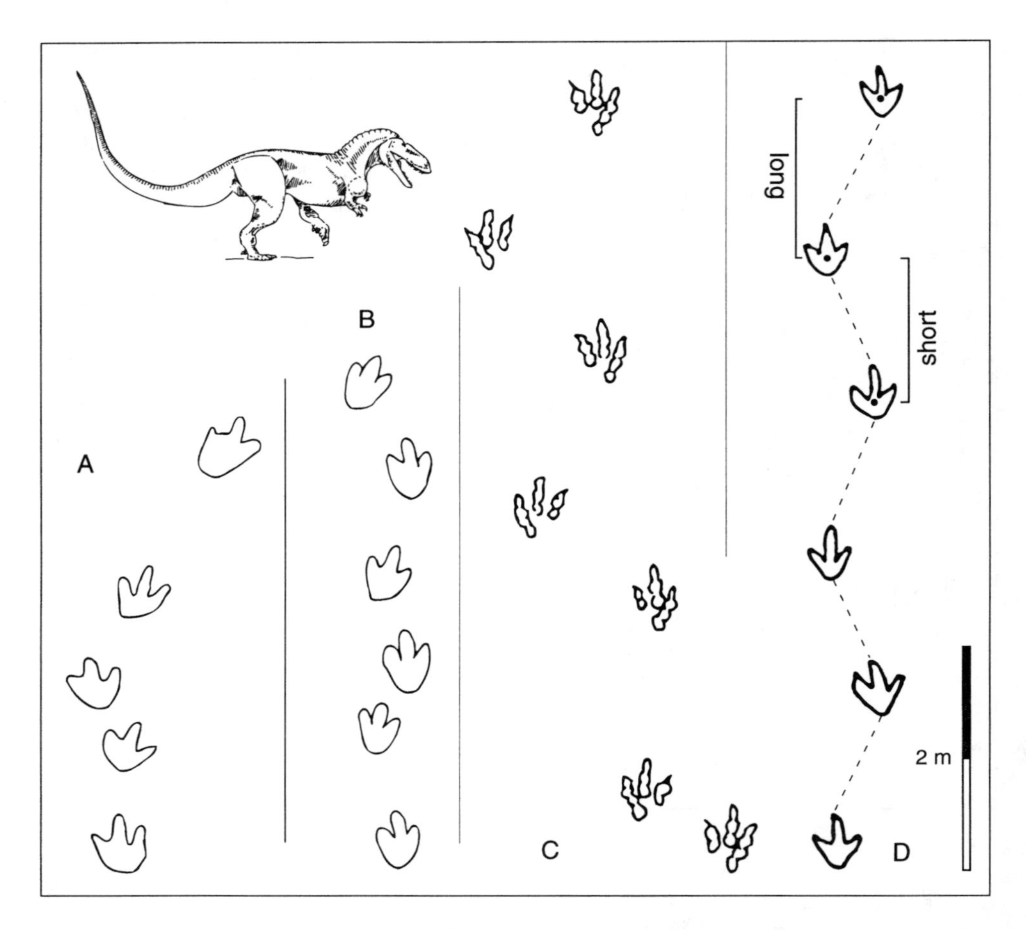

FIGURE 7.7. Pronounced irregularity has been noted in trackways attributed to *Megalosauripus,* as examples from North America (A and B), Asia (C), and Portugal (D) show.

whether the giant Cretaceous carnivore *Tyrannosaurus rex* was a predator or a scavenger. Perhaps the debate can now be extended to address the relationship between the megalosaurids and brontosaurs.

SAUROPODS ON THE RISE: GERMANY, IBERIA, AND SWITZERLAND

The Late Jurassic has been called the golden age of sauropods, based largely on the high diversity of skeletal remains unearthed, particularly in North America. The track record clearly supports this conclusion. In Europe, where the bone record of brontosaurs is not spectacular, the track record is quite impressive. Based on a recent compilation,[23] there is one site in Germany and there are two in Spain, sixteen in Portugal, and eleven in Switzerland. This total of 30 sites represents two thirds of the grand total of 46 Late Jurassic brontosaur tracksites currently known worldwide. So the Late Jurassic European sauropod track record could be said to be the best currently known, at least in terms of quantity of material available.

In terms of quality, the sauropod tracks record is variable. Addressing first the smaller sites, in Germany and Spain, we encounter tracks of dubious quality that have nevertheless been named formally. Tracks from Barkhausen, a site in Germany that is accessible to the public, have been given the name *Elephantopides barkhausensis* and were the first named from Europe in 1974.[12] Although it is generally accepted that the tracks are sauropodan in origin, the tracks as originally illustrated did not display enough detail to demonstrate this affinity with absolute certainty, and it was suggested that the name is dubious (a *nomen dubium*).[24] This is because the description was inadequate, indicating only a large quadrupedal animal. Our recent investigations of the site show that the tracks are those of fairly narrow-gauge sauropods that were quite small, evidently with a relatively small manus. It could be argued that although the tracks are elephantine in appearance (figure 7.8), the name *Elephantopides* is inappropriate because there is clearly no evidence for elephants or their relatives in the Jurassic. Such a name may confuse laypersons without a good understanding of evolution and geologic time.

The potential confusion does not end here. In their 1974 study, Kaever and de Lapparent[12] indicated that the sauropod trackways were trending in one direction (up the face of the outcrop) when in fact they trend in the opposite direction, down the face. Because the published maps are incomplete and misleading,[12] we include our own sketch map (figure 7.9) showing at least seven relatively small sauropods moving in approximately the same direction, with two theropods (megalosaurs) moving in different directions. The name *Megalosauropus* is incorrect (as noted above). The megalosaur tracks (63 cm long) are almost twice the size of the sauropod hind feet (32 to 38 cm).

The next brontosaurian track type named from the Jurassic of Europe, in 1984, was *Gigantosauropus asturiensis,* from Asturias in Spain.[25] According to Hans Mensink and Dorothee Mertmann, the authors of this ichnospecies, the trackmaker was a large megalosaurid with a footprint length of 1.35 m. This would be a truly gigantic theropod (figure 7.8) with almost twice the foot length of the examples cited in the previous section. As pointed out by Australian ichnologist Tony Thulborn,[26] the track is evidently that of a sauropod, a conclusion with which we agree.[18] *Gigantosauropus* occurs in conjunction with a three-toed track named *Hispanosauropus hauboldi,* which measures 52 cm long by 36 cm wide. In our opinion this track might be a megalosaurid track of the type we assign to *Megalosauripus* (figure 7.6). Tracks similar to *Hispanosauropus* also have been reported by Spanish researchers from several outcrops in the region.[27–30] Such an abundance of megalosaurid tracks in Spain is not surprising in light of their abundance in rocks of the same age in Portugal, as well as in North America and Asia.

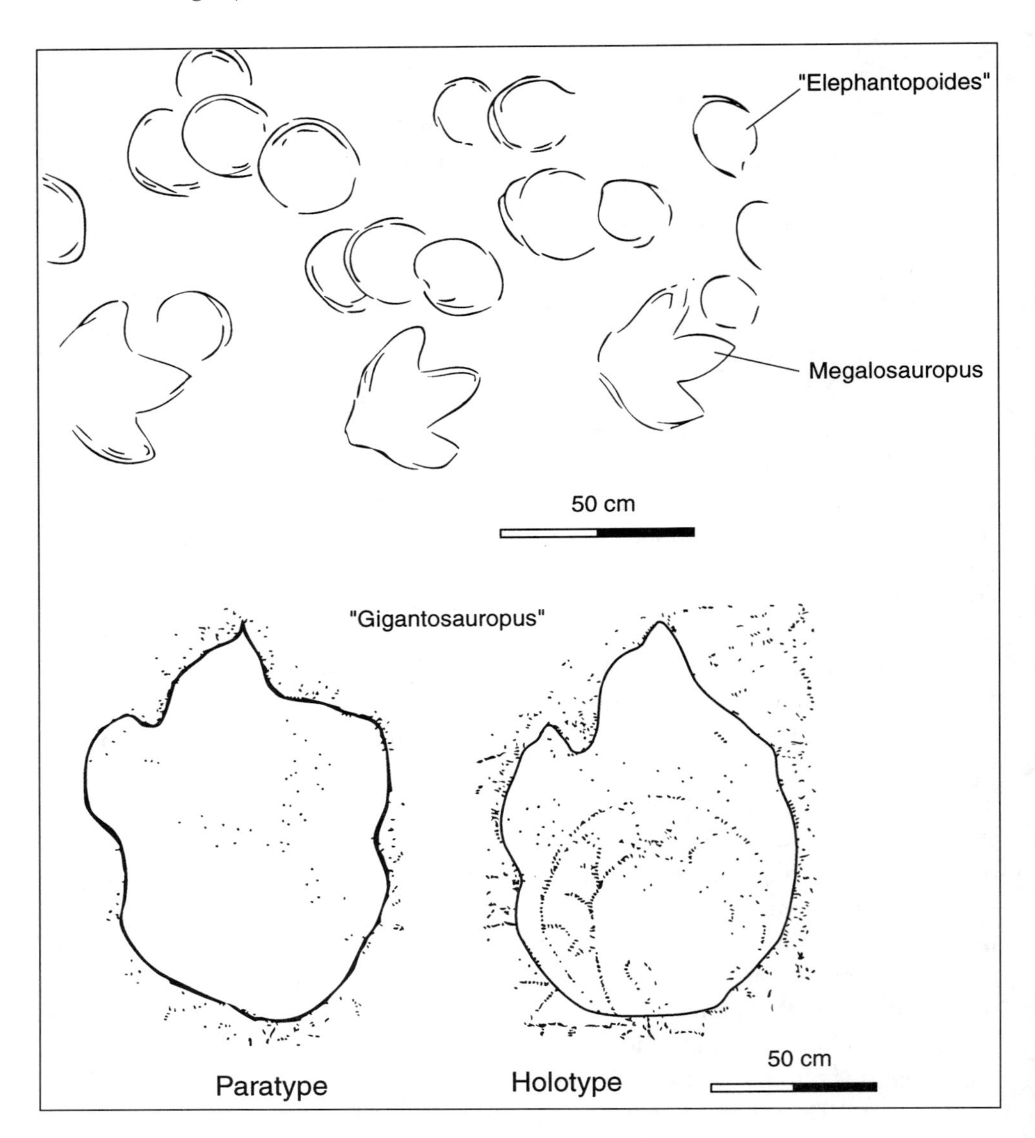

FIGURE 7.8. Top: "Elephantine" sauropod tracks from Barkhausen have been named *Elephantopides barkhausensis.* Tridactyl tracks have been incorrectly attributed to *Megalosauropus.*[12] Below: *Gigantosauropus,* a sauropod track, incorrectly attributed to a megalosaurid dinosaur.[25]

As we shall soon see, much of the sauropod track record of Europe is found in the Upper Jurassic limestones of Portugal, with important sites also found in Switzerland. The Upper Jurassic is divided into three stages: Oxfordian, Kimmeridgian, and Tithonian (or Portlandian), each spanning an average of about 7 million years. Recent work has demonstrated that the sauropod track record is ap-

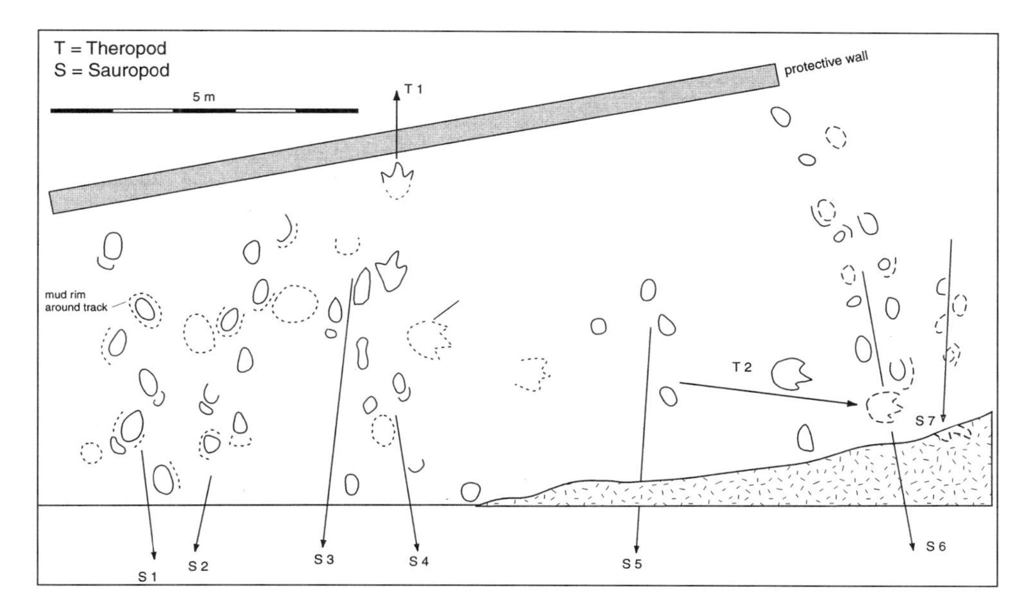

FIGURE 7.9. A map of the Barkhausen tracksite reveals at least seven parallel trackways of small, narrow-gauge sauropods and two much larger "megalosaurid" trackways. The name *Megalosauropus,*[12] however, is technically incorrect.

parently different in each one of these stages. This is perhaps what one might expect given that the average longevity of dinosaur genera has been estimated at about 7 million years.[31] What the track record appears to show is that sauropod tracks are rare in the Oxfordian, either because they really were scarce at this time or because conditions were not particularly favorable for the preservation of tracks. In the Kimmeridgian, however, tracks become very abundant, and we find a mixture of different track types, both wide gauge (*Brontopodus* type) and narrow gauge (*Parabrontopodus* type).[24] In the Portlandian, however, there is an apparent shortage of narrow-gauge types. A preliminary interpretation of the track record suggests that, in Europe, sauropods reached their peak of abundance or diversity in the middle stage of the Upper Jurassic. In the sections that follow, we shall first examine the Kimmeridgian track record before moving on to discuss Portlandian track assemblages.

BABY BRONTOSAURS

The Upper Jurassic rocks of Portugal belong to two distinctive sedimentary facies, namely, limestones deposited in a carbonate platform setting and mudstones and sandstones deposited in a fluvial–floodplain sedimentary environment. Tracks are found in both environments but are far more abundant in the carbonates because they weather to expose large surfaces of bedding plane. Spectacular exposures of these limestones are found in the area south of Lisbon, especially in the vicinity of Cabo Espichel Monastery. There are also many inland outcrops of limestone, especially in the many active quarries that exploit this local natural resource. It appears that brontosaur tracks are by far the most common in this limestone facies, with tracks of bipedal dinosaurs being comparatively rare. In order to find much in the way of other footprints, it is necessary to examine the nonlimestone facies north of Lisbon.

We shall begin our tour of the brontosaur-rich tracks layers in the Avelino quarry south of Lisbon, where the rocks are of Kimmeridgian age. Although a small quarry, it reveals a single surface with five crisscrossing trackways of different sizes (figure 7.10). Although described only in 1993,[32] these tracks were the first reported from Europe that show clear definition of both hind and front footprints. This is perhaps a little surprising when we consider that the tracks may be underprints.[32] They also include one that represents the smallest sauropod trackmaker known from Europe with a hind foot only 30 cm long, no larger than an adult human. (Although the smallest reported to date, it is only a little smaller than the Barkhausen tracks, i.e., foot lengths of 32 to 38 cm.) The significance of such small trackways is that in general the skeletal record of sauropods is heavily biased toward large animals.[2] The track record therefore provides useful information on the size, distribution, and behavior of juveniles that is not readily available at most sites where skeletal remains occur. At the Avelino site, for example, the tracks tell us that small, medium-sized, and large individuals mingled, whereas at Barkhausen, only small individuals appear to have traveled as a group. All the Avelino trackways also appear to be of the narrow-gauge, *Parabrontopodus* type. Although this suggests individuals of the same species, they do not appear to have been moving in the same direction. The Avelino tracksite reveals one other noteworthy trackway, that of a large individual whose front footprints are much more clearly impressed than those of the back feet. As we shall see, there have been many discoveries of trackways in which only the manus footprints are preserved, or in which they are much more clearly preserved than the pes traces (see figure 6.6).

We can move quickly from A to Z, or from Avelino quarry to Zambujal quarry. At Zambujal the main track-bearing layer was until recently situated on a vertical surface (figure 7.11). In order to study this surface, we used a "cherry-picker" crane of the type used by telephone and electrical companies for work on high poles and wires. Unfortunately, this wall collapsed recently,[20] destroying the ichnologic record. Among the casualties were the large 77-cm-long megalosaur track mentioned in the previous section (figures 7.5 and 7.6). Also lost was the only sauropod trackway known from this site.[33] This trackway was another example of one in which only the manus tracks were well preserved (figure 7.12) More importantly, the manus tracks showed clear traces of all five digits and were in this respect quite unique. As noted in the previous chapter, Middle Jurassic tracks from the Fatima site reveal the presence of a well-developed "thumb" (i.e., digit I claw). The Zam-

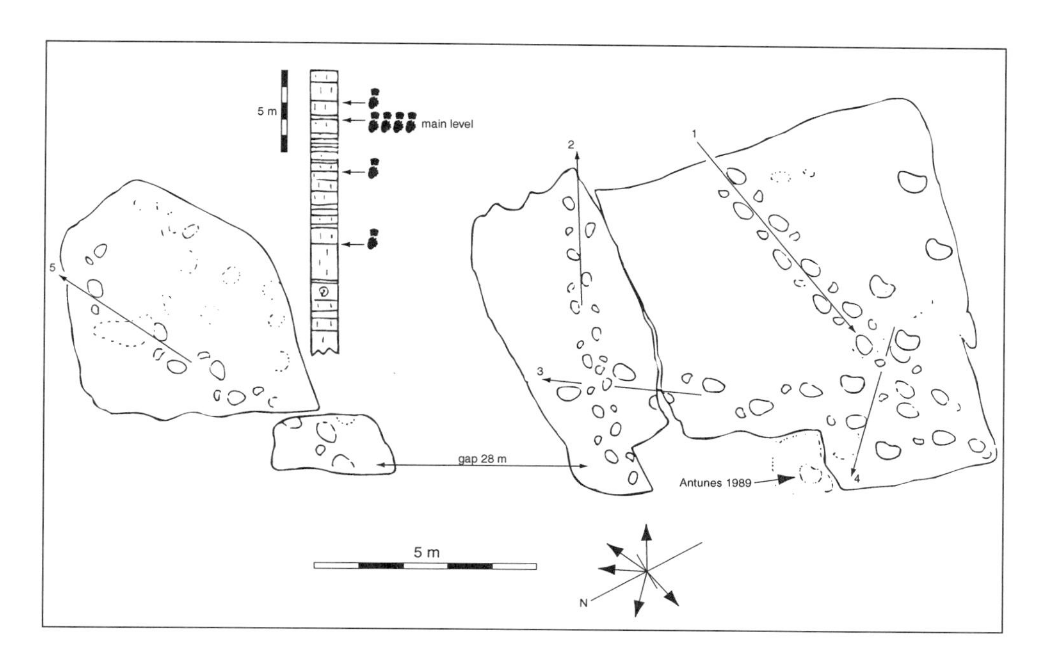

FIGURE 7.10. Map of the Late Jurassic Avelino site in Portugal. Trackway number 2 represents the smallest sauropod trackmaker known from Europe (after Lockley and dos Santos[32]). Photo of main outcropping (below).

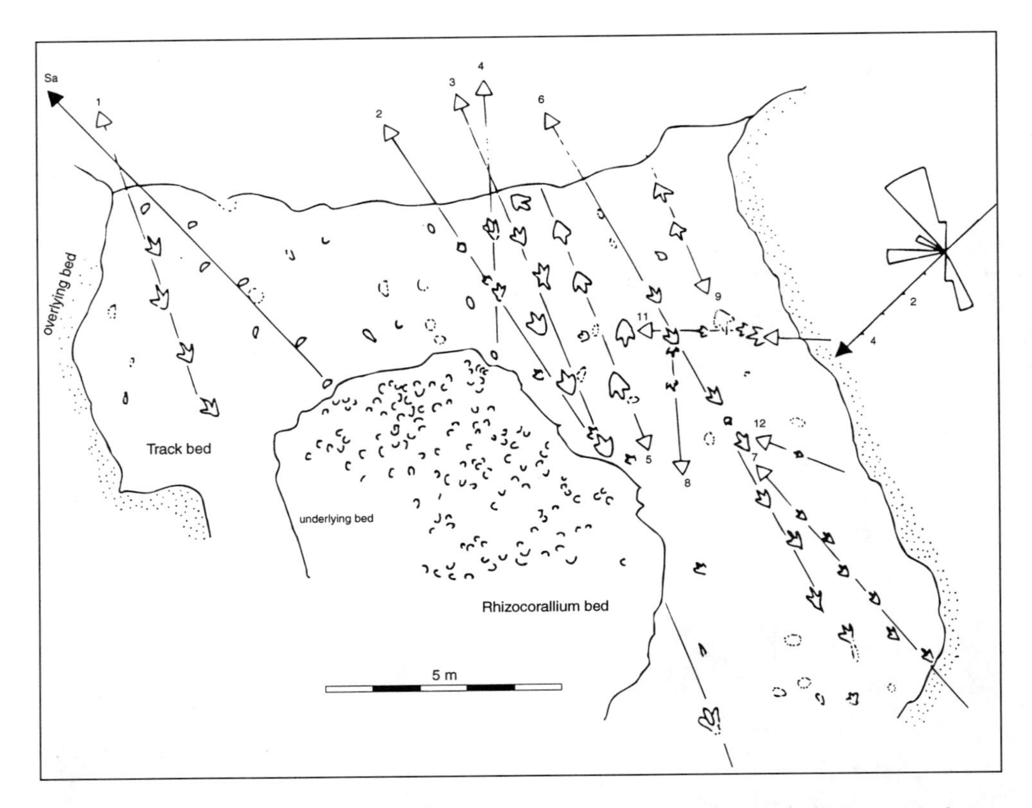

FIGURE 7.11. Map of the Late Jurassic Zambujal site in Portugal showing theropod trackways, including *"Megalosauripus,"* and a single sauropod trackway (see figure 7.12 for details; after Lockley et al.[19]). This surface was destroyed by a collapse of the vertical wall (after dos Santos et al.[20]).

bujal tracks show this thumb or claw also, but in addition reveal smaller protrusions indicating the positions of digits II through V. It may be technically incorrect to refer to these as digit impressions because they are not associated with the plantar surface, or sole, of the foot. Instead they are located more as corrugations on the anterior wall of the footprints. For this reason, they could be interpreted as the traces of the vertically oriented metacarpal bones rather than the impressions of protruding toes. Regardless of the interpretation, the tracks were unique among sauropod footprints and their loss is lamentable for ichnology. Fortunately, however, one of the best-preserved footprints was replicated for the National Museum. A photographic and cartographic record of the track-bearing surface also has been preserved (figures 7.11 and 7.12).

As we shall see when we examine younger Late Jurassic deposits in the following section, manus-only and manus-dominated sauropod trackways are abundant in Portugal. When first discovered, at a site in the Cretaceous of Texas, such trackways were incorrectly interpreted as evidence of swimming behavior.[34] Sub-

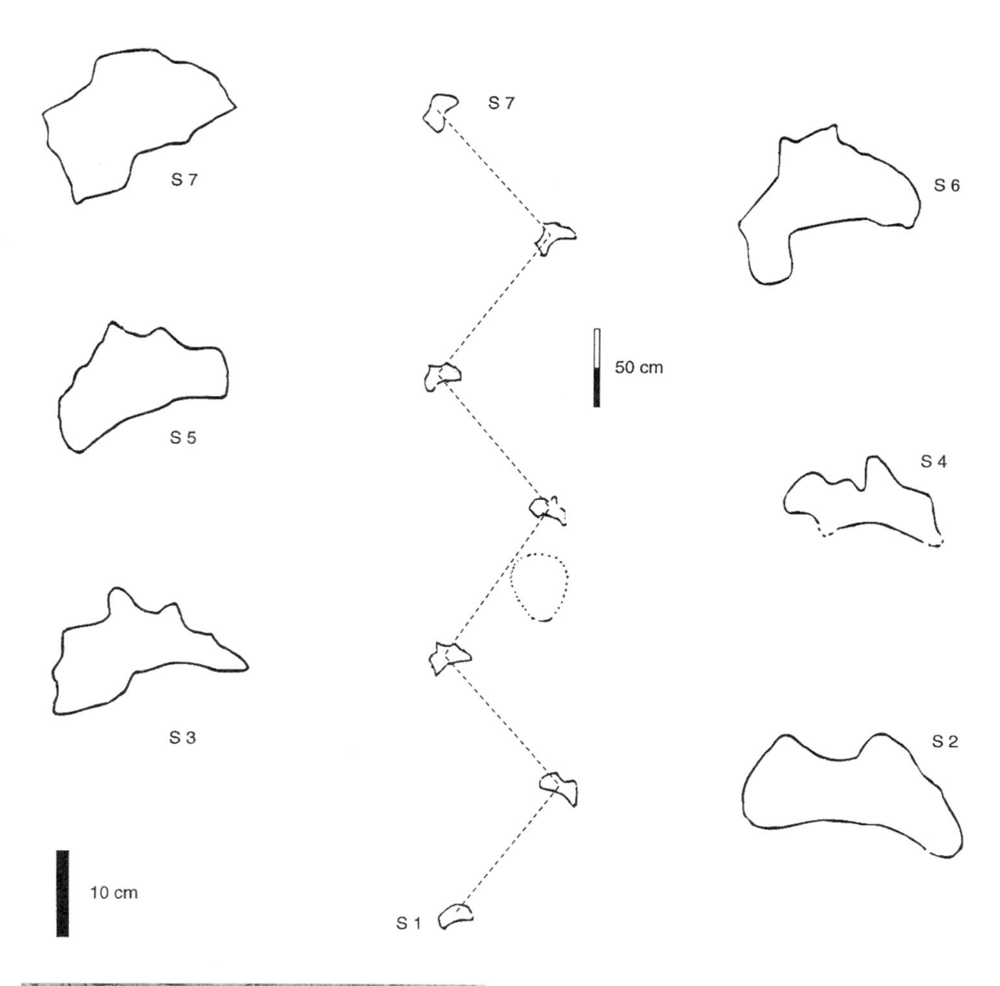

FIGURE 7.12. Sauropod trackway dominated by manus tracks that show traces of all five digits. Late Jurassic Zambujal site in Portugal[33] with photo of track S3 (left manus). This surface was destroyed by a collapse of the vertical wall (see figure 7.11).

sequent work has shown that these are underprints of sauropods that were walking normally on an overlying layer. Such evidence reveals that the hind and front feet of sauropods sometimes sank to different depths, depending on such variable factors as the condition of the substrate, the weight distribution of sauropods over their front and rear quarters, and the size difference (heteropody) between manus and pes.

SOCIAL SAUROPODS

When we move from Avelino and Zambujal to the spectacular cliff exposures at Cabo Espichel, we also move up the stratigraphic section from the Kimmeridgian to the Portlandian. Tracks from this site had long been the source of a local legend which holds that the footprints in the rock were left by the mule ridden by the Virgin Mary as she rode up from the sea. A shrine to the mule tracks (Pegadas de Mula) is situated beside the old Cabo Espichel Monastery. As recent studies have shown, in addition to the obvious brontosaur tracks, limestone surfaces in this region also contain horseshoe-shaped traces that are attributed to burrowing invertebrates, possibly shrimp-like arthropods.[19] Such traces have been named *Rhizocorallium* by invertebrate ichnologists. In other parts of the world, such horseshoe-shaped traces have been mistaken for the tracks of hoofed mammals, such as horses and their relatives, and have provided fuel for creationists hoping to use such features as evidence that paleontologists have made fundamental mistakes in their interpretation of the evolutionary record.

Returning to the subject of real dinosaur tracks, we find that despite their prominence, the tracks of Cabo Espichel had never been studied seriously. A preliminary study in 1976 by Miguel Antunes[35] gave only brief descriptions of a few of the more accessible tracks, found mainly in Cretaceous deposits on the north side of Lagosterios Bay, across the bay from the Cabo Espichel headland. In a study conducted mainly in 1993 with our colleague Vanda Faria dos Santos, we mapped the broad sweep of Late Jurassic (Portlandian) surfaces exposed on the south side of Lagosterios Bay immediately beneath the walls of the shrine and monastery. This endeavor required the use of ropes and climbing equipment loaned to us by the Portuguese Speleological Society. Although not a technically difficult climb, it is dangerous to attempt to approach these tracks without safety ropes. Seabirds and wind loosen small rocks from above, and in places the rock is too smooth for safe footing, especially for the researcher who needs free hands for measuring and note taking.

The result of our survey of the Cabo Espichel exposures was to establish the presence of at least eight different track-bearing layers (figure 7.13), most revealing only a few sauropod trackways. By far the most important layer was designated number 2, in ascending stratigraphic sequence.[36] This layer reveals evidence of a group of seven subadult dinosaurs heading in the same southeasterly direction (figure 7.14). The trackways reveal animals that were all about the same size (foot

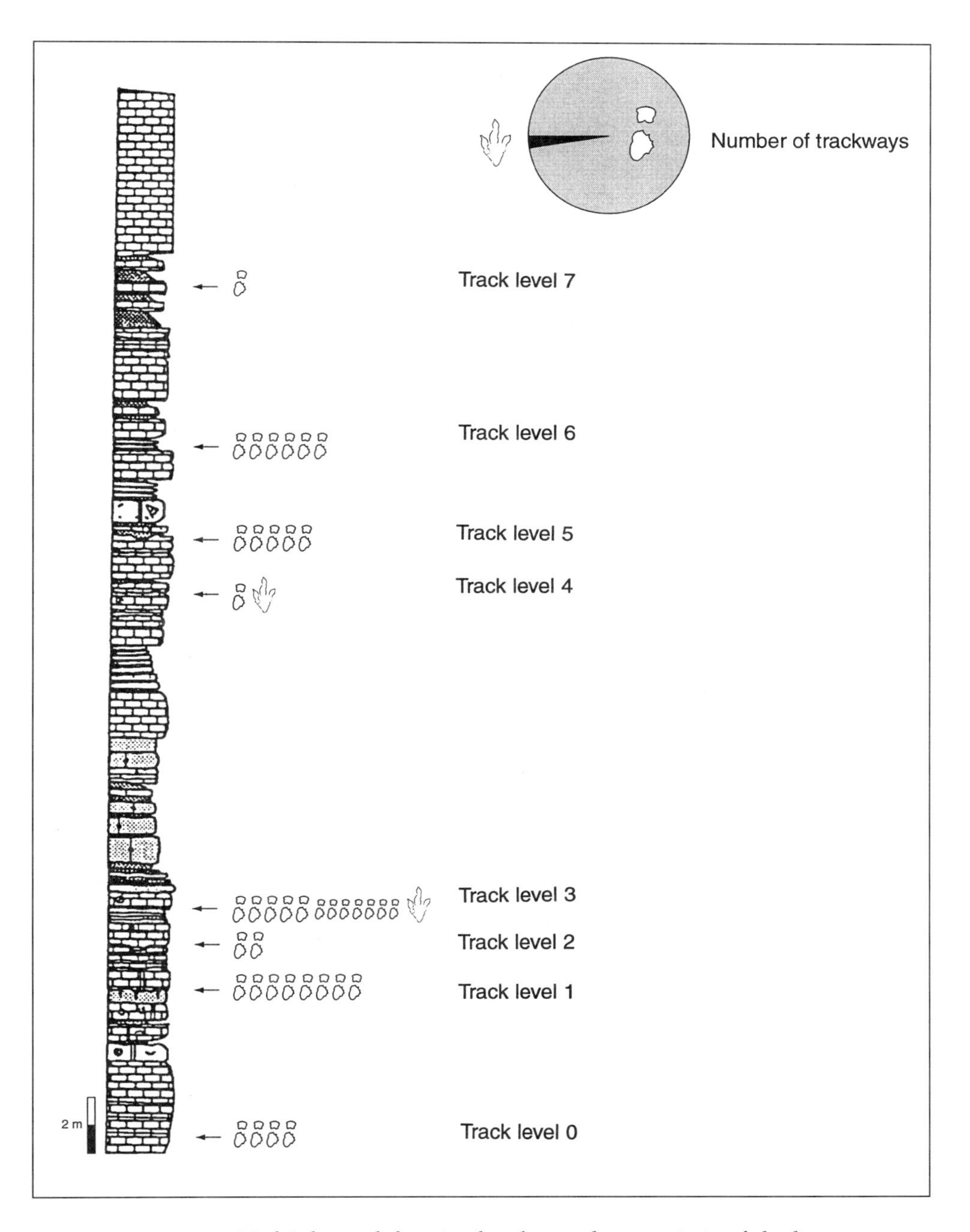

FIGURE 7.13. Multiple track-bearing levels are characteristic of the limestone sequences at Cabo Espichel, where about 40 sauropod trackways have been recognized from at least eight stratigraphic levels. *After Lockley et al.*[36]

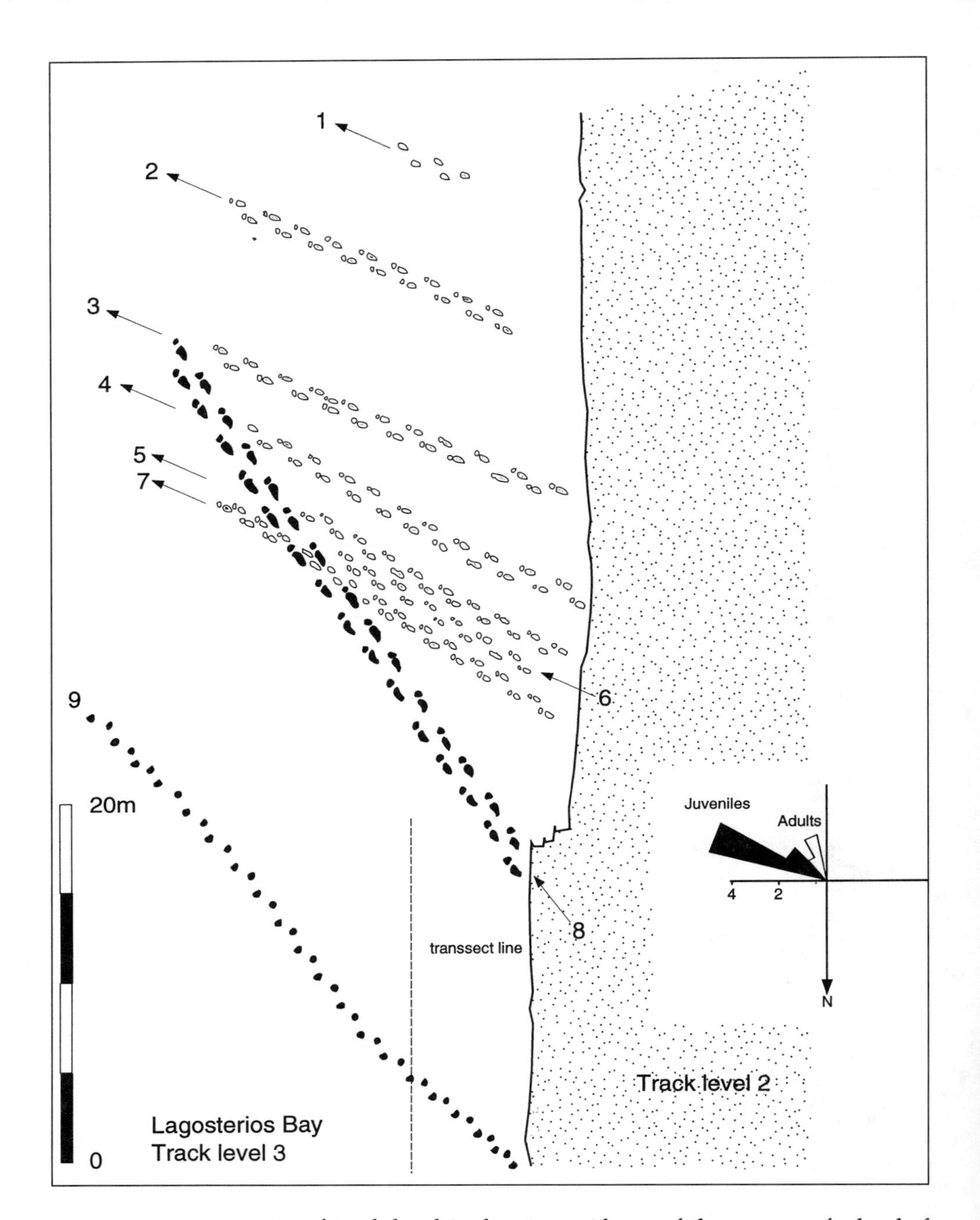

FIGURE 7.14. Map of track level 2, showing evidence of the passage of a herd of seven subadult sauropods, Upper Jurassic, Cabo Espichel, Portugal.[36] Compare with figure 1.7. Artwork at beginning of this chapter is based on this trackway map.

length between 38 and 46 cm), with trackways that are more or less regularly spaced. Such regular "intertrackway spacing" is taken as an indication that the animals were moving as a group in much the same way that soldiers march and birds fly in formation. In short, such evidence strongly suggests gregarious or social behavior.

Based on the size of these sauropod footprints in this group of seven, we can infer animals that were a little larger than the Avelino and Barkhausen juveniles but considerably smaller than the largest tracks (foot length 73 cm) found on this surface, which belong to "adults" heading in a slightly different direction. Also in contrast to the Avelino and Barkhausen sauropods, the Cabo Espichel sauropods represent the wide-gauge (*Brontopodus*) type, which, as noted above, became dominant in the Portlandian.

As a footnote to our discussion of the track-bearing limestones of Portugal, we should note that traditionally the sedimentary sequence has been regarded as representing a shallow marine, shelf, or platform environment. The fossils, which comprise most of the rock, are mainly composed of marine invertebrates and some marine vertebrate remains, such as fish and turtle fragments, and numerous burrows of marine invertebrates. Such fossil and sedimentologic evidence would not normally suggest to geologists that the platform was periodically exposed subaerially. However, the track evidence shows that parts of the platform were periodically exposed. Indeed, we might go so far as to say that if the tracks were not preserved on numerous layers, we would not have definite evidence of the episodic rise and fall (or emergence and submergence) of the platform.

THE SWISS MEGATRACKSITE

From our temporary accommodations in the small port of Sesimbra, not far from Cabo Espichel, we watched large barges exporting limestone. Some of the rock eventually finds its way to Switzerland. It seems ridiculous to send limestone from Portugal to Switzerland, especially when one considers that Switzerland is blessed with large supplies of exactly the same type of rock. But the incentive for this export trade is dictated by the large differential in price between Swiss and Portuguese building stone.

Switzerland is famous for its limestones, especially those from the Jura Mountains, which give their name to the Jurassic System and the corresponding time period. Like those from Portugal, they are traditionally regarded as marine. Thus, one would not expect them to contain much evidence of terrestrial animals such as dinosaurs. On the contrary, they are known for yielding the remains of marine species such as sea turtles. One such Swiss deposit rich in marine vertebrate fossils is the Solothurn Turtle limestone exposed in various quarries and natural outcroppings in the vicinity of the town of Solothurn. During the course of investigations of this limestone at a quarry near the small town of Lommiswil, the junior author

FIGURE 7.15. Photograph of the track-bearing surface at Lommiswil, in the famous Jura Mountains of Switzerland.[37] Compare with map in figure 1.6.

of this volume identified what turned out to be the first Late Jurassic dinosaur tracks reported from Switzerland.[37]

Given the attention that geologists have focused on quarries such as the one at Lommiswil, it is surprising that tracks are so easily overlooked, for, as this discovery shows, footprints are both large and abundant. At Lommiswil, for example, individual tracks are up to a meter or more in diameter, and one trackway can be traced for about 90 m, making it one of the longest in Europe (figure 7.15). The tracks are not so well preserved that they reveal clear details of manus and pes outlines and toe impressions in most cases. The general trackway configuration, however, reveals that the trackmaker was of the wide-gauge variety.

The Lommiswil site has proved something of a Rosetta stone for Swiss dinosaur ichnology. Several subsequent tracksite discoveries reveal that they are associated with the same stratigraphic levels near the top of the Reuchenette Formation. This means, in short, that the Lommiswil site is part of a much larger megatracksite complex. Megatracksites have been defined as regionally extensive track-bearing surfaces or layers, often covering geographic areas of several hundred or several thousand square kilometers. Such megatracksites, also dubbed "dinosaur freeways," are best known from the western United States. The Swiss megatracksite was the first identified in Europe.[38]

THE DINOSAUR DISCO: AN ANCIENT STOMPING GROUND

One of the Swiss national pastimes is rock climbing, and it is through this activity that a number of dinosaur tracksites have been found. In fact, most of the dinosaur tracksites found in Switzerland were found by the junior author of this volume as a result of combining geologic curiosity with rock-climbing excursions. The second largest dinosaur tracksite in Switzerland, in terms of surface area and number of footprints, is found in the Gorge of Moutier in the Jura Mountains. We discovered this site simply by deciding that it was important to inspect large surfaces of Late Jurassic limestone that had not previously been looked at with an eye to their footprint-bearing potential; sure enough, a casual inspection soon revealed tracks. But the site is not in a remote location. Quite the contrary; it is a well-known face where rock climbing is taught and has been seen close up by many pairs of eyes, but evidently by none that were looking for dinosaur tracks.

In fairness to the countless climbers and hikers who have climbed and skirted this large surface, many of the tracks are not obvious, and the site can be best described as a stomping ground where there is evidence of various degrees of trampling or dinoturbation. Such crossroads, congregation points, or "dinosaur discos," as the local press labeled the site, are interesting paleontologically but messy from an orderly scientific perspective, owing to the confusion of footprints.

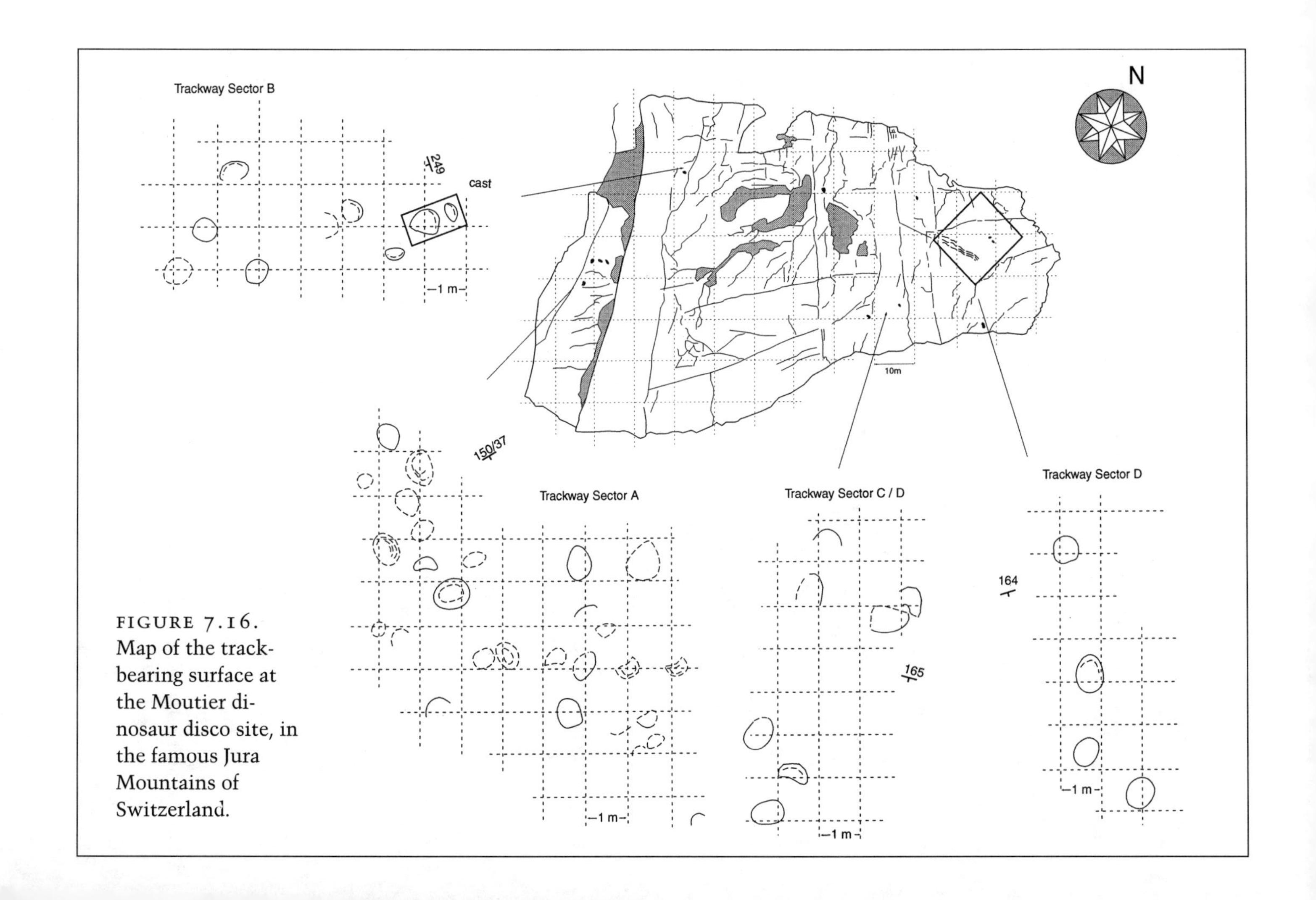

FIGURE 7.16. Map of the track-bearing surface at the Moutier dinosaur disco site, in the famous Jura Mountains of Switzerland.

In order to document the Moutier dinosaur disco, we adopted the most basic scientific procedure used at most dinosaur tracksites: we made a map of the surface. But this was not the usual low-budget compass and tape measure project. In characteristic Swiss style, we went hi-tech with helicopters and the latest computer surveying technology. While we trackers climbed on the surface and marked individual footprints onto photographs of the surface taken from a previous helicopter flyby, it was impossible to remove the distortion in the photographs without accurate surveying methods. Therefore, our colleagues, using theodolites, surveyed in a half dozen fixed points around the perimeter of the site. The next step was to fly over again, using a helicopter without back rotors, to reduce vibration, and shoot 28 × 28 cm format digital photographs. These were then transformed into what is called a "digital elevation model," producing a map image with no distortion and the ability to resolve the surface topography with a resolution of a few centimeters.

The result of our combined high-tech and climb-on-the-surface survey was the recognition of about 200 footprints distributed more or less randomly across a steeply inclined surface of about 6000 m^2 (figure 7.16). Only a few tracks were sufficiently clear to warrant casting with rubber and fiberglass, and only a few allowed us to distinguish large hind footprints (pes tracks) from small front footprints (manus tracks). Despite the latest technology, the Moutier site does not easily reveal its secrets, and we are still hard pressed to say much about the number of sauropod trackmakers that were in the area, the type (wide or narrow gauge), the size (although some were large), or the direction of travel of these sauropods. Work is still in progress on the track-bearing surface in an attempt to explain why some tracks are very deep (40 cm) and others so shallow as to be barely perceptible. For now we must accept that it is a stomping ground or disco site, where the spoor left behind is messy and hard to interpret unambiguously.

While writing this book, new discoveries have revealed several new surfaces not far from Moutier that also appear to correlate well with each other. These indicate another extensive track-bearing unit, or megatracksite, near the base of the Reuchenette Formation, that is, at a level lower than the aforementioned upper Reuchenette megatracksite. Some of these discoveries include a 6-meter segment of theropod trackway in fresh outcrops along a new road that will connect northern Switzerland and France. There also have been finds of dinosaur tracks at levels above the Reuchenette Formation. Thus, the classic Jura region is full of tracks that were previously not known to exist.

SMALLER SPOOR

It is easy to get the impression that most of the Upper Jurassic track record, especially in limestone facies, consists almost exclusively of sauropod tracks. As we shall see, a closer look reveals that this is not so. If we look in the right places, we can find the spoor of small dinosaurs, turtles, and pterosaurs. The most promising area for small dinosaur tracks appears to be on the Iberian Peninsula, so let us first return to Portugal. North of Lisbon, instead of finding large surfaces of limestone, the sedimentary facies consists of mudstones and sandstones deposited on ancient floodplains. These deposits are similar in some respects to bone-rich dinosaur beds found in the Late Jurassic of North America and East Africa, and, like these deposits, have also yielded the skeletal remains of sauropods and other well-known dinosaurs. The track record, however, is quite intriguing and different from that found south of Lisbon.

Tracks are preserved almost exclusively as natural casts on the underside of fluvial sandstone units in fine-grained floodplain deposits. Many of these track-bearing surfaces are quite small and first appear in coastal rock exposures as overhangs of sandstone, undercut by the sea, and subsequently become fallen blocks on the beach. Depending on their location, they are subject to erosion by waves and periodic burial under the sand. About a half dozen track-bearing slabs have been reported, and although some have been preserved in local museums, a few have been destroyed by erosion, and all that remains are photographs and sketch maps. Most of these sites reveal the presence of small three-toed footprints, some no larger than 7 to 10 cm in length, others larger, in the range of 25 to 33 cm. These appear to be evidence of small and medium-sized theropod dinosaurs, smaller than the large to gigantic megalosaur footprints described at the beginning of this chapter. This area therefore holds promise of providing insight into the track record of small bipedal dinosaurs whose footprints have yet to be found in the limestones south of Lisbon.

Only one set of tracks, from a locality at San Martinho do Porto,[39] has been named as *Dinehichnus* (figure 7.17). This name derives from the "Dineh," the indigenous name by which the Navajo people know themselves. In the Navajo Nation region of the so-called four corners area (Utah, Colorado, Arizona, and New Mexico), 16 similar parallel trackways were found, indicating a social group of small to medium-sized bipeds, in the size range of turkeys and emus. The North American forms were named *Dinehichnus socialis*[39] to indicate their inferred tendency to flock together in gregarious or social groups. Based on their foot shape and its similarity to certain tracks in the Cretaceous, the trackmakers are thought to be ornithopods, not theropods. It is hoped that further study of the Late Jurassic outcrops in Portugal will reveal more tracks of this type and provide further insight into the spoor of smaller dinosaurs.

From Portugal we can turn our attention to Spain, where there are several Late Jurassic sites from which the spoor of small dinosaurs has been reported. As we shall see in the following chapter, the La Rioja region of north central Spain, long known for its fine wines, also has become famous in the past decade or so for an abundance of dinosaur tracks. Although many of these are Cretaceous in age, some have been assigned a Late Jurassic age.

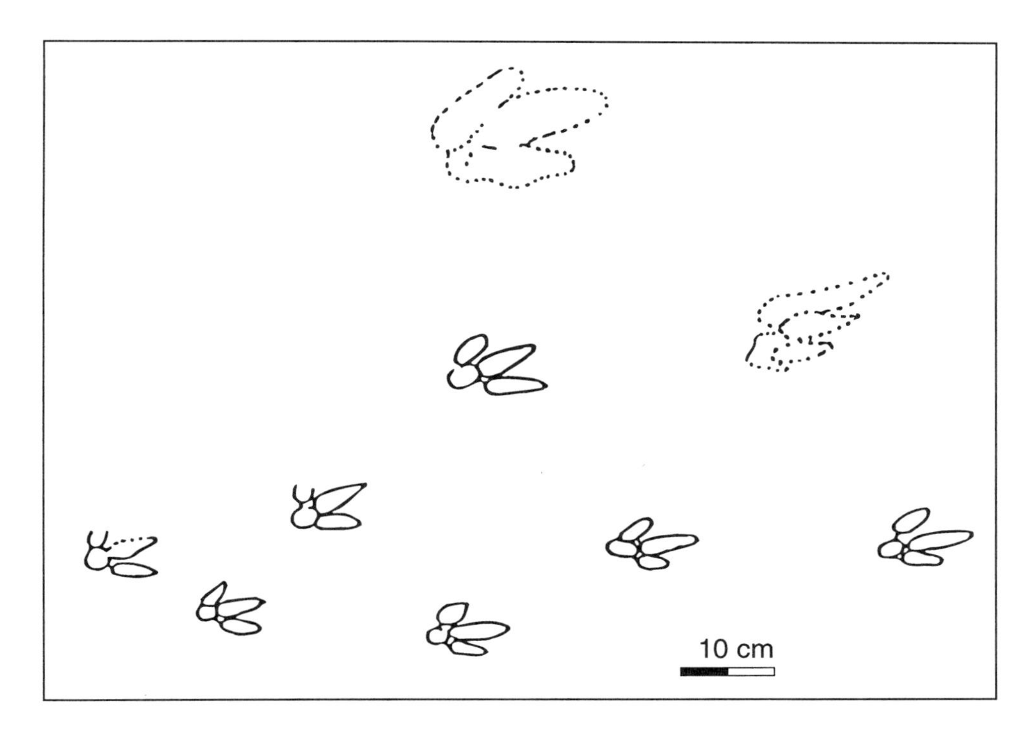

FIGURE 7.17. *Dinehichnus* represents an ornithopod dinosaur with strong gregarious tendencies, from the San Martinho do Porto site. *Redrawn from Lockley et al.*[39]

TURTLES AND HOPPING DINOSAURS

What is the difference between the trackway of a hopping dinosaur and a large marine turtle? The answer to this seemingly bizarre question is that the difference is not obvious enough to all ichnologists to result in happy agreement. Tracks that we consider to be those of walking turtles, from the lithographic limestones of Cerin, near Lyons, France (figure 7.18,) are apparently not controversial, but those that we consider to be evidence of swimming turtles (figure 7.18) have proved controversial, and according to some researchers were made by hopping dinosaurs. We can probably agree by inspection of the two trackway types that they are significantly different. But were they made by entirely different animals, or do they simply represent different behaviors? The names given the two trackways would suggest a biologic distinction. *Chelonichnium cerinense* (meaning "turtle track from Cerin") is the label designating the "walking" trail, and *Saltosauropus latus* (meaning "wide trackway of a hopping saurian" or dinosaur) is the name given to what we consider the swim tracks.

We have not been directly involved in the debate over hopping dinosaurs versus swimming dinosaurs, although the evidence, presented below, compels us to take the side of the turtles. The story began in 1984, when a group of French pale-

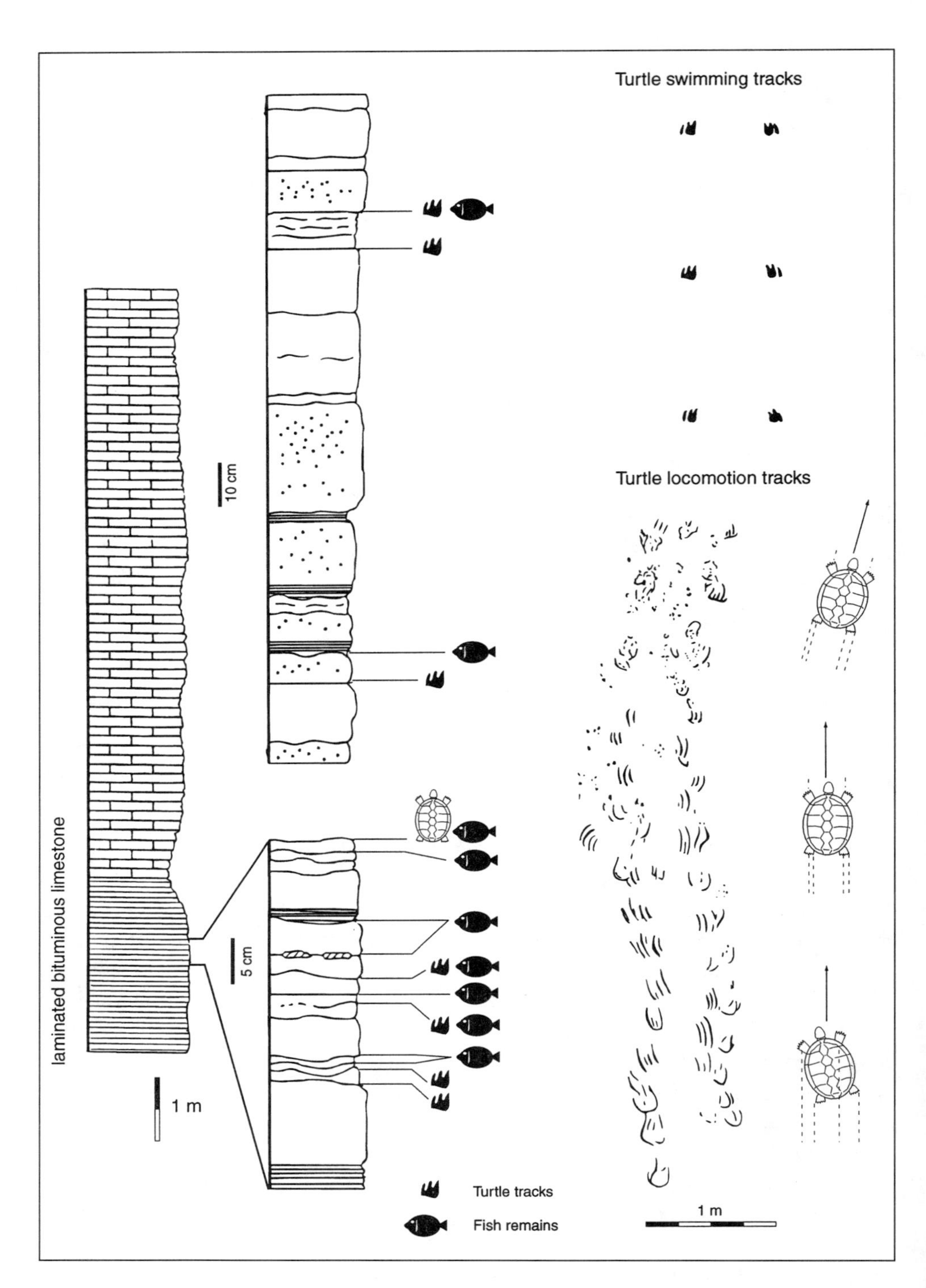

FIGURE 7.18. *Chelonichnium* tracks from the Upper Jurassic lithographic limestones of Cerin, France, were made by large marine turtles, although some have been incorrectly assigned to hopping dinosaurs. Note the trackways of the "walking" turtle and those that indicate swimming. Also note the multiple stratigraphic levels with tracks (see Bernier et al.[40–42]).

ontologists[40,41] reported and named *Saltosauropus.* The trackways consist of pairs of large, widely spaced, three-toed tracks, about a meter apart, and separated by variable distances of as much as 3 m. There can be no doubt that they are quite different from normal trackways of walking dinosaurs. Instead, they resemble the tracks of modern turtles. Although these researchers attributed the tracks to hopping dinosaurs, at about the same time they also described *Chelonichnium* as the trackway of a giant turtle.[42] The alert tracker may note the similarity between this ichnogenus label and *Chelichnus* from the Permian, once thought to be a turtle track (see chapter 2).

Several people questioned the hopping dinosaurs interpretation, notably Australian ichnologist Tony Thulborn,[26] who suggested that a hopping dinosaur would leave closely spaced pairs of tracks, as seen in the trackways of kangaroos or hopping birds. He suggested instead that the widely spaced pairs of tracks are best explained as the traces of a large sea turtle swimming near the bottom, touching the substrate with synchronous strokes of its flippers. We agree with Thulborn's interpretation[43,44] and note that the reconstruction of the environment in which the tracks were found, as that of a tropical marine lagoon, is the perfect habitat for marine turtles.

The idea of hopping dinosaurs was not entirely new in 1984, having been first suggested by M. A. Raath in 1972[45] to explain two similar theropod dinosaur tracks, which he attributed to *Syntarsus,* found side by side in Lower Jurassic deposits of Rhodesia. He suggested that it might have "had a kangaroo-like saltatory gait, using both hind legs together." We are skeptical about this interpretation because the pair of prints was isolated and not part of a trackway. Moreover, no similar paired tracks have ever been found despite the abundance of this type of Lower Jurassic track (see chapter 5). Another report of the trackway of a very small hopping dinosaur, from desert sand dune deposits in South America,[46,47] has been reinterpreted as the trackway of a hopping mammal. So, despite two other suggested examples, convincing evidence for hopping dinosaurs remains scant at best and quite possibly nonexistent.

Once a sensible and pleasing interpretation has been put forward (Thulborn's turtle model), it is easy to see the flaws in the alternate (hopping dinosaur) hypothesis. It should be noted, however, that fossil footprints attributed, either rightly or wrongly, to swimming vertebrates have often proved controversial. The simple reason for this is that such trackways are uncommon, and often incomplete and irregular. Trackways that really represent swimming vertebrates by definition reveal evidence of animals that were not touching the substrate with all parts of their feet in the way that they do when walking. So their tracks are incomplete, often with irregular spacing patterns and unusual features such as long toe scrape marks, all of which can be attributed in some way to the effects of buoyancy on the trackmaker.

Before we leave the subject of turtle tracks and move on to the subject of pterosaur tracks in the next section, it is historically honest to draw attention to a few additional problems of interpretation that have surrounded the investigation of purported turtle tracks. Although turtles—as well as that other group of common aquatic vertebrates, the crocodiles—are among the most common vertebrate remains in the body fossil record, their tracks are surprisingly rare. Part of the reason for this is that, as aquatic creatures, their remains were easily incorporated into stream channels, ponds, and other subaqueous deposits. By contrast, they rarely made trackways by walking on substrates that were subsequently preserved in the track record. Most reports of turtle and crocodile footprints come from Upper Jurassic and Lower Cretaceous freshwater deposits in western North America.[44,48] For various reasons these tracks also have proven quite controversial. Crocodile trackways have perhaps proved the most controversial, in some cases having been confused with the footprints of ornithischian dinosaurs[48] and in other cases having been confused with those of pterosaurs.[49] The footprints of true pterosaurs also have been confused with those of crocodiles.[50]

We have already noted the confusion that exists in the interpretation of Late Jurassic turtle tracks from the lithographic limestones of Cerin.[26,40–44] We can trace the discussion of trackways in lithographic limestones back to the mid-nineteenth century, when trackways from the famous *Archaeopteryx* and pterosaur-bearing lithographic limestones of Solnhofen were attributed to pterosaurs,[51] small dinosaurs, and also to *Archaeopteryx.* Such interpretations were shown to be incorrect when Kenneth Caster, a famous American student of vertebrate traces, showed that these were the trackways of horseshoe crabs, also known as "limulids," that were sometimes found literally dead in their tracks.[52] Regular scrape marks were also attributed to turtles, but these too were shown to be of invertebrate origin, having been produced by ammonite shells that were drifting and bouncing along the bottom.

SPOOR OF THE PTEROSAUR

We have just seen that purported pterosaur footprints from the Solnhofen limestone turned out to be the tracks of horseshoe crabs. However, this does not mean that real pterosaur tracks do not exist. In fact, recent discoveries by Jean-Michel Mazin and colleagues reveal the first Late Jurassic pterosaur tracks (figure 7.19) known from Europe.[53,54] Moreover, these tracks occur in sublithographic limestones, thereby giving us some insight into the habitats frequently by pterosaurs. It seems that in the region of present-day Crayssac (Lot, France), during the Late Jurassic, a shallow gulf known as Golfe Charantais extended into southwestern France between land masses in the region of the Iberian Peninsula and central and northwest France. On the substrates of this gulf, sometimes emergent above sea level, we find mud cracks, raindrop impressions, and the tracks and traces of invertebrates and vertebrates, including dinosaurs and pterosaurs.

Before describing the Crayssac pterosaur tracks it is worth discussing the controversy that surrounded, and still surrounds, the interpretation of pterosaur footprints. Leaving aside the incorrectly interpreted horseshoe crab tracks from Soln-

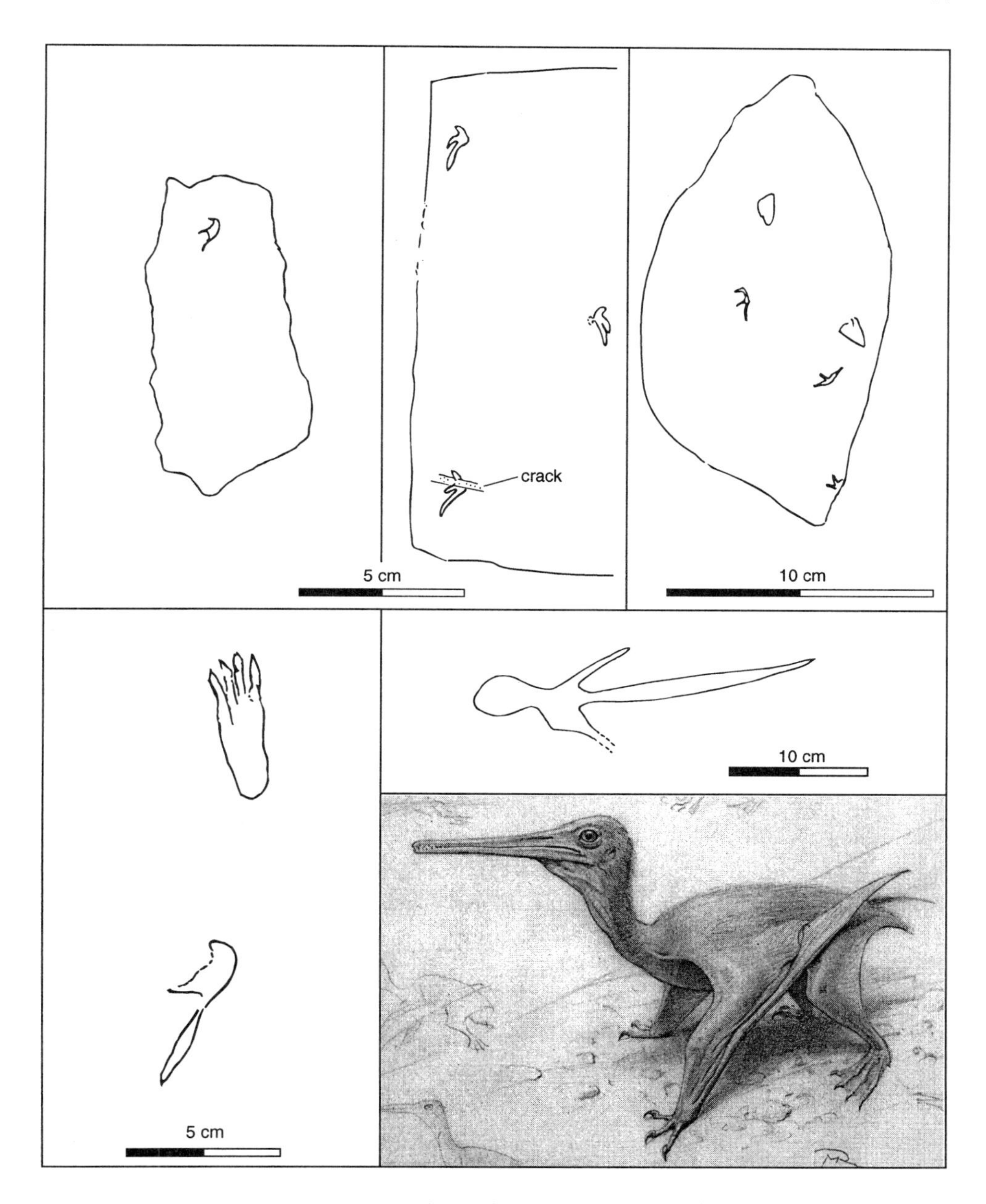

FIGURE 7.19. Pterosaur tracks from the Upper Jurassic of Crayssac, France. Tracings courtesy of Mr. Michel Dutrueux.[53,54] Inset shows pterosaur in quadrupedal pose. *(After Reichel, unpublished.)*

hofen, the first convincing report of pterosaur footprints was a trackway from the Late Jurassic of Arizona named *Pteraichnus* by William Stokes in 1957.[55] Although *Pteraichnus* literally means "pterosaur trace," the validity of this interpretation was thrown into doubt in 1984 when Kevin Padian and Paul Olsen[50] reinterpreted the trackway as crocodilian in origin. At that time no other *Pteraichnus* trackways

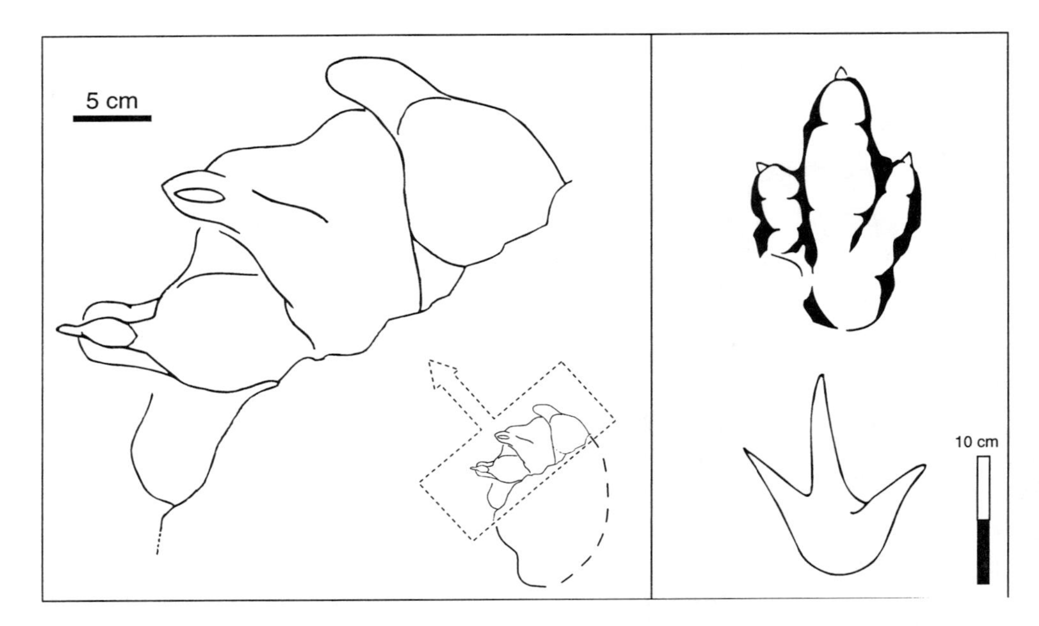

FIGURE 7.20. Brontosaur tracks[61] and theropod tracks[54] were recently reported from the pterosaur track-bearing beds near Crayssac, France. Note the claw impressions on the partial brontosaur track (left).

had been found, and no further studies were conducted to support or refute the hypotheses of Stokes, Padian, and Olsen. This situation changed dramatically in 1995, when dozens of *Pteraichnus* trackways came to light at many localities in the western United States, leading to a multi-authored paper on the pterosaurian origin of *Pteraichnus*.[56] Subsequently, many other researchers joined the rush to confirm that *Pteraichnus* and *Pteraichnus*-like tracks from the Jurassic and Cretaceous of North America, Europe, and Asia are pterosaurian in origin.[57–59]

Among these researchers, the contributions of Mazin and colleagues in 1995[53] and 1997[54] have been pivotal. The Crayssac tracks consist primarily of small footprints (from 15 to 45 mm in length), indicating animals ranging in size from sparrows to very small sea gulls. Many of these diminutive tracks are very well preserved. Prior to the discovery of *Pteraichnus* at Crayssac, there was still uncertainty about the interpretation of the tridactyl manus impression. Stokes had suggested that the elongate posterior trace was made by the wing finger (digit IV). The problem with this interpretation is that it suggests that digit I did not touch the ground. The alternative hypothesis—that the tridactyl impression represents digits I, II, and III—requires that digit IV be rotated inward toward the body. The French ichnites show this to be the case, supporting this latter hypothesis (figure 7.19). Several of the Crayssac trackways also show the manus tracks situated far from the trackway midline and well outside the pes footprints. Such a configuration matches certain predictions of what pterosaur trackways should look like[60] and establishes that they could not possibly have been made by the short forelimbs of crocodiles.

As we shall see in the following chapter, although the Crayssac site is the only Late Jurassic pterosaur tracksite currently known in Europe, there are several Cretaceous pterosaur tracksites now known. Such ongoing discoveries of previously unknown footprint assemblages, representing major vertebrate groups, seem to underscore the considerable potential of the European track record to reveal new and unexpected ichnofaunas.

The presence of hitherto undiscovered ichnofaunas in the Late Jurassic is again demonstrated by another discovery near Cahors, not far from Crayssac. At this site, Brigitte Lange-Badré and colleagues reported the first brontosaur tracks (figure 7.20) from the Upper Jurassic of France.[61] Although not abundant or well-preserved, owing to the small size of exposures in limestone quarries from which the tracks originate, it is becoming clear that tracks are not uncommon at this site. In addition to pterosaur and brontosaur tracks, a few well-preserved small theropod (figure 7.20) tracks have also been reported. Like the Late Jurassic limestones of Portugal and Switzerland, which are proving to be a rich source of dinosaur tracks, the limestone facies of southern France is also proving productive, not just for brontosaur tracks, which predominate in these other regions, but also for the tracks of smaller vertebrates, which appear to be well preserved in the lithographic limestone facies.

A NOTE ON THE *BRONTOPODUS* ICHNOFACIES AND OTHER CARBONATE ICHNOFACIES

In 1994, we collaborated with Adrian Hunt to define the *Brontopodus* ichnofacies.[62] Our concept of vertebrate ichnofacies evolved quite simply from the observation that particular track assemblages are found recurrently in distinct sedimentary facies. For example, sauropod tracks are found almost exclusively at low paleolatitudes, within 30 degrees of the equator,[23] and in most cases are associated with limestone substrates.

In principle, if a particular sedimentary facies repeatedly reveals a certain type of track, it must reflect an enduring relationship between particular trackmakers and their preferred habitats. Alternatively, different tracksites in a particular sedimentary facies would reveal a random selection of tracks if no relationships existed. But this is not the case. We identified the recurrent association of brontosaur tracks with platform carbonate facies in the Late Jurassic of Portugal and Switzerland, as well as the Early Cretaceous of Texas, as examples of the *Brontopodus* ichnofacies. Here it should be noted that other sporadic occurrences of brontosaur tracks are not automatically considered as examples of the *Brontopodus* ichnofacies. An ichnofacies cannot be defined if there are insufficient recurrent examples of the same track assemblages in the same facies. It should also be noted that the three examples cited here are different expressions of the *Brontopodus* ichnofacies.

In all three regions the dominant sauropod trackways are the wide-gauge type (ichnogenus *Brontopodus*), although in Portugal at least one site in the carbonate facies is dominated by the narrow-gauge variety (ichnogenus *Parabrontopodus*). Given that this latter ichnogenus was named[24] after the *Brontopodus* ichnofacies had been defined, the appropriate historical perspective is that there is at least one defined example of the *Brontopodus* ichnofacies in which a *Parabrontopodus*-dominated assemblage has been recorded. By the same token, the presence of theropod tracks at particular sites within the broader *Brontopodus* ichnofacies does not invalidate the ichnofacies concept.

We can take the examples of the Cerin and Crayssac lithographic limestones to further illustrate the ichnofacies principle. For example, at Cerin there are at least eight stratigraphic levels with turtle tracks (ichnogenus *Chelonichnium*) occurring in laminated bituminous limestone and lithographic limestones (figure 7.18). Thus, we can infer a recurrent association of turtle tracks with these two lagoonal facies that we can characterize as the *Chelonichnium* ichnofacies. Similarly, at the Crayssac tracksite, *Pteraichnus* occurs at six stratigraphic levels in a sequence of platy laminated limestones and sublithographic limestones, again suggesting a *Pteraichnus* ichnofacies. Given that *Pteraichnus* from somewhat older deposits in North America often occur in noncarbonate facies, we must be aware that the Crayssac *Pteraichnus* ichnofacies is not identical to its expression elsewhere. We can extend this to a general rule that an ichnofacies may be expressed in various subtly different ways in different areas and different facies. Similarly ichnofacies may overlap. For example, in the Late Jurassic carbonate facies of Europe we find expressions of the *Brontopodus* ichnofacies and local *Pteraichnus* ichnofacies that are evidently not completely unrelated. Future work should help shed light on the distribution and interrelationships of these and other as yet undefined ichnofacies.

THE FIRST ANKYLOSAUR TRACKS

What were once thought to be the youngest fossil footprints known from the Jurassic, reported by Paul Ensom in the late 1980s, may also be the earliest known examples of ankylosaur tracks from Europe. The tracks in question were found only about 2 m below the Jurassic-Cretaceous boundary, as then defined, in the Purbeck Limestone Formation, near Langton Matravers, in Dorset, England.[63,64] The precise stratigraphic location of the Jurassic-Cretaceous boundary is not universally agreed upon by paleontologists and stratigraphers. Nevertheless, in Dorset, most authorities place it near the so-called Cinder Bed within the Purbeck Limestone Formation. Since Ensom described these tracks as Jurassic, the boundary has been "pushed back" or "down," making the tracks Cretaceous in age.[65,66] This shifting of the Jurassic-Cretaceous boundary does not make the tracks any older, although it does change the label we apply. Given that they were originally described as Jurassic footprints, and that more ankylosaur tracks are mentioned in chapter 8, we

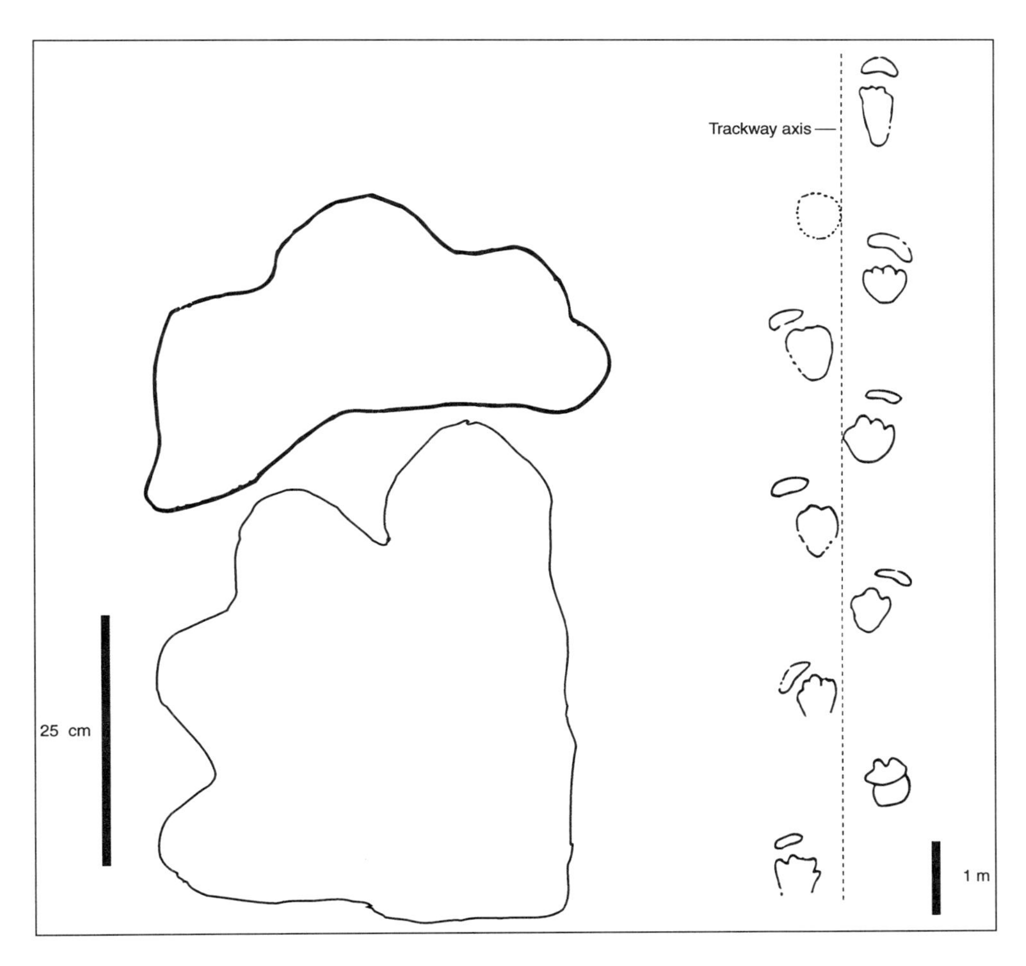

FIGURE 7.21. Ankylosaur tracks close to the Jurassic-Cretaceous boundary in the Purbeck Limestone Formation of Dorset, England. *After Ensom,[64] and Wright.[66]*

include a reference to these tracks in this chapter. It is surprising that more ankylosaur tracks are not known from deposits of Late Jurassic age, because their skeletal remains are reasonably well known in rocks of this age.

It is not just the age of the tracks that makes them controversial or potentially controversial (moving the boundary down a few meters has put them in the Cretaceous), but also the fact that researchers do not universally agree on their origin. Ensom, who originally described the tracks when working for the Dorchester Museum, suggested that they might have been made by sauropods.[64] In 1991 the senior author disagreed with this interpretation and proposed an ankylosaur origin.[43] At that time only a few ankylosaur tracks were known (see chapter 8). Recent discoveries, mainly from the Cretaceous of North America, reveal that purported ankylosaur tracks from this region are very similar in morphology to the Purbeck specimen (figure 7.21). It is worth noting here that these footprints are not much older than the skeletal remains of *Hylaeosaurus* from the overlying Wealden beds. *Hylaeosaurus* is one of the most historically famous of all dinosaurs, having been described in 1832 and included, with *Megalosaurus* and *Iguanodon,* in Sir Richard Owen's original description of the Dinosauria. Thus, after a century and a half we may have discovered the tracks of one of Britain's best-known dinosaurs or one of its close relatives.

BIBLIOGRAPHIC NOTES

1. Buckland, W. 1824.
2. Weishampel, D. B., P. Dodson, and H. Osmolska. 1990.
3. Lapparent, A. F. de, G. Zbyszewski, F. Moitinho de Almeida, and O. Viega Ferreira. 1951.
4. Lapparent, A. F. de and G. Zbyszewsky. 1957.
5. Gomes, J. P. 1915–16.
6. Lessertisseur, J. 1955.
7. Lockley, M. G. and A. P. Hunt. 1994.
8. Abel, O. 1935.
9. Nopsca, F. von. 1923.
10. Colbert, E. H. and D. Merrilees. 1967.
11. Haubold, H. 1971.
12. Kaever, M. and A. F. de Lapparent. 1974.
13. Gabuniya, L. K. and V. Kurbatov. 1982.
14. Calkin, B. J. 1968.
15. Casanovas Cladellas, M. L. and J-V. Santafe Llopis. 1971.
16. Casanovas Cladellas, M. L. and J-V. Santafe Llopis. 1974.
17. Lockley, M. G., C. A. Meyer, and V. F. dos Santos. 1996.
18. Lockley, M. G., C. A. Meyer, and V. F. dos Santos. In press.
19. Lockley, M. G., V. Novikov, V. F. dos Santos, L. A. Nessov, and G. Forney. 1994.
20. Santos, V. F. dos, G. Carvalho, and C. Marques da Silva. 1995.
21. Lockley, M. G., A. P. Hunt, J. J. Moratalla, and M. Matsukawa. 1994.
22. Bakker, R. T. 1996.
23. Lockley, M. G., C. A. Meyer, A. P. Hunt, and S. G. Lucas. 1994.
24. Lockley, M. G., J. O. Farlow, and C. A. Meyer. 1994.
25. Mensink, H. and D. Mertmann. 1984.
26. Thulborn, R. A. 1990.
27. Valenzuela, M., J. C. Garcia Ramos, and C. Suarez de Centi. 1986.
28. Valenzuela, M., J. C. Garcia Ramos, and C. Suarez de Centi. 1988.
29. Garcia Ramos, J. C and M. Valenzuela. 1977a.
30. Garcia Ramos, J. C and M. Valenzuela. 1977b.
31. Dodson, P. 1990.

32. Lockley, M. G. and V. F. dos Santos. 1993.
33. Lockley, M. G., V. F. dos Santos, M. M. Ramahlo, and A. Galopim. 1993.
34. Bird, R. T. 1944.
35. Antunes, M. T. 1976.
36. Lockley, M. G., C. A. Meyer, and V. F. dos Santos. 1994.
37. Meyer, C. A. 1993.
38. Meyer, C. A. 1997.
39. Lockley, M. G., V. F. dos Santos, C. A. Meyer, and A. P. Hunt. 1998.
40. Bernier, P., G. Barale, J-P. Bourseau, E. Buffetaut, G. Demathieu, C. Gaillard, J-C. Gall, and S. Wenz. 1984.
41. Bernier, P. 1985.
42. Bernier, P., G. Barale, J-P. Bourseau, E. Buffetaut, G. Demathieu, C. Gaillard, and J-C. Gall. 1982.
43. Lockley, M. G. 1991.
44. Foster, J. R., M. G. Lockley, and J. Brockett. In press.
45. Raath, M. A. 1972.
46. Leonardi, G. 1994.
47. Rainforth, E. C. and M. G. Lockley. 1996.
48. Foster, J. R. and M. G. Lockley. 1997.
49. Bennett, S. C. 1993.
50. Padian, K. and P. E. Olsen. 1984.
51. Oppel, A. 1862.
52. Caster, K. E. 1938.
53. Mazin, J-M., P. Hantzpergue, G. Lafaurie, and P. Vignaud. 1995.
54. Mazin, J. M., P. Hantzpergue, J. P. Bassollet, G. Lafaurie, and P. Vignaud. 1997.
55. Stokes, W. L. 1957.
56. Lockley, M. G., T. J. Logue, J. J. Moratalla, A. P. Hunt, R. J. Schultz, and J. W. Robinson. 1995.
57. Unwin, D. A. 1997.
58. Bennett, S. C. 1997.
59. Lockley, M. G., M. Huh, S-K. Lim, S-Y. Yang, S. S. Chun, and D. Unwin. 1997.
60. Unwin, D. A. 1989.
61. Lange-Badré, B., M. Dutrieux, J. Feyt, and G. Maury. 1996.
62. Lockley, M. G., A. P. Hunt, and C. A. Meyer. 1994.
63. Ensom, P. 1987.
64. Ensom, P. 1988.
65. Feist, M., R. D. Lake, and C. Wood. 1995.
66. Wright, J. 1996.

A megalosaurid dinosaur. Considerable debate surrounds the concept of megalosaurid dinosaurs and their tracks and whether they are found in the Jurassic and Cretaceous, or only in the Jurassic. *Courtesy of Mauricio Anton.*

CHAPTER 8

THE AGE OF *IGUANODON*: EARLY CRETACEOUS

ARCHOSAURS IN THE AIR (PTEROSAURIAN GIANTS)

As hinted in the previous chapter, the recent spate of pterosaur track discoveries has not only improved our knowledge of Late Jurassic ichnites, but has also shed new light on the Cretaceous footprint record. This new illumination of the pterosaur track record is well illustrated by a study of the ichnogenus *Purbeckopus*, which, as the name suggests, originates from the Purbeck Limestone Formation of Dorset, England.[1] When originally described in 1963 by Justin Delair, the tracks were named *Purbeckopus pentadactylus* and attributed to an unknown reptile, possibly a crocodile. Since that time, several studies[2,3] have concluded that the tracks are tetradactyl, like the pterosaur track *Pteraichnus*, and not pentadactyl. This would seem to rule out crocodiles. Nevertheless, the crocodilian interpretation was reinforced by the claims of Padian and Olsen[4] that *Pteraichnus* was crocodilian in origin (see chapter 7).

Because the pendulum of opinion on the origin of *Pteraichnus* has swung back firmly in favor of a pterosaurian interpretation, it became necessary to reexamine *Purbeckopus*. This restudy[3] yielded a total of three slabs that reveal six pes footprints ranging in length from 18.7 to 22.5 cm. These dimensions make these the second largest pterosaur tracks currently known, with the largest (24 to 33 cm) recently reported from the Cretaceous of South Korea.[5] The restudy of *Purbeckopus* also reveals the presence of manus tracks that show the typical pterosaurian (i.e., *Pteraichnus*-like) morphology (figure 8.1) and confirms that the trackmakers were quadrupedal.

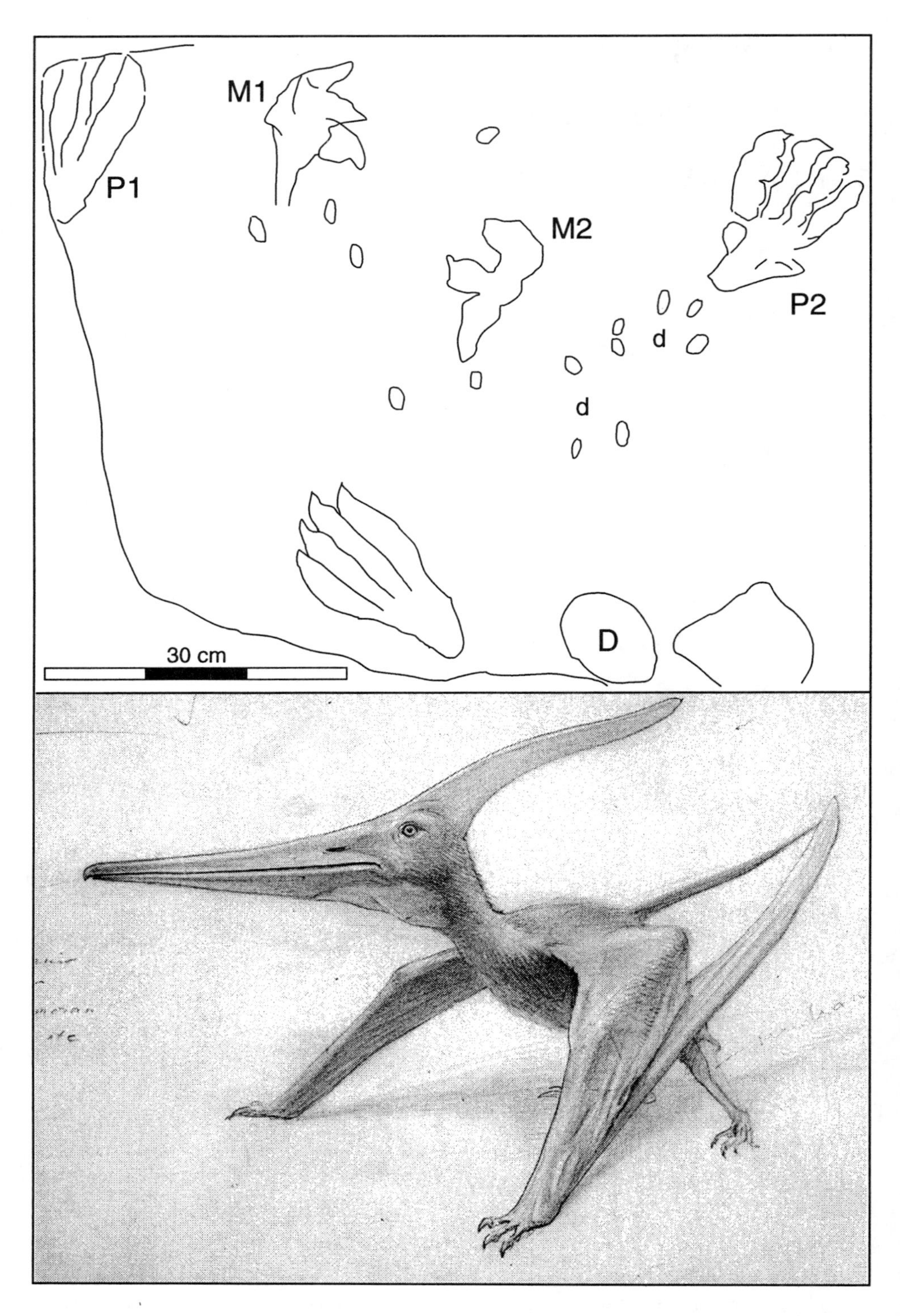

FIGURE 8.1. *Purbeckopus* from the Early Cretaceous of England is currently the largest pterosaurian track known from Europe. Line drawing of slab with manus (*M*) and pes traces (*P*), and possible beak prod marks. *Top: After Wright et al.*[3] *Bottom: After Reichel, unpublished.*

Purbeckopus therefore represents a large Early Cretaceous pterosaur that inhabited lagoonal and intertidal habitats represented by the Purbeck Limestone Formation.[6] Geologic and paleontologic evidence places the tracks in the "intermarine member," which is rich in faunal remains and represents lagoonal intertidal to supratidal habitats where there was a dilution of sea water by fresh water.[3] Based on footprint size, it appears that the *Purbeckopus* trackmaker may have had a wingspan of about 5 or 6 m, making it larger than any pterosaurs known from the skeletal record at this time. Associated with the manus and pes tracks are "small round to oval shaped depressions"[3] that may represent beak traces or prod marks made during feeding activity. Thus, we have a picture of large pterosaurs, presumably pterodactyloids, based on the trace of metatarsal V (figure 8.1),[3] inhabiting coastal environments where there was a mixing of marine and freshwater influence and a generally favorable habitat for the proliferation of diverse vertebrate life.

EUROPE'S EARLY BIRDS

It is not only pterosaur footprints that have been the subject of renewed interest in the Mesozoic track record. Bird tracks also have commanded increasing attention in recent years,[7,8] with a large number of discoveries having been made in Asia. In combination with discoveries from North America and elsewhere, it is becoming apparent that the tracks of waterbirds, or shorebirds, are more common than previously supposed in middle and Late Cretaceous deposits. As yet, however, no bird tracks are known prior to the earliest Cretaceous. Such a stratigraphic distribution conforms well with the bone record, which shows that birds were becoming very diversified as a group by Early Cretaceous times.

The European track record is interesting in this regard because the only significant bird track assemblage appears also to be the oldest currently known. The tracks in question originate from the Wealden Beds of the Oncala Group in north central Spain and have been assigned a Berriassian age, placing them very close to the Jurassic-Cretaceous boundary.[9] As we shall see in subsequent sections, the Early Cretaceous track record in this region, known geologically as the Cameros Basin and geographically as the famous wine-producing region of La Rioja, is extensive and pivotal in our understanding of the vertebrate footprint record.

The bird track locality, in the vicinity of Villar del Rio, was discovered in 1991 and subsequently described by Fuentes Vidarte.[9] He recorded more than 250 distinctive footprints, which he named *Archaeornithipus meijidei* (meaning "ancient bird footprint" and named in honor of M. Meijide Calvo, M. Meijide Fuentes, and F. Meijide Fuentes, who discovered the footprints and assisted with field research). The tracks are remarkable among Cretaceous bird footprints for their large size (average length 12 cm, range 7.5 to 16.6 cm); without exception, all other bird footprints described to date are less than 7.5 to 8.0 cm in length, and many are considerably smaller. Although the skeptic might suggest that these are dinosaur tracks,

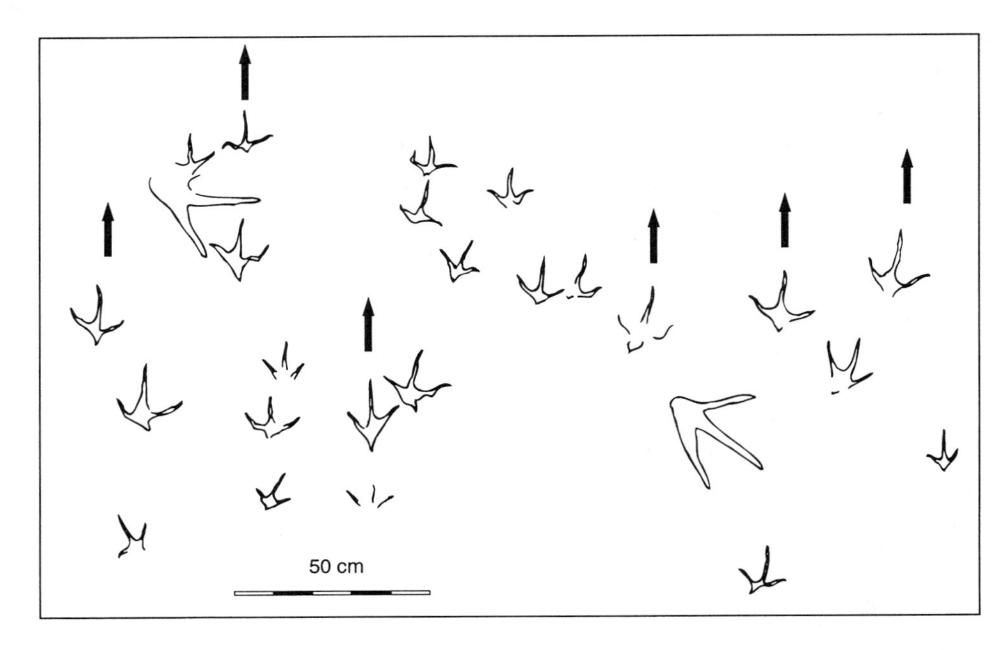

FIGURE 8.2. *Archaeornithipus,* Europe's oldest bird footprints, from the earliest Cretaceous, La Rioja region, Spain. Note the preferred orientation trend. *Modified after Fuentes Vidarte.*[9]

this interpretation is not supported by the footprint morphology, which is typically avian. For example the footprints reveal very slender digit impressions with sharp, tapered extremities, lack of pronounced digital pads, wide digit divarication angles (up to 150 degrees between II and IV) and, in some cases, a pronounced angular articulation or "joint" associated with digit IV (figure 8.2). Such tracks, although unique in the Cretaceous record, are like all other Cretaceous bird footprints, characteristic of shorebirds or waders except that they are larger.

Another remarkable feature of the *Archaeornithipus* track assemblages is that they are virtually all oriented in the same direction, toward the north and northeast. Most bird track assemblages reveal random trackway orientations and high footprint densities, which are often characteristic of modern shoreline habitats where birds move restlessly in many directions as they feed. However, flocks of birds may move more or less in unison in a single direction along a shoreline, if only for short distances. Perhaps the Villar del Rio site indicates such behavior; sedimentologic studies in the region indicate that the paleoenvironment was one of a large alluvial plain with fluvial channels connecting to a permanent lake. Runoff at times may have been intermittent or ephemeral.

Virtually all known Cretaceous bird track sites are associated with fresh water rather than marine shorelines. Also, as noted, all bird tracks from such habitats indicate shorebirds, waders, or waterbirds. By contrast, the growing record of skeletal remains of Cretaceous birds, most of which come from freshwater lakes, consists mainly of small perching birds (passerines and their songbird-like relatives) and other groups not classified as shorebirds. Given the proven potential of lakes as graveyards for birds, it is a little surprising that we have yet to find any Early Cretaceous skeletal remains of shorebirds. Indeed, it is only because we have tracks dating to the earliest Cretaceous that we can extrapolate the origin of shorebirds back so far.[7] This is a significant example of the utility of footprints in shedding light on the timing of the origin of a large and well-known group. Until adequate skeletal remains are found, it will not be known whether these shorebird trackmakers were closely related to modern varieties in a biological or phylogenetic sense, or whether they were ecologic equivalents that were much more distantly related.

THE AGE OF *IGUANODON*

Despite the importance of recent discoveries of pterosaur and bird tracks in the earliest Cretaceous strata of Europe, the best-known footprints from this epoch are those attributed to *Iguanodon,* or iguanodontid dinosaurs. For a multiplicity of reasons, *Iguanodon* tracks provide a variety of challenges to the twenty-first-century ichnologist striving to understand the Early Cretaceous track record. Some of the more important challenges posed by *Iguanodon* track assemblages are as follows:

1. What are the correct ichnogenus and ichnospecies names to use for these tracks?
2. Were the tracks made by representatives of the genus *Iguanodon* only, and if so, which species?
3. Did these animals walk bipedally, quadrupedally, or by using both modes of progression?
4. Does the track evidence indicate gregarious behavior?
5. What is the age range of various *Iguanodon* track types in the Lower Cretaceous, and are any trends discernible in terms of geographic and stratigraphic distribution?
6. What other tracks are associated with *Iguanodon* footprints in various track assemblages?

Other questions may well occur to other ichnologists. In the context of the European track record, however, these appear to be significant questions. In the context of the European track record, the following general points are noteworthy. Most so-called *Iguanodon* tracks come from southern England, northern Germany, or the Cameros Basin in Spain. The former regions have produced many isolated track casts, and a few trackways preserved as impressions, from strata that are reasonably well dated. Tracks from Spain are often preserved in trackways, but they are less common and their precise age is not always known with certainty. Given these considerations, we shall examine, in sequence, the six questions posed above.

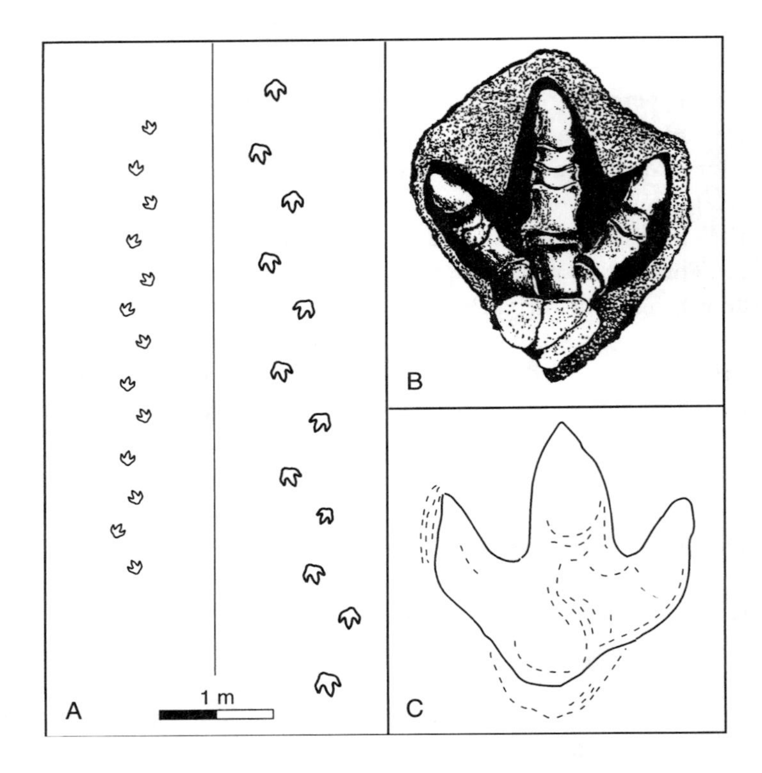

FIGURE 8.3. *Iguanodontipus* (see Sarjeant et al.[12]) showing two trackways, individual tracks with matching *Iguanodon* foot skeleton (A), and a photo of one of the first tracks ever collected in England in 1846 (B and C). *After Tagart.*[10]

1. What Are the Correct Ichnogenus and Ichnospecies Names to Use for These Tracks?

It is axiomatic in ichnology that one does not apply the name of a genus or species to a track, even if there is strong or compelling evidence that the track in question was made by that genus or species (see chapter 1). This rule does not preclude casual reference to *Iguanodon* tracks; it just means that the concept is not formal and therefore not supported by a detailed description of the morphology of the tracks in question. *Iguanodon* tracks provide a classic case study of this exact situation. They were first reported in England in 1846 (figure 8.3)[10] and by 1860 had been correctly attributed to *Iguanodon.*[11] Thus, they were the first dinosaur tracks ever to be attributed to a particular genus of dinosaur. Perhaps because of this early success in identification, and the common habit of simply referring to *Iguanodon* tracks, no further serious study was undertaken. As a result, no ichnogenus name or type specimen was established until recently, when the name *Iguanodontipus*[12] was introduced (figure 8.3). The name *Iguanodonichnus* had been introduced in 1968 to describe tracks from South America,[13] but as subsequent studies[14] have revealed, the tracks are of sauropod origin and are not those of iguanodontids. This name therefore is now considered a *nomen dubium*[12] that cannot be applied to the typical tridactyl tracks found in Europe. The name "*Iguanodonipus*"* also was coined in an unpublished thesis[15] but was not introduced formally into the ichnologic literature.

2. Were the Tracks Made by Representatives of the Genus Iguanodon *Only, and If So, Which Species?*

This question is related to the first, which seeks to establish formal names for any track types that show a distinctive morphology. The question was first seriously addressed by Luis Dollo in 1905,[16] when he placed the foot of *Iguanodon bernissartensis* into a track attributed to *Iguanodon* in what must be the first attempt to match these tracks with a particular species of the genus *Iguanodon* (figure 8.3). It is now well established that the genus *Iguanodon,* originally based on skeletal material from England and Belgium, comprises several distinct species, including larger, robust forms such as *Iguanodon bernissartensis* and smaller, gracile forms such as *Iguanodon atherfieldensis.*[17] The obvious question then arises: Can *Iguanodontipus* be correlated with one of these species? The answer remains ambiguous, but as we shall see in subsequent sections, there is some indication that larger and smaller iguanodontid footprints can be differentiated. Alternatively, we

* The use of quotation marks around Latinized ichogenus names indicates that they originate in an unpublished thesis or have since been used in the literature or formally redescribed in papers in press at the time of writing.

also must be aware that the foot skeletons and feet of different species were sufficiently similar to make ready distinction and differentiation of footprints a challenge. Ultimately, progress depends to a significant degree on looking carefully at large samples and distinguishing distinctive track types whenever possible. What we can say is that despite the presence of various small, gracile ornithopods, such as *Hypsilophodon,* in the Lower Cretaceous of England and Europe, there are no large *Iguanodon*-sized ornithopods that were as abundant or as widespread as *Iguanodon,* so there is a high probability that most tracks really were made by representatives of the genus *Iguanodon.*

3. Did These Animals Walk Bipedally, Quadrupedally, or by Using Both Modes of Progression?

We can begin to answer this question. There are two sites in Spain, at Cabezón de Caméros[18] and Regumiel de la Sierra,[19] one in England,[20] and one in Germany, where trackways of large quadrupedal ornithopods (iguanodontids) are known (figure 8.4). Most other iguanodontid tracks from Europe, including all those known from England, appear to represent bipedal animals. We must bear in mind here that although the presence of manus tracks is proof of quadrupedal progression, the absence of manus tracks has three possible interpretations (i.e., the animal was progressing bipedally, or progressing quadrupedally, then overprinting its manus tracks with its pes footprints,[21] or leaving only shallow manus tracks that were not preserved). If some species, notably larger forms, were "obligatory" or "habitual" quadrupeds, whereas others were "obligatory" or "habitual" bipeds, it appears probable that there is more than one species represented in the European footprint record, as outlined below. Again, however, we must not lose sight of the alternative explanation, that some iguanodontids may have been "facultative" or "optional" bipeds capable of switching into or out of quadrupedal progression at will.

Here it is interesting to note that the front footprints of *Iguanodon* trackways from Early Cretaceous deposits are placed well away from the trackway midline in some cases. This is in contrast to a more central positioning in Late Cretaceous ornithopods. The extent to which this is a significant trend through time is not clear and requires further investigation.

4. Does the Trackway Evidence Indicate Gregarious Behavior?

There is already abundant evidence from ornithopod trackways around the world that this group exhibited gregarious tendencies. We have already seen (see chapter 7) that the small ornithopod track *Dinehichnus* from the Upper Jurassic of Portugal, and similar tracks from the United States and Spain,[22] indicate gregarious behavior, but these are not attributed to iguanodontids. Similarly, trackways of large ornithopods (probably iguanodontids) from Canada and Korea[23] show parallel ori-

FIGURE 8.4. Trackways of quadrupedal iguanodontids from Spain. A: Ceradiccas site.[57] B: Regumiel de la Sierra site.[19] C: Cabezón de Caméros site.[18]

entations indicative of herding behavior. Direct trackway evidence for gregarious behavior among European iguanodontids is somewhat sparse. However, several parallel trackways have been recorded at Lower Cretaceous sites in England,[24] so it appears that European trackmakers were no exception to this general tendency among large Early Cretaceous ornithopods (figure 8.5).

In 1973 Justin Delair and A. B. Lander[25] illustrated three parallel trackways from the so-called Roach Stone at Suttles Quarry, Herston, near Swanage, which were described as those of *Megalosaurus* by Alan Charig and Robert Newman, but are clearly of iguanodontid origin.[6] These trackways indicate progression toward the southwest, and there is evidence of two more trackways with a similar trend on an underlying surface (the so-called Sugar bed). This could be an example of a repeated preferred direction of travel, as has been recorded for Cretaceous ornithopod tracks from multiple track-bearing layers in South Korea.[23] One trackway was excavated for the British Museum (see Appendix). Newman also reported on an iguanodontid trackway from the freestone bed at Queensground quarry (number 9) that was excavated for the Hunterian Museum in Glasgow (see Appendix). This particular specimen is actually now known to be a theropod trackway segment.[6] Three additional parallel trackways of iguanodontids, with an east-southeasterly trend, were uncovered in the Roach Stone (pink bed). Purported megalosaur trackways were excavated from the Roach Stone at Lock's Quarry, Acton, and a segment of one of these was acquired by the Royal Scottish Museum, Edinburgh (see Appendix). Originally, in 1967, two interwoven trackways were uncovered at this latter site, both with a similar north-northwesterly trend. They are also probably of iguanodontid affinity.

In 1982 Delair summarized several of his previous papers in his article "Multiple Dinosaur Trackways from the Isle of Purbeck,"[24] but erring on the side of caution, he failed to reach the conclusion that the evidence is suggestive of gregarious behavior. His reluctance to mention this topic may in part be the result of his belief that some of the trackways were of megalosaurid origin. It is our belief that all the tracks are of iguanodontid origin, a conclusion supported by our own analyses of Jurassic megalosaurid tracks (see chapter 7) and Delair's own observation that all the tracks are similar in size and pace length. Based on the examples given—from Lock's Quarry, Suttle's Quarry (two levels), Queensground, and Warbarrow Bay—we can conclude that there are at least five examples of groups of two or more parallel iguanodontid trackways and that individual trackways from three of these locations have been excavated for public display at major museums, even though in some cases they are incorrectly characterized as megalosaurid tracks. Such evidence may not be indicative of large herds, although the areas of trackway exposure were small in all cases, but it is suggestive of some gregarious behavior, and we should bear in mind that many of the parallel trackway groupings may have been much larger. Such patterns are consistent with the evidence of multiple parallel trackways of ornithopods from many Lower Cretaceous trackway sites from other regions, notably in North America and South Korea.[26]

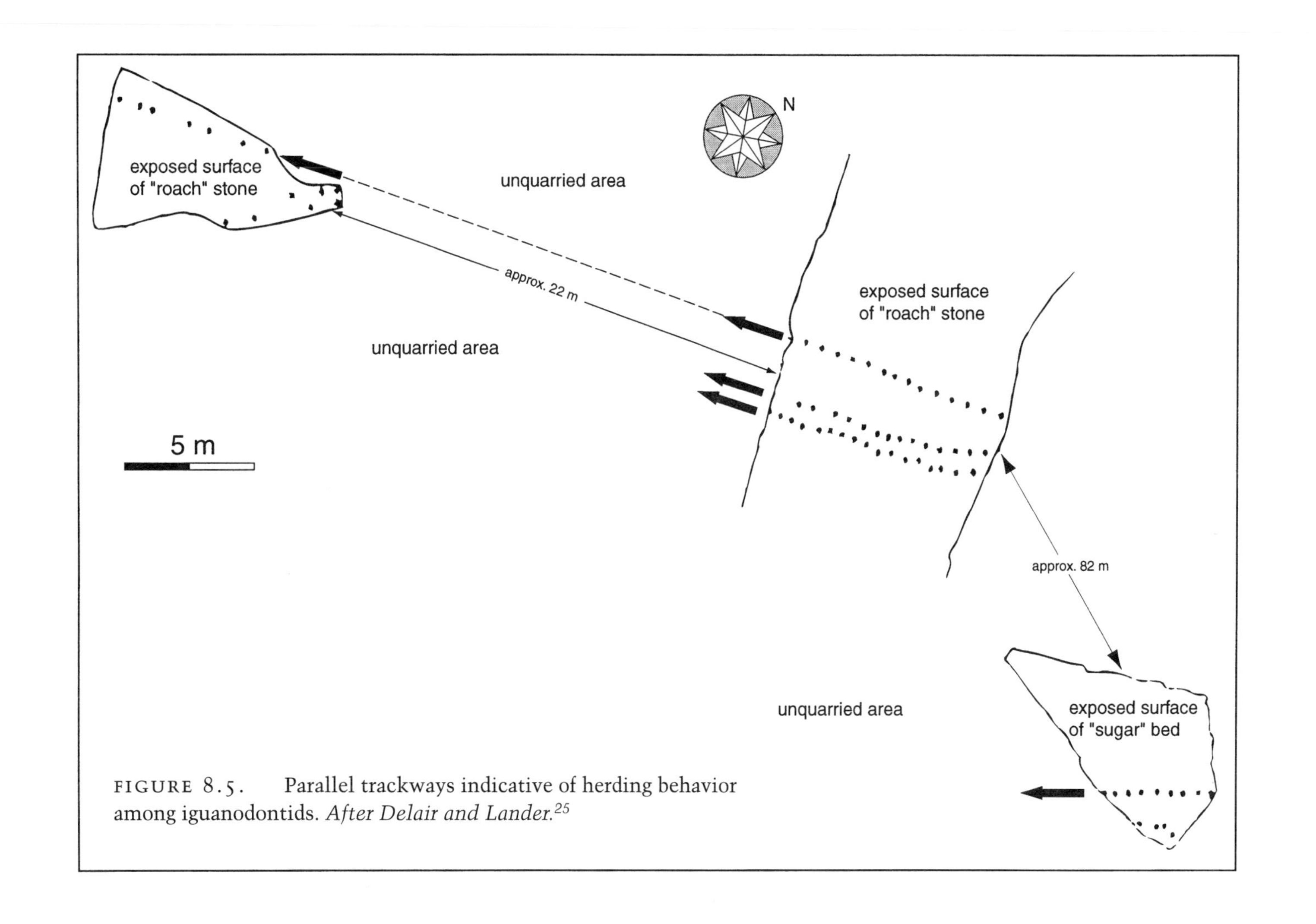

FIGURE 8.5. Parallel trackways indicative of herding behavior among iguanodontids. *After Delair and Lander.*[25]

5. *What Is the Age Range of Various* Iguanodon *Track Types in the Lower Cretaceous, and Are any Trends Discernible in Terms of Geographic and Stratigraphic Distribution?*

This is not a question that can be answered either fully or unambiguously at present. However, it is an important question to ask, because it is only by understanding tracks in their chronostratigraphic, biostratigraphic, and paleogeographic context that we can hope to get a complete picture of their distribution. It should be noted first that *Iguanodon* and other iguanodontids are particularly abundant in Western Europe, with more scattered representation in other parts of the world, such as Africa, Australia, and North America. In general, track-bearing Lower Cretaceous strata are imprecisely dated, so that various so-called *Iguanodon* or *Iguanodon*-like tracks may be anywhere from Berriasian (145 to 138 million years) to Aptian or Albian in age (110 to 98 million years). It should be noted, however, that David Norman[17] has shown that *Iguanodon* species from Early Wealden deposits (Berriasian-Valanginian) are different from later Wealden species (Hauterivian-Barremian-Aptian). So, providing that foot morphology changed, we might expect that track assemblages would change over a long period.

Given the track stratigraphy emerging in the Jurassic, where track faunas change every stage (5 to 10 million years on average), it is unlikely that the entire Lower Cretaceous, represented by six stages spanning almost 50 million years, would display precisely the same ichnofauna throughout. We therefore predict that future refinements on dating of Lower Cretaceous strata, coupled with further track discoveries and improved ichnotaxonomic differentiation of distinctive track types, will lead to the recognition of several different Lower Cretaceous ichnofaunas. Another general characteristic of iguanodontid tracks that is of interest is size. Several of the parallel Isle of Purbeck trackways reported in the previous section represent quite small individuals, with foot widths in the range of 15 to 20 cm, whereas some of the largest tracks measure about 50 cm in length and width. Whether such differences in size are randomly distributed in space and time or are a reflection of evolutionary or taxonomic changes (size increases) in the trackmaking fauna can only be determined by a thorough study of all the available data in their proper geologic context. Such topics are considered further in the sections that follow, but it is worth noting that this 15 to 50 cm size range is typical of Lower and middle Cretaceous tracks from North America.[27]

6. *What Other Tracks Are Associated with* Iguanodon *Footprints in Various Track Assemblages?*

As we shall see in the sections that follow, *Iguanodon* tracks are sometimes associated with tracks of other dinosaurs such as theropods, ankylosaurs, noniguanodontid ornithopods, and occasional sauropods, and with tracks of birds and pterosaurs. Again, once the ichnotaxonomic composition, stratigraphic distribution, and relative abundance of such tracks are more precisely described, it should be possible to define characteristic ichnofacies and so say something about the ecologic composition of trackmaking faunas. This topic is addressed in the sections that follow.

IGUANODON AND CONAN DOYLE'S LOST WORLD

S. H. Beckles, who was the first to write extensively on footprints from the Wealden in the 1850s,[28–30] adopted the terminology of Edward Hitchcock and applied the label *Ornithoidichnites* ("bird traces"), admitting that he did not know what had made the tracks. He stated in 1854 that he adopted the term *Ornithoidichnites* "provisionally and most cautiously" and continued, "although the evidence seems to connect the footprints with the class Aves, yet I am not aware that it is such as positively to exclude animals of different organization. . . . I shall therefore leave to future inquiry the solution of the interesting problem of the natural affinities of these wonderful bipedal forms."[30] This paper is the source of the illustration of trackways in figure 8.3. It is against this background that we can look at the next, early-twentieth-century, chapter in the interpretation of *Iguanodon* tracks.

According to Dana Batory and William Sarjeant,[31] "the discovery in 1909 of *Iguanodon* footprints in the Wealden Bed at Crowborough, Sussex, excited the attention of Sir Arthur Conan Doyle and served as a stimulus to his writing of 'The Lost World'." Given that a deluxe 1912 edition of this famous adventure story has a cover decorated with stylized *Iguanodon* footprints (some say they are theropod tracks), the assessment of Batory and Sarjeant is well supported. The footprint discovered at Crowborough, and actually illustrated in the aforementioned deluxe edition, was just one of several dozen discoveries of small track assemblages or isolated footprint impressions and casts found in southern England (Sussex, Surrey, Isle of Wight, and Dorset) between 1846 and the present time.

It is interesting to look back almost a century and try to recreate the excitement and mystery experienced by Conan Doyle as he contemplated the significance of *Iguanodon* footprints found on his doorstep. When introducing the footprints, he emphasizes their tridactyl shape and so has his character Lord John Roxton declare that the trackmaker must have been "the father of all birds." The paleontologically astute Professor Challenger, predecessor of such heroes as Indiana Jones, replies that these are not bird tracks but those of reptiles, specifically dinosaurs. This exchange precisely mirrors the nineteenth-century shift in paleontologic interpretation of three-toed tracks from giant birds, early in the century, to dinosaurs by the end of the century. Similarly, when Professor Summerlee, another expedition member, identifies the actual trackmakers as "*Iguanodons*" whose footmarks were found "all over Kent, and in Sussex" at a time when "the South of England was alive with them," Conan Doyle is again providing as accurate an interpretation of the tracks as was then available from the paleontologic literature.

Despite the accuracy of Conan Doyle's paleontologic knowledge, "The Lost World" was nevertheless a work of fiction and may have influenced scientists inadvertently. For example, he set his lost world on a high, inaccessible jungle plateau in South America. It is therefore interesting to note that in 1968,[13] South American ichnologists interpreted tracks from this continent as those of *Iguanodon* (*Iguanodonichnus*) despite a lack of evidence for this genus in this region. This seems to be a case of a work of fiction influencing one of science. In this case, the work of fiction has proved more enduring and accurate, in terms of correlation between track and trackmaker, than the footprint study.

It is outside the scope of this book to enter into a detailed analysis of British iguanodontid footprints. As already mentioned, the ichnogenus *Iguanodontipus* has been introduced to fulfill a long overdue need for a proper ichnotaxonomic label for these tracks.[12] Joanna Wright has recently completed a Ph.D. thesis[6] on the subject of Lower Cretaceous (Purbeck and Wealden) footprints, and we consider it premature to summarize her conclusions. Further study is required to establish the extent to which all large ornithopod footprints from these formations are assignable to *Iguanodontipus.* Studies by Ken Woodhams and John Hines revealed the presence of "traces of hide pattern in the form of clearly separated hemispherical bumps, or tubercles, 3–4 mm in diameter" on natural casts of the heel.[32] Such discoveries are associated with tracks that consistently reveal other "fine preservation of detail," such as "hoof-like terminations," to the robust, padded digit impressions (figure 8.6). It is interesting to note that these tracks, although clearly of iguanodontid type, are much longer than they are wide. They can be contrasted with much less elongate iguanodontid tracks described from Spain.

Woodhams and Hines[32] also reported "several theropod footprint forms . . . described for the first time." Some of these, we believe, are iguanodontid tracks, but others, including those illustrated here (figure 8.6), are quite distinct and probably of theropod origin. One variety was described as "birdlike" and is slender toed, with wide digit divarication angles and well-defined phalangeal pad impressions. Based on the large number of reports of iguanodontid tracks and the apparent rarity of theropod tracks, it was, until recently, easy to be persuaded that the Wealden track assemblages of England were heavily dominated by ornithopod tracks. This conclusion is reinforced by the observation that so-called *Megalosaurus* tracks, often reported from the Purbeck Limestone, are in fact iguanodontid tracks. As outlined in the previous chapter, megalosaurid tracks are quite distinctive and, based on recent studies, evidently confined to the Late Jurassic. As we shall see in the next section, however, it appears that in other parts of Europe a predominance of iguanodontid tracks in Lower Cretaceous strata is not evident, and the picture is also changing in England now that Woodhams and Hines, in a second paper, have reported additional theropod tracks.[33]

It seems, therefore, that the evidence for theropod tracks in the Wealden has been accumulating steadily. In 1993, Parkes[34] described theropod tracks from the Ashdown beds at Fairlight Cove, near Hastings, Sussex, and Woodhams and

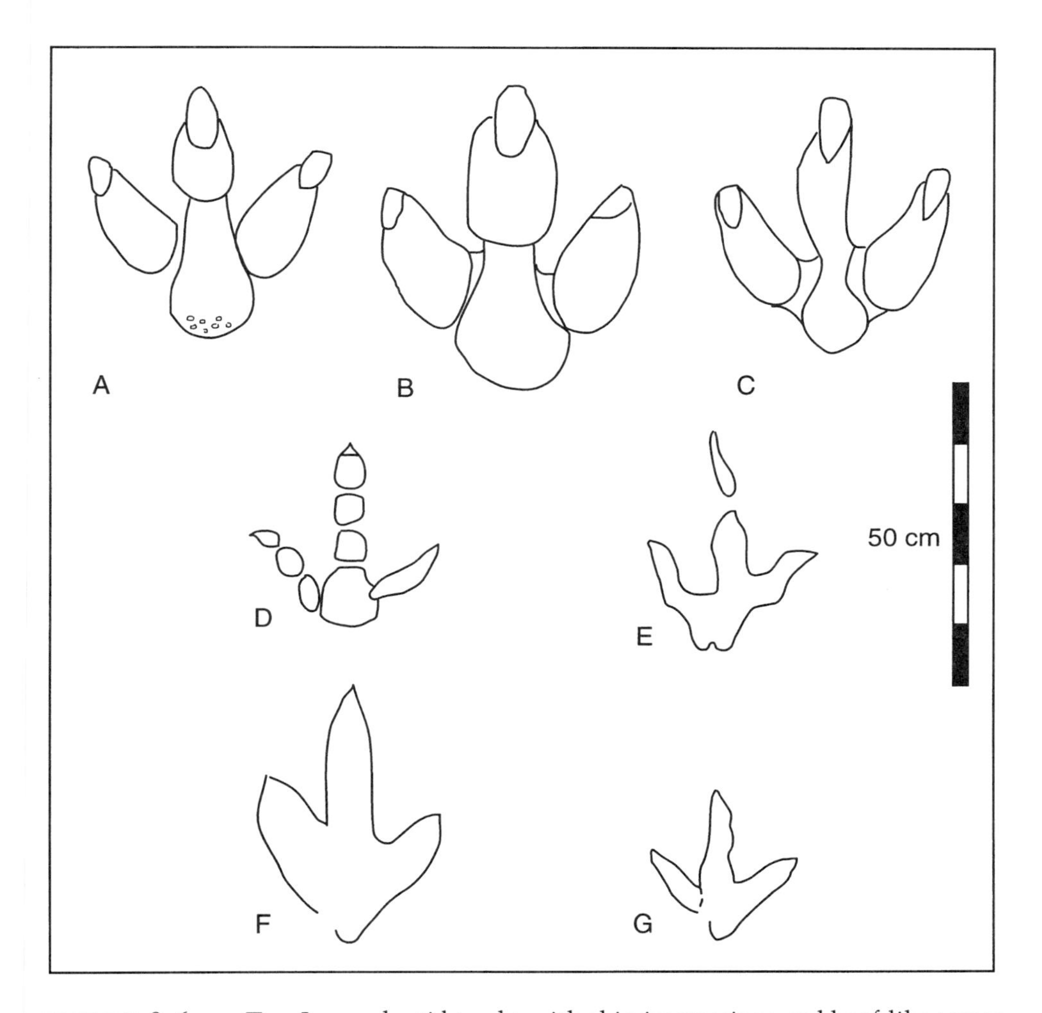

FIGURE 8.6. Top: Iguanodontid tracks with skin impressions and hoof-like traces (A–C). Bottom: Theropod tracks (D and E) from the Hastings beds, Sussex, England. *After Woodhams and Hines.*[32] (F and G) Theropod tracks. *After Parkes.*[34]

Hines[33] again documented a variety of new theropod footprint discoveries, not all of which are convincingly attributed to theropods. New reports of theropod track assemblages are also emerging from the Isle of Wight.[5,35] The result of these theropod track discoveries is that the perception of the Wealden ichnofauna as iguanodontid dominated is changing. Most of the new track theropod discoveries represent gracile, slender-toed forms of moderate size. If we were to have made a preliminary attempt at a census of dinosaur tracks from the Purbeck and Wealden of England 20 years ago, we would have found only reports of *Iguanodon* and "megalosaur" tracks, the latter probably misidentified in most or all cases. Moreover, there would have been little in the way of useful description of how these two track types differed in any significant morphologic details. Today, however, it is clear that there are a variety of distinctive theropod tracks known from horizons that yield abundant iguanodontid footprints, and that in some cases we may find the tracks of other dinosaurs, such as sauropods,[6,36] ankylosaurs (see chapter 7), and even other vertebrates such as pterosaurs.[3]

LA RIOJA

In a review of the Cameros Basin dinosaur trackways of La Rioja Province, in north-central Spain, Joaquin Moratalla and José Sanz reported that "most of the Cameros Basin trackway ichnorecord has been produced by theropods (80%). Ornithopods are represented by 16% and, finally, sauropod tracks are scarce, representing about 4%."[37] Thus, it appears superficially that the Spanish track record in the Lower Cretaceous is very different from the English ichnofaunas outlined above. We should note, however, that the Cameros Basin sedimentary sequence is approximately 9 km thick, covering a stratigraphic interval from latest Jurassic (probably Tithonian) to late in the Early Cretaceous (Aptian), a span of about 40 million years. The summary statistics just cited do not apply consistently to the ratio of track types at different levels throughout this thick sequence. In addition, it is not possible at present to correlate precisely between Spain and England in such a way as to demonstrate the age relationships between track assemblages in the two regions. Some attempts have been made to make correlations between deposits that contain skeletal remains,[38,39] but such correlations have not been explored for track-bearing sections.

In addition to stressing the need for stratigraphic correlations within basins such as the Cameros Basin and across Europe in general, we also draw attention to the fact that because the Cameros Basin is so rich in tracks, it has sometimes been referred to, incorrectly, as a megatracksite; the term "megatracksite"[40–43] refers to an area where tracks are found repeatedly, and predictably, in a single unit of strata, or in a thin stratigraphic package of beds that has a wide lateral, or regional, distribution. Based on present knowledge, the Cameros Basin could be described as a large area in which dinosaur tracks are abundant at many different stratigraphic levels. Future detailed stratigraphic and sedimentologic studies may show that some important track-bearing beds are laterally extensive, in which case one or more megatracksites might be revealed within the basin. At present, however, no such evidence has been published, and it is not possible to describe the whole basin as a megatracksite without radically altering our concept of megatracksites.

It is also important to point out that the Cameros Basin track record is not composed entirely of dinosaur tracks. In addition to the bird tracks[9] reported from the lower part of the sequence (Oncala Group) at Villar del Rio, pterosaur tracks are also known from at least two locations,[44] in one case with associated turtle

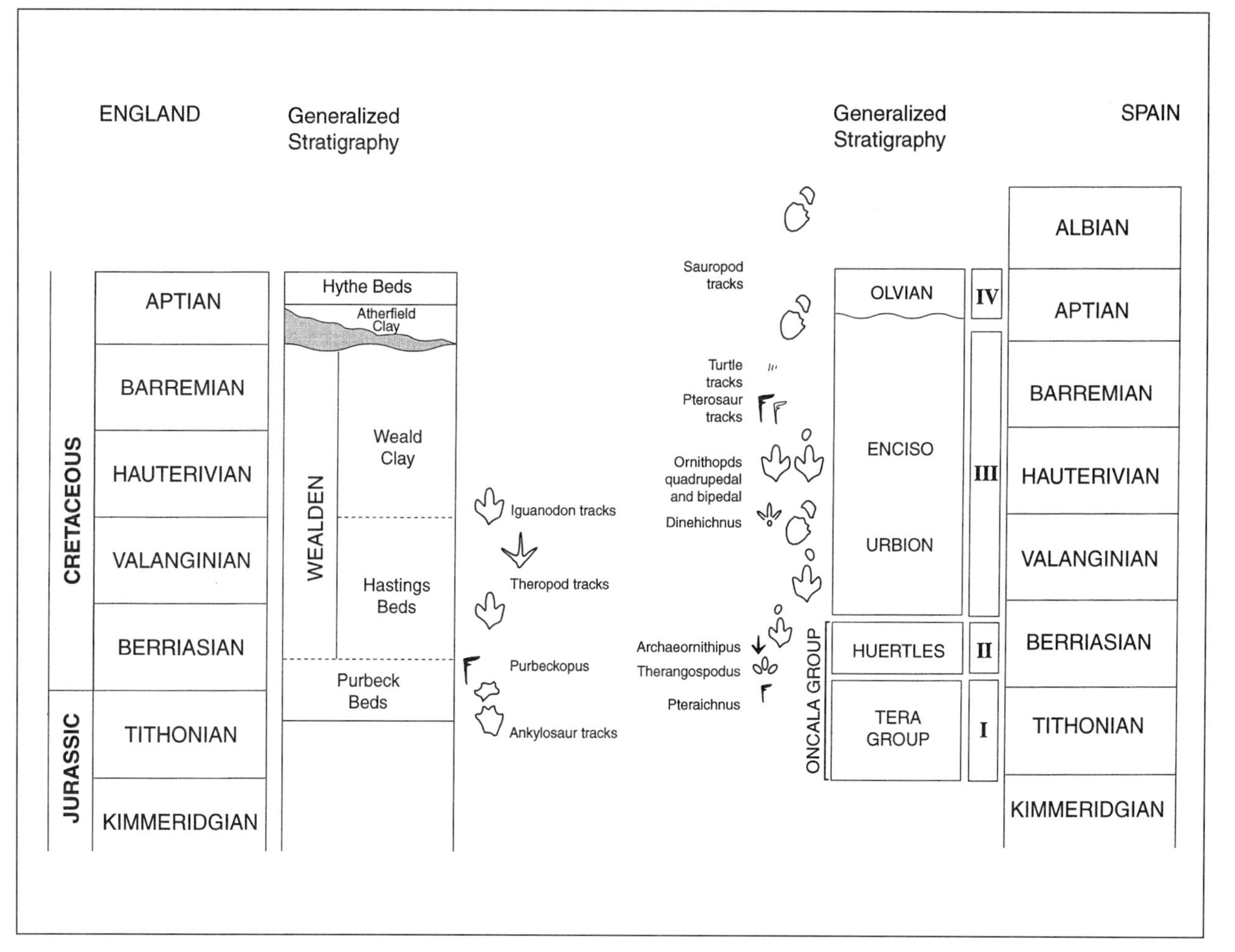

FIGURE 8.7. Simplified stratigraphic sections for the Lower Cretaceous track-bearing sequence of England and the Cameros Basin of Spain, showing the main track types represented. Note the slightly greater variety of tracks known from Spain, owing in part to better exposures over a larger area.

tracks[15,45] and additional examples of possible bird tracks (figure 8.7). Some of these are described in the section that follows. In the meantime we attempt to describe the distribution of the three main dinosaur track types (theropod, sauropod, and ornithopod) in relation to the general stratigraphic sequence.

Theropod Tracks

We have already noted that the Cameros Basin stratigraphic section spans the latest Jurassic to the upper part of the Lower Cretaceous. As noted in chapter 7, especially on the Iberian Peninsula, there are a variety of theropod track types known from the Upper Jurassic, including large forms such as *Megalosauripus* and a variety of smaller, unnamed forms. Although in theory we are not concerned with Jurassic tracks in this chapter, it is not possible to precisely determine the age of tracks close to the Jurassic-Cretaceous boundary in many cases. The theropod track "*Therangospodus oncalensis*" (figure 8.8), first reported by Joaquin Moratalla[15] and later formally described,[46] provides a good example. It is a distinctive footprint type lacking clear creases that separate the digital pads. It originates from the Oncala Group, and so may be about the same age as the bird tracks (*Archaeornithipus*) described by Fuentes Vidarte.[9] Only one continuous trackway and a couple of isolated footprints, from the type section (Fuentesalvo), two trackways from Santa Cruz de Yanguas, and one trackway and other footprints from Bretun are currently known. According to Moratalla, the age of the "*Therangospodus*"-bearing beds is possibly Berriasian within the Late Jurassic (Tithonian) to Lower Cretaceous (Berriasian) Huerteles "Alloformation" of the Oncala Group.[15,47] So the tracks could be Late Jurassic in age because elsewhere "*Therangospodus*" is associated with Late Jurassic *Megalosauripus*. But this possibility is not proven because "*Therangospodus*" might have had a longer age range, extending into the Lower Cretaceous. A large track from Bretun is 43 cm long and could be evidence of *Megalosauripus* in association with "*Therangospodus*" in Spain. But such a co-occurrence, if validated by further discovery, still does not precisely date the age of these supposedly Lower Cretaceous (Berriasian) tracks.

Moratalla[15] noted that in the Oncala Group (?Tithonian-Berriasian) "*Therangospodus*" is actually dominant in the overall paleoecologic composition of the ichnofauna. By contrast, "*Therangospodus*" is virtually absent in the overlying Urbion and Enciso Groups, where the ichnofauna is dominated by large iguanodontid footprints and much larger theropod tracks (ichnogenus? *Buckeburgichnus*). The appearance of abundant iguanodontid tracks is often seen close to the Jurassic-Cretaceous boundary, again indicating the possibility of a pre-Cretaceous age for the "*Therangospodus*"-dominated assemblages. Further studies are required to determine the age of these beds, but it is nevertheless clear that the older ichnofaunas of the Oncala Group are different from those of the younger groups.

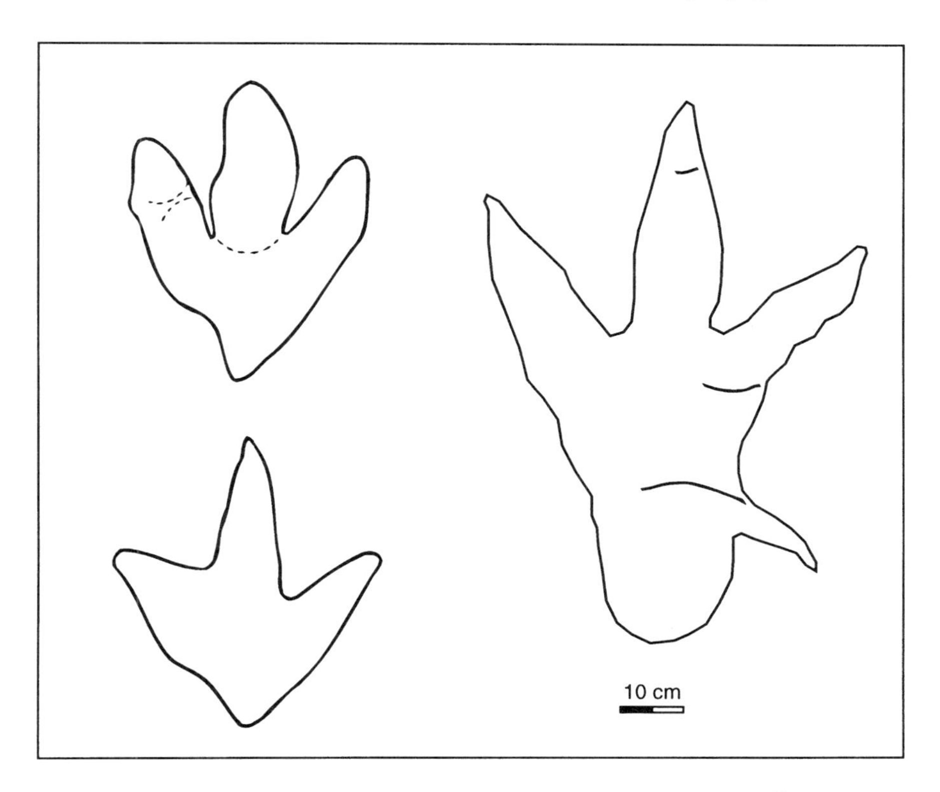

FIGURE 8.8. *Therangospodus oncalensis* (top left) after Moratalla[15] and Lockley et al.[46] with *Buckeburgichnus maximus* for comparison (right) after Kuhn[50] and (lower left) after Moratalla.[15] Note lack of well-defined digital pads. (Note that *Buckeburgichnus* from Germany is currently being restudied.)

In beds that are the same age as those containing "*Therangospodus*," inside a corral or stock enclosure known as the Coral de la Peña in the village of Bretun, there occurs, abundantly, another variety of theropod track, marked by its small size (length:width 24:21 cm) and slender digit impressions. This type was first reported by Luis Aguirrezabala and Luis Viera in 1980[48] and was later labeled as "*Filichnites*" by Moratalla,[15] and it appears to be an ichnotaxon not known from higher in the stratigraphic succession. Thus, for want of further ichnotaxonomic resolution, at present we can speak of an assemblage containing theropod tracks labeled "*Therangospodus*" and "*Filichnites*," with at least one other large track type that occurs close to the Jurassic-Cretaceous boundary in parts of the Cameros Basin.

Higher in the stratigraphic sequence, in the Enciso Group, at such well-known sites as Los Cayos,[49] we find abundant large theropod footprints that have been assigned to the ichnospecies *Buckeburgichnus maximus*.[15,37] This ichnogenus was named by Oskar Kuhn in 1958 to describe large theropod tracks from the Lower

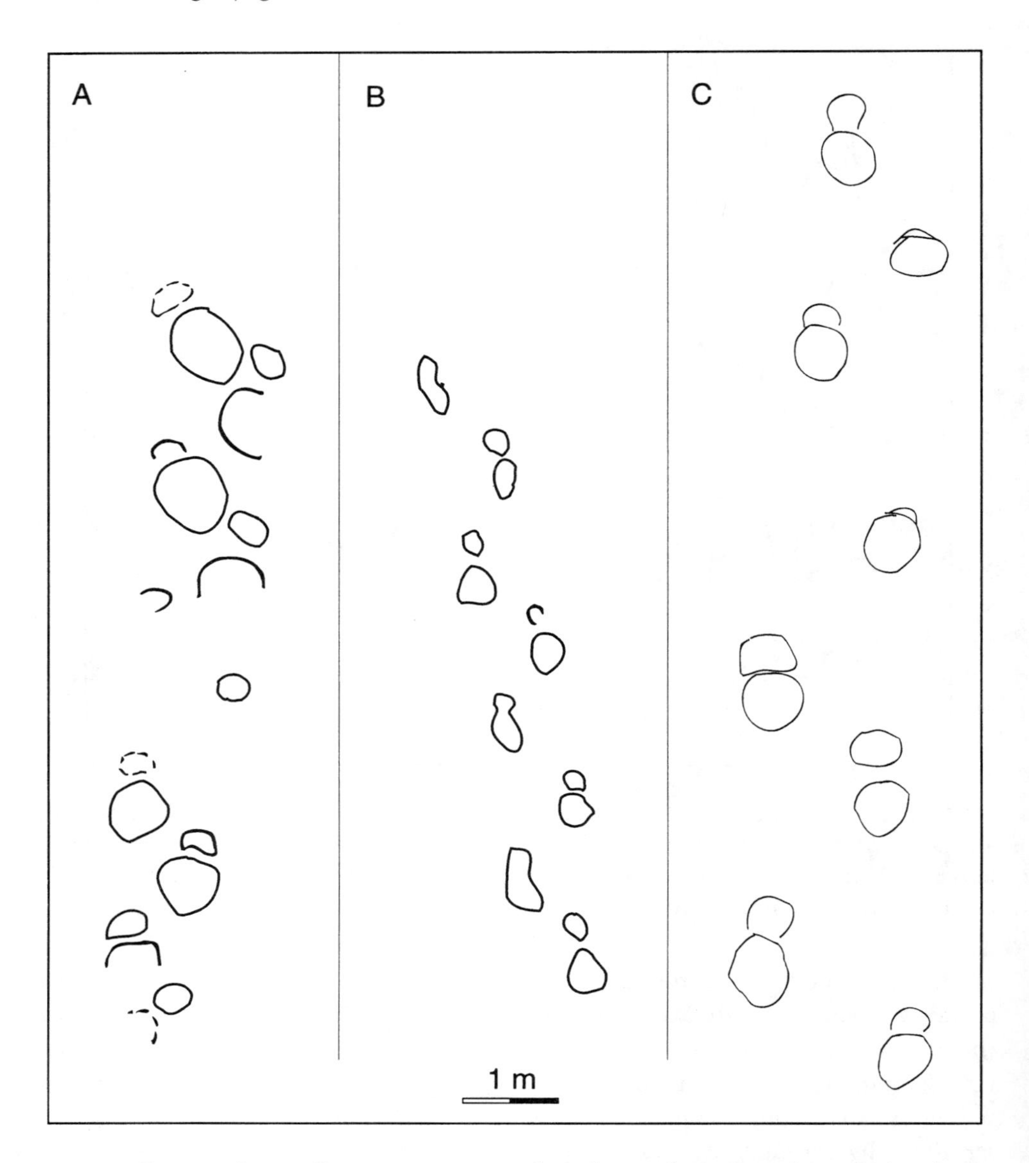

FIGURE 8.9. Lower Cretaceous sauropod trackways from Spain (A and B) and Germany (C). A: Based on the Valdemurillo site.[52] B: Based on the El Sobaquillo site.[53] C: Based on the Münchehagen site.[56] Trackway A may be Jurassic in age.

Cretaceous of Germany that are characterized by a distinctive hallux (figure 8.8; see also chapter 7). They are also apparently characterized by a lack of well-defined creases separating the digital pad impressions,[51] a feature that is conspicuous by its absence in the track assemblage at Los Cayos and certain other tracksites in the Enciso Group. Thus we arrive at the preliminary conclusion that theropod tracks are different at lower and higher levels within the Cameros Basin sequence, although we must admit that details of the lithostratigraphic (vertical) and chronologic age of these different ichnofaunas remains hazy.

Sauropod Tracks

Although in 1994 only 4 sauropod tracksites were documented,[52] by 1997[53] it had been reported that at least 10 sauropod trackway sites were known from the Lower Cretaceous of La Rioja Province. Although Moratalla and Sanz reported that sauropod tracks make up only about 4% of the Cameros Basin ichnofauna, this is probably an underestimate, owing to the recent increase in discoveries. Most of the sauropod tracksites do not reveal particularly well preserved trackways, and based on published information, it is not easy to determine their precise stratigraphic position. However, several of the trackways that have been described originate from near the base of or from other levels within the Enciso Group, also referred to as "depositional sequence III" (figure 8.7). They are therefore estimated to be of late Berriasian to early Aptian age. A trackway from near the base of this group, at the Valdemurillo locality,[52] is the narrow-gauge (*Parabrontopodus*) type with small manus (figure 8.9), whereas at other sites and levels—for example, at Valdecevillo and El Sobaquillo[53]—we find the *Brontopodus* type, which is by far the most common in the Cretaceous.[54] The youngest sauropod trackway known from the Lower Cretaceous of Spain originates from the Monte Grande Formation (Albian age) from outside the Cameros Basin, near Bilbao.[52]

Sauropod trackways are not particularly common in the Lower Cretaceous of Europe, although they are known from isolated sites in Portugal (Praia Grande site of Aptian age),[55] Germany (Buckeberg Formation of Berriasian age),[56] and England (Purbeck Formation of Berriasian age).[6,36] Of these various sites, the most important and most accessible to visitors is probably the German site at Münchehagen (see Appendix). The tracks at this site, named *Rotundichnus münchehagensis* by Alfred Hendricks, are the wide-gauge (*Brontopodus*) type[56] and are preserved as seven extensive trackways suggestive of herding behavior (figure 8.10). Note that they also display a relatively large manus.

Ornithopod Tracks

As noted above, ornithopod tracks are purported to make up about 16% of the Cameros Basin. This figure may be reduced slightly owing to the aforementioned increase in reports of sauropod tracksites. What then can be said about the character and distribution of ornithopod tracks in this sequence? Among the oldest tracks known are a set of ornithopod tracks from the Berriasian Villar del Arzobispo Formation at the Las Cerradicas tracksite, near Galve. Although this tracksite is outside the Cameros Basin, to the southeast, it is nonetheless important in showing the earliest known example of quadrupedal locomotion among or-

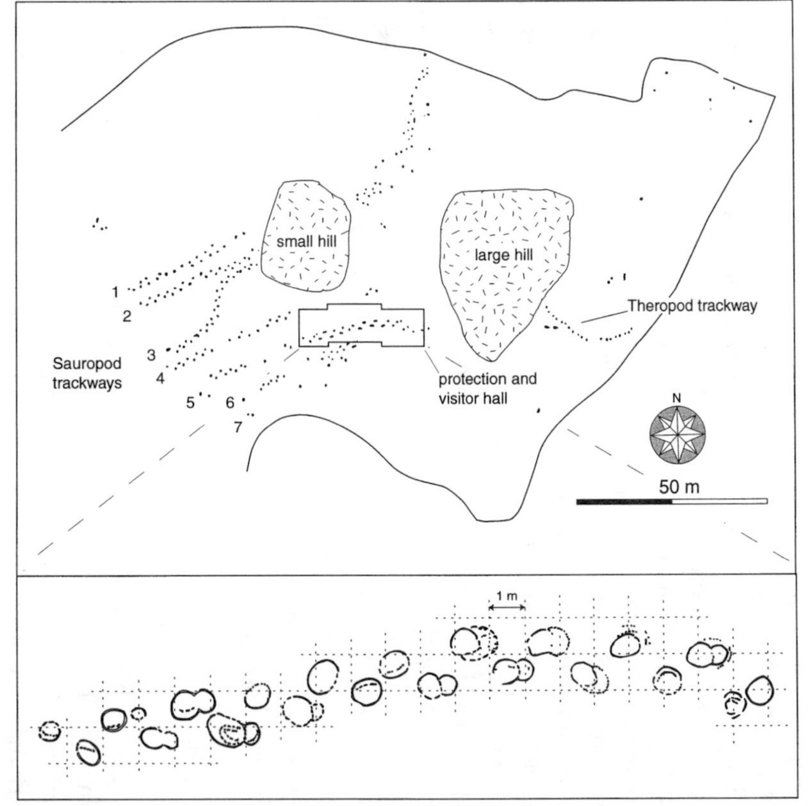

FIGURE 8.10. Map of part of the Münchehagen site showing multiple sauropod trackways with detail of sauropod manus-pes set attributed to *Rotundichnus münchehagensis*. *After Hendricks.*[56]

nithopods. Under the title "Growing Evidence for Quadrupedal Ornithopods," Félix Pérez-Lorente and colleagues[57] described a trackway with pes prints 23 cm long and 23 cm wide, with a small oval manus print situated anterolaterally (figure 8.4A). They noted that there are now about 10 sites with quadrupedal ornithopod trackways from the Lower Cretaceous of Spain, North America, and South America. They identified the Las Cerradicas trackmaker as a probable iguanodontid. The Spanish sites from Regumiel de La Sierra, in the western part of the Cameros Basin, and Cabezón de Caméros (also in the Cameros Basin), reveal trackways, in both cases attributed to iguanodontids, with pes prints averaging about 60 cm in length and width (figures 8.4 B, C).

Given that there are also many purported iguanodontid tracksites in Spain where the trackmakers were moving bipedally, we can conclude that iguanodontids of different sizes adopted both quadrupedal and bipedal gaits. This conclusion is interesting in light of the observations by Cambridge paleontologist and *Iguanodon* expert David Norman[17] that larger, robust forms of *Iguanodon* may have walked quadrupedally, whereas smaller, gracile forms were probably bipedal. Evidence from large sites in North America suggests that both small and large forms were optional bipeds, so the same may have been true for iguanodontids on the Iberian peninsula. Common iguanodontid pes casts from England might be unrepresentative. If the English trackways were more extensively exposed on large surfaces, we might find among them more evidence for quadrupedal progression. Recent finds in Germany show several examples of quadrupedal progression.

Ornithopod tracks from Spain, according to Moratalla,[15] are mainly attributed to iguanodontids, and he named the ichnotaxa "*Iguanodonipus*" and "*Brachyiguanodonipus*," respectively, for a gracile, long-footed trackmaker and a robust, short, wide-footed trackmaker. The former ichnogenus label was applied to the trackway of a quadrupedal iguanodontid from Cabezón de Caméros, but the name was not formally published (see earlier footnote).[18] Instead the name *Iguanidontipus* (with a "t") was suggested.[12]

The Valdevajes (or Valdebrajos) site, also in the Lower Cretaceous of the La Rioja region, also has produced parallel trackway evidence for small, probably gregarious ornithopods tentatively assigned to hypsilophodontids, possibly *Hypsilophon*. The site was first described in 1985 by Luis Aguirrezabala and colleagues,[58] who recorded nine parallel trackways that they attributed to *Hypsilophon* or a similar trackmaker (figure 8.11). In 1991 Casanovas-Cladellas and colleagues[59] also studied the site and attributed the tracks to theropods, without adequate reference to the earlier work. This oversight was criticized by subsequent authors, and again in 1992,[60,61] the opinion that the trackmakers were hypsilophodontids was reiterated. During this debate no name was given to the tracks, but in 1998 it was pointed out that the tracks are very similar to the Upper Jurassic ichnogenus *Dinehichnus*, known from Portugal and Utah, and tentatively attributed to a dryosaurid.[22]

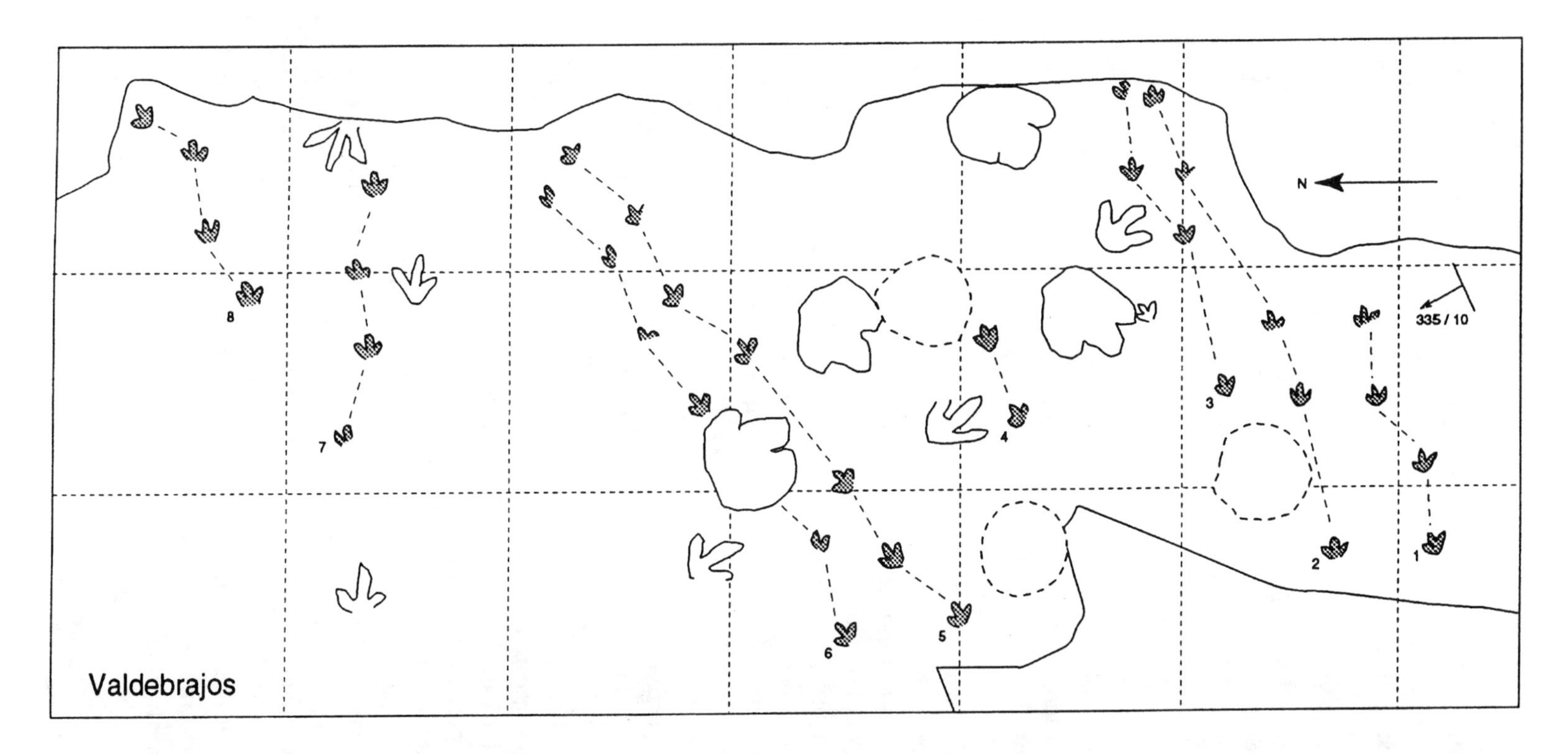

FIGURE 8.11. Controversial ornithopod tracks, possibly made by *Hypsilophon,* an unknown hypsilophodontid or a dryosaurid, from the Valdebrajos site, La Rioja, provide evidence of gregarious behavior.

MORE SPOOR OF THE PTEROSAUR

Lower Cretaceous deposits on the Iberian Peninsula have yielded a number of pterosaurian tracksites, as well as at least one site where other nondinosaurian vertebrate tracks have been reported. The first site, at Santa Cruz de Yanguas, reveals pterosaur tracks (compare with *Pteraichnus*) in beds of the same age as those containing "*Therangospodus.*"[46] The tracks at this site form part of a single trackway in which manus footprints predominate (figure 8.12). Such a configuration has been explained by the fact that pterosaurs had front quarters that were much heavier than their hind quarters, so they often produced trackways in which the front footprints are deeper than the hind footprints.[44] In size and shape the trackway is similar to *Pteraichnus.* Although originally referred to as Berriasian in age,[15,44] this ichnogenus is widely distributed in North America in Oxfordian-Kimmeridgian beds that also contain "*Therangospodus*" and *Megalosauripus.*[44,46,62] With few exceptions,[63] most other Cretaceous pterosaur tracks are much larger that those found at Santa Cruz de Yanguas.

An example of a much larger, probably pterosaur track is that reported from the lithographic limestones at Las Hoyas, in the Province of Cuenca. These limestones represent lake deposits Hauterivian to early Barremian in age and are famous for producing articulated remains of birds, small dinosaurs, crocodilians, and other well-preserved vertebrates. Such subaqueous deposits are not traditionally regarded as a good place to look for tracks. Despite this poor ichnologic potential, the Las Hoyas limestones have produced a number of interesting but problematic footprints that have yet to be fully and correctly interpreted to everyone's satisfaction. Study on the trackways began in 1993 when a trackway and several large tracks of large vertebrates were discovered.[64] The trackway (figure 8.13) has an elongate tetradactyl or pentadactyl pes (20 cm long by 14 cm wide) and a small manus with at least three digits, situated very close to, and slightly on the inside of, the front of the pes impression. The trackway is moderately wide (32 to 33 cm), with a pace angulation of 130 degrees, and has been interpreted tentatively as that of a crocodile, with a hip to shoulder (glenoacetabular) length of about 80 cm and an overall body length of about 3 m. It might, however, be a pterosaur trackway.

The other footprints are different in character, consisting of large three-toed tracks (25 to 30 cm long) that have so far been found only in isolation (figure 8.13). Such tracks bear some resemblance to the large three-toed turtle tracks found in the Upper Jurassic lithographic limestones of Cerin (see chapter 7), and this was the first interpretation given.[64] In 1994 and 1995, however, tracks of pterosaurs began to be identified abundantly all over the world. As a result of their becoming better known, it became evident that large pterosaur tracks were known at least in the Cretaceous. As a result, the purported turtle tracks were reinterpreted as pterosaurian in origin.[44] Such a revised interpretation perhaps makes some sense in light of the fact that huge marine turtles of the size indicated by the Cerin tracks (see chapter 7) are not known from terrestrial lake settings. It is also possible that

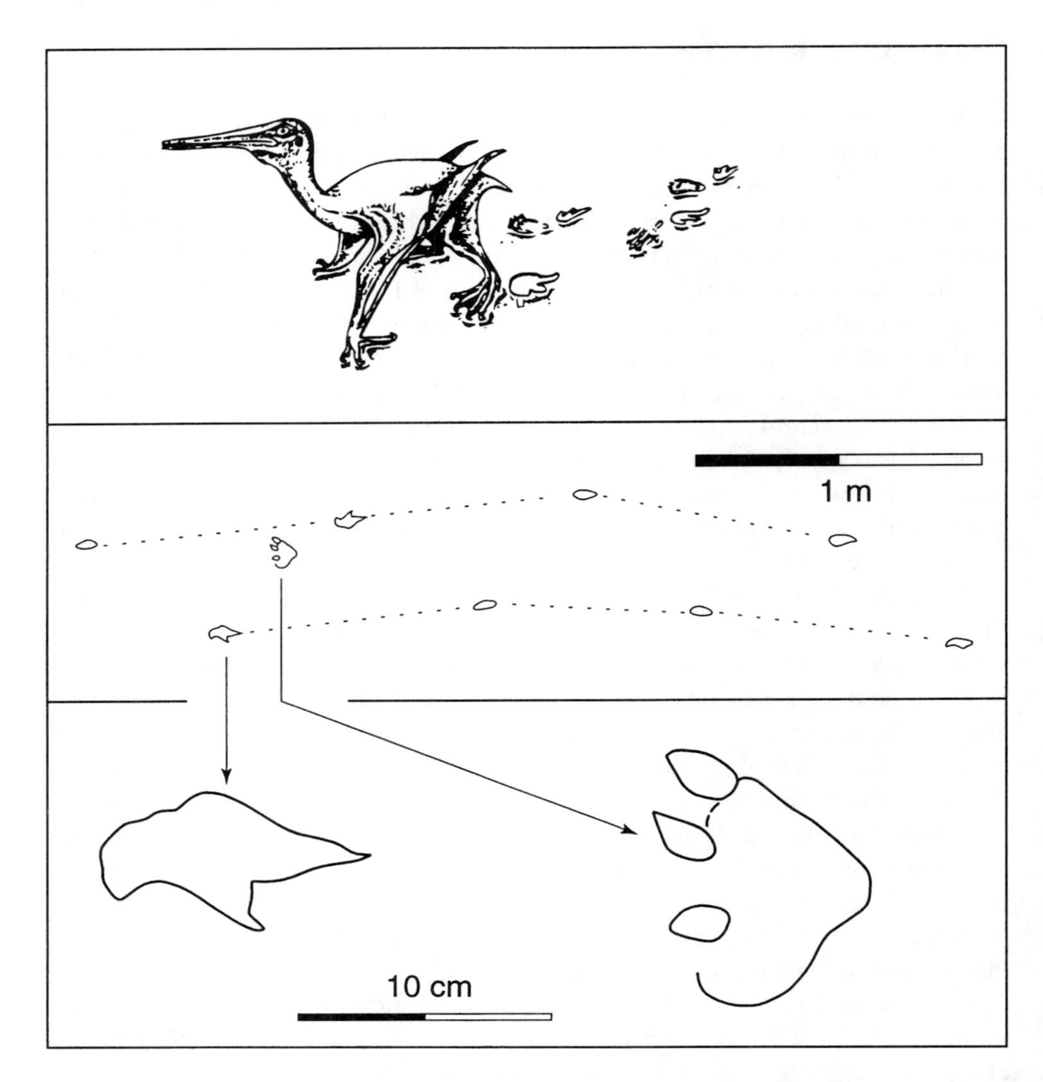

FIGURE 8.12. A pterosaur trackway (compare with *Pteraichnus*) from Santa Cruz de Yanguas is dominated by front footprints. *Redrawn from Lockley et al.*[44]

the crocodilian trackway is pterosaurian in origin. In fact, in the history of debate on the subject of pterosaur tracks, they have been confused with crocodile tracks on more than one occasion.[3,4,44] For the time being, however, we do not attempt to reinterpret the purported crocodilian trackway for several reasons. First, it appears somewhat narrower than large pterosaur trackways of comparable size; second, the position of the manus anterior to, and slightly inside, the pes impression is not characteristic of pterosaur trackways; third, the trackway is not well preserved. One of the most interesting features of this track record is that despite its limitations, it indicates the presence of large vertebrates not known from the body fossil record.

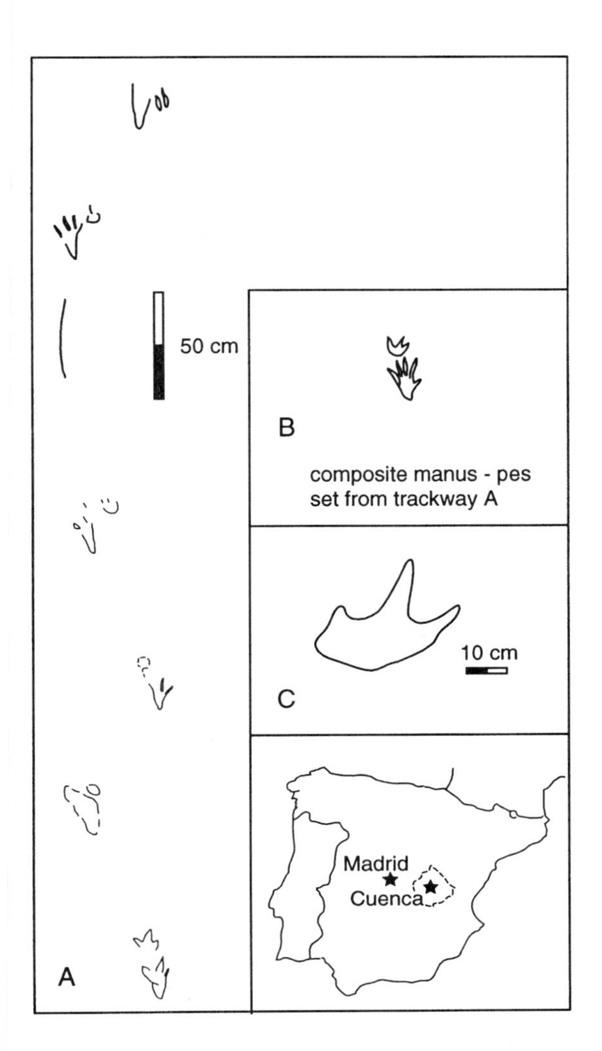

FIGURE 8.13. Probable large crocodilian or pterosaurian trackway (A) and pterosaurian track (C) from the Las Hoyas lithographic limestones, Province of Cuenca, demonstrate an interesting track record in deposits otherwise famous for body fossils of birds and other small vertebrates. Crocodilian trackway (A) with detail (B) of composite manus-pes set. Large pterosaur manus track (C). *From Moratalla et al.*[64]

A third nondinosaurian vertebrate track locality of significance is found in the Enciso Group at Los Cayos, La Rioja Province. In the vicinity of the large, publicly accessible, roofed site where abundant theropod tracks (?*Buckeburgichnus*) have been identified, there are smaller sites where one can find tracks that have been tentatively attributed to pterosaurs, turtles, and birds (figure 8.14). The pterosaurian tracks consist only of three small manus impressions, about 10 cm long, that are quite similar to *Pteraichnus*.[44] The turtle tracks, reported in 1993 by Moratalla, are small, 3 to 4 cm long and wide, and are among the first turtle tracks ever reported from Mesozoic freshwater deposits.[15,45] The presence of the skeletal remains of turtles in this region supports the interpretation of these tracks. The purported bird tracks are controversial. Part of the problem surrounding their interpretation is that they are represented by only one or two partial tracks that are poorly and incompletely preserved. Despite uncertainly about the interpretation of

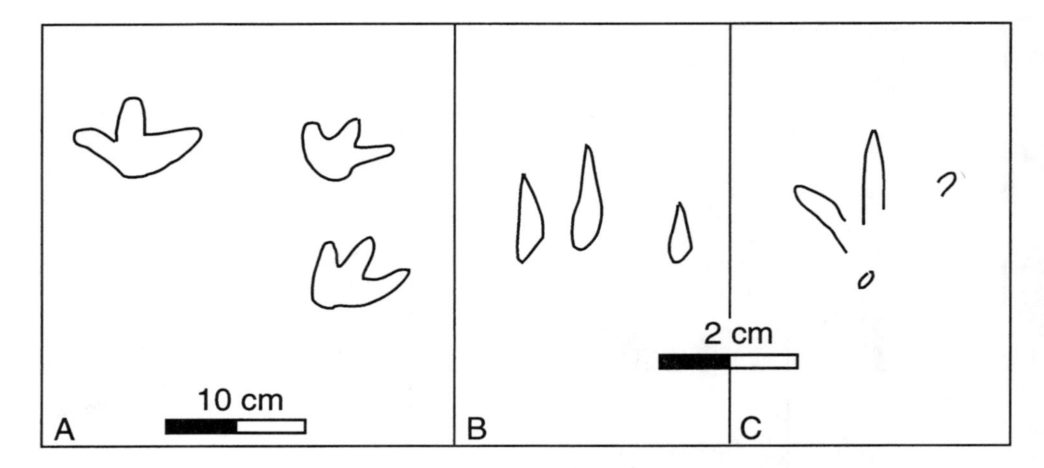

FIGURE 8.14. Small vertebrate tracks from Los Cayos, Enciso Group, La Rioja Province, include footprints attributed to pterosaurs (A), turtles (B), and possibly birds (C). *Redrawn from Moratalla*[15] *and Moratalla et al.*[45]

these bird tracks, there is no known paleoecologic or evolutionary reason not to predict the occurrence of bird tracks in this fluvio-lacustrine environment. In fact, the presence of tracks of all the nondinosaurian groups mentioned in this section (crocodiles, turtles, birds, and pterosaurs) adds richness to the vertebrate track record by confirming the presence and activity of small and medium-sized aquatic and flying vertebrates alongside the more strictly terrestrial dinosaurs.

FARTHER ALONG THE TRAIL OF THE ELUSIVE ANKYLOSAUR

We have seen that the oldest trackway evidence of ankylosaurs is that found from close to the Jurassic-Cretaceous boundary in the Purbeck Limestone of Dorset (see chapter 7). This trackway was originally described as a sauropod trackway from the Jurassic but later was interpreted as an ankylosaurian trackway. It was then rejuvenated, or made younger, when the Jurassic-Cretaceous boundary was lowered a few meters. One trackway that has not proved quite so controversial to date is *Metatetrapous valdensis*,[65] formally described by Hartmut Haubold on the basis of a trackway from the Wealden Beds of Germany. As shown in figure 8.15, and as described by Haubold, *Metatetrapous valdensis* is characterized by a large, relatively elongate tetradactyl, plantigrade pes footprint and a manus track with at least three toe impressions. In both manus and pes the toes appear about the same length. The stride is about 140 cm and the pace angulation is about 130 degrees. The trackway name originates from the work of Baron von Nopsca in 1923. As suggested by Haubold, *Metatetrapous* is similar to *Tetrapodosaurus borealis* from the Aptian-Albian of western Canada,[66] which is now considered to be ankylosaurian in origin.[67] Studies in Canada have revealed that ankylosaur tracks are abundant toward

FIGURE 8.15. A: *Metatetrapous valdensis*[65] from the Lower Cretaceous of Germany. B: *Tetrapodosaurus borealis*[66] from the Lower Cretaceous of Canada. Both are attributed to ankylosaurs.

the end of the Early Cretaceous. We might predict that more ankylosaur trackways will be found in the future in Europe.

DALMATIAN DINOSAURS

In April 1995 the new state of Croatia, born from the collapse of the Socialist Yugoslavia, was still a country at war. I had a grant, from the Dinosaur Society, to study Cretaceous dinosaur footprints on the Brioni Islands in southwest Istria. The main island is a National Park and the date of my visit had been approved by the Assistant Director of the Park, who knew the aim of my visit. The first day I mapped, without problems, a site at Ploce with the help of a student. The day after we planned to visit the new site of Pljisevac, but when we arrived, we discovered that going further was not possible, as that part of the island had fallen under military control during previous months and was delimited by a fence. We walked along the coast, as we found no fence or warning signs, and so reached the wide surfaces with the tracks.

While busy mapping the tracks, a captain with a platoon of armed soldiers arrived and told us that we were under arrest. They brought us to the base and handed us over to the police. We were not aware that at the time the Croatian army was preparing the final offensive against Serbian rebels, and that the soldiers were therefore very suspicious and considered us as spies. The police interrogated each one of us separately (with some problems of linguistic understanding) and Luisa, the student, burst into tears. Other problems arose and we remained under suspicion in the policemen's eyes. They confiscated our passports and placed us under house arrest. The day after we were quickly tried. Fortunately I was able to demonstrate that we were in Croatia only to study footprints and the judge believed me, but we had to pay just the trial expenses. I have been warned that if they find me in a military zone again without permission I will be put in a jail. One year later I was back on the island with my friend Igor Vlahovic, geologist of the *Geoloski Zavod* and sergeant in the Croatian army. With Igor I finished the mapping.

Letter from Fabio Dalla Vecchia to the authors,
February 1998

Studying dinosaurs on the Dalmatian coast of what is now Croatia can be a hazardous business. In one area one has to dive for fossils because they are eroding from strata exposed below sea level. Nevertheless, sauropod vertebrae and fragments of femora and ribs of Berriasian age have been recovered. More hazardous, it seems, in some cases is the attempt to map dinosaur tracks on dry land, as indicated by the reports of Italian ichnologist Fabio Dalla Vecchia, who was arrested by the authorities while trying to map tracks in Istria.

The story of dinosaur tracks research in Croatia began on Brioni/Brijuni islands in 1925 when the German geologist Adolf Bachofen-Echt[68] described three-toed dinosaur tracks and attributed them to *Iguanodon* (figure 8.16). Bachofen-Echt also suggested that certain short, wide tracks (about 14 cm long by 15 cm wide) were made by turtles, a suggestion apparently encouraged by the eccentric paleontologic genius of Baron von Nopsca.[69] In 1993, Dalla Vecchia and colleagues reinterpreted the three-toed "*Iguanodon*" tracks as those of theropods,[70,71] an identification with which we agree. They also noted that they could not relocate the

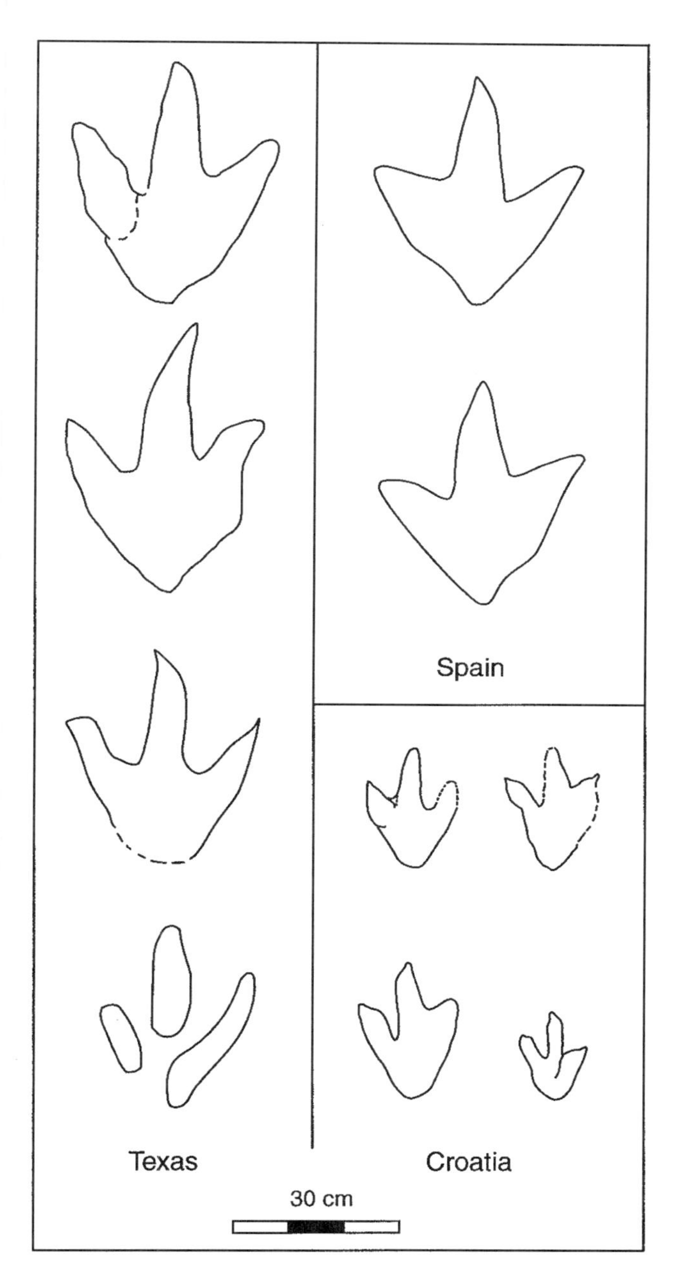

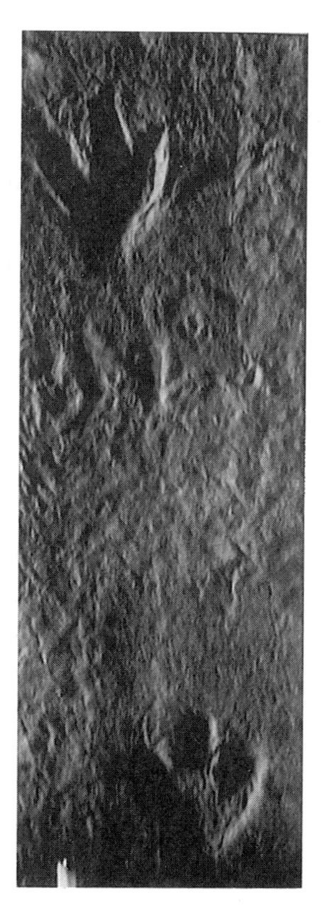

FIGURE 8.16. Purported *Iguanodon* tracks from the Lower Cretaceous of Croatia are now interpreted as the footprints of theropods.[70–72] These tracks appear dwarfed in comparison with tracks from rocks of the same age and type in North America. Photo (right) shows segment of theropod trackway from Brioni Island.

purported turtle tracks, and that they might have been the traces of very small sauropods. Recent work by Dalla Vecchia and colleagues revealed at least a half dozen Early Cretaceous (late Hauterivian to middle Albian) sites with theropod tracks, most of which are small to medium-sized.[72] Together with the footprint evidence of very small sauropods, this raises the intriguing possibility of what are known as "dwarfed faunas," which are often characteristic of island settings (see chapter 9), for the paleogeography of this region indicates that it was an island archipelago during the Cretaceous.

The footprints observed in Bachofen-Echt's original study are late Lower Cretaceous in age (middle Albian), but dinosaur tracks have been reported from a number of different stratigraphic levels in Lower Cretaceous limestones exposed along the Dalmatian coast of Istria. The oldest known dinosaur tracks are possible sauropod tracks from Berriassian deposits at Fantazija Quarry, near Rovinj. These were originally observed as inexplicable features or oddities during the course of a sedimentologic study, but later interpreted by us as the destructive effects of sauropods trampling certain types of carbonate sediment, causing more or less circular features associated with "breccia-like intrusions" into underlying layers.[55] It is interesting to note that such features were evidently not interpreted as footprints simply because sedimentologists were not looking for tracks; they did not have the so-called footprint "search image" in mind.

Theropod and sauropod footprints of late Albian age have been reported from near the mouth of the Quieto and Mirna rivers.[70–72] Track-bearing surfaces at this site are characterized by a random distribution of isolated manus and pes tracks and considerable bioturbation or dinoturbation. There is some suggestion that the theropod tracks are similar to theropod tracks from the Albian of Texas, although in general they appear to be smaller. In addition, the co-occurrence of theropod and sauropod tracks in platform limestones is reminiscent of the *Brontopodus* ichnofacies (see chapter 7). As noted in chapter 7, these sauropod track occurrences also fit the paleogeographic model that associates their distribution with tropical and subtropical latitudes (less than 30 degrees), as well as with the carbonate substrates that are concentrated in these regions.

ARCTIC DINOSAURS

> As the geologic knowledge of all parts of the world increases, the discovery of prints are being more and more frequently reported. The Spitzbergen locality will be the twenty-seventh or the twenty eighth.
>
> Albert de Lapparent, 1960[73]

So far the most northerly examples of European dinosaur footprints mentioned in this volume have been Jurassic tracks from Yorkshire and Scotland (see chapter 5). However, a few sites have been found much farther to the north in Spitzbergen, Norway. The first of these sites was discovered during the field trip that preceded the XXIth International Geological Congress, when 36 geologists toured Svalbard (western Spitzbergen) from July 30 to August 14, 1960. The discovery made by Albert de Lapparent and Robert Lafitte, on August 3, 1960, during a brief disembarkation from the ship *Valkyrien,* in the fjord known as Isfjorden, was described

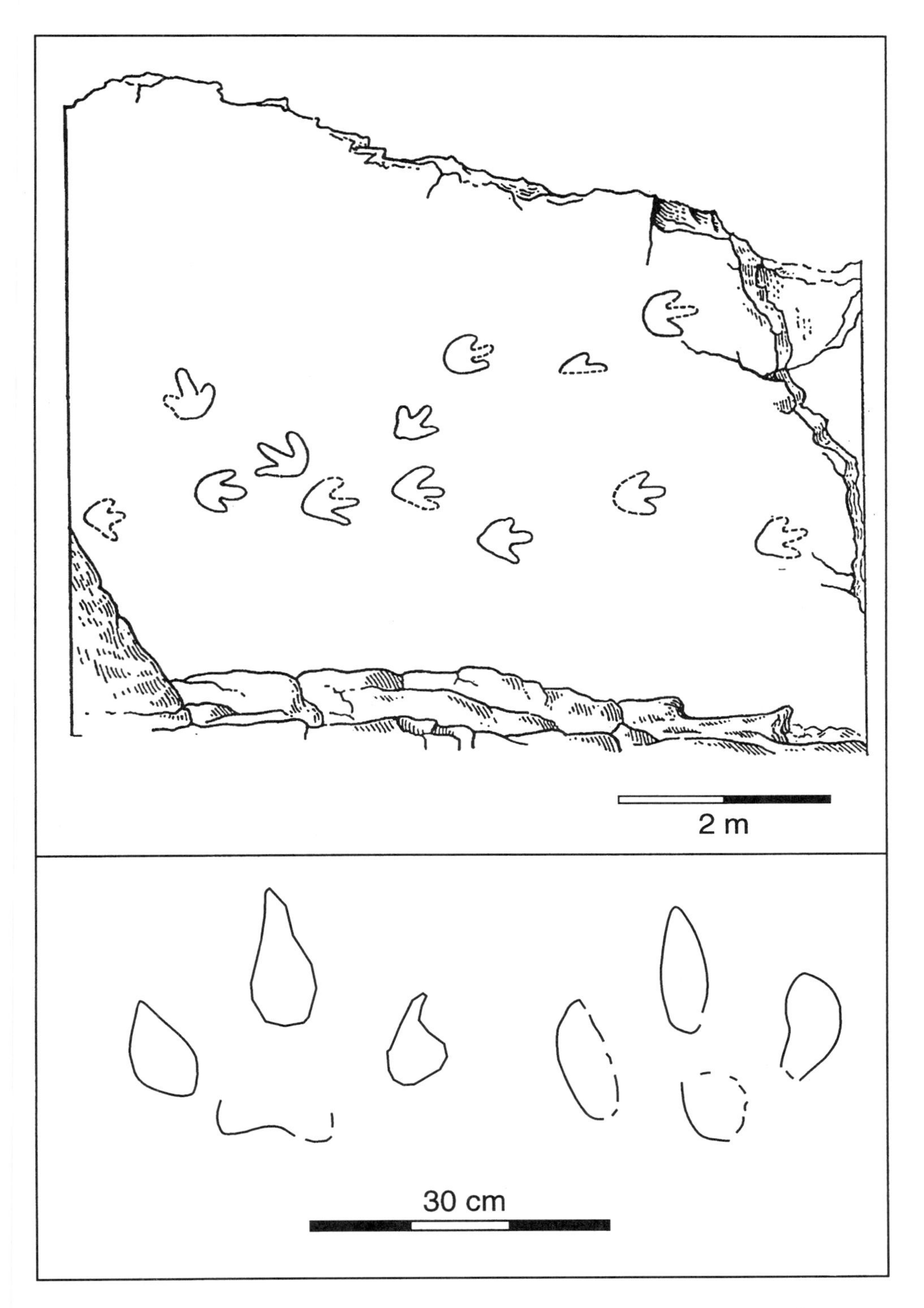

FIGURE 8.17. Theropod and iguanodontid tracks from two localities in the Lower Cretaceous of Spitzbergen.[73,78] Top: Probable theropod ichnites in situ. Below: Probable iguanodontid tracks.

by field trip leader Natasha Heintz[74] as "the peak of the day and the whole excursion." As described by de Lapparent, he and Lafitte found themselves "at the foot of a high sandstone" on which they "suddenly saw some very large footprints probably those of a giant bipedal dinosaur" (figure 8.17).[73] The importance of the track discovery and the impression it made on field trip participants is quaintly and artistically summarized by the collage of fossils drawn by Anatol Heintz and reproduced here as an endpiece to the chapter (figure 8.18)

What is known of the stratigraphy in this region indicates that the tracks are in the Lower Cretaceous Festningsodden sandstone of Hauterivian or Barremian age. There are about a dozen tracks preserved in three trackway segments (figure 8.17). Trackways have strides ranging from about 1.5 to more than 2 m, and individual footprints about 68 cm long and 60 cm wide, although some isolated tracks are smaller. The longest trackway shows a lengthening (accelerating) stride from 1.50 m to more than 2 m. De Lapparent reported that "we first believed them to belong to the carnosauria. This was also the opinion of Joseph T. Gregory."[73] He then went on to argue that the lack of claw impressions and the rounded shape of the toe impressions suggest that the tracks should be interpreted as those of *Iguanodon.* He then argued that the foot of *Iguanodon bernissartensis* is about 50 cm long, and that with flesh and skin taken into consideration, it would match the Festningen footprints quite well. We argue that the large size, elongate nature of the foot, and slight outward rotation of the foot axis leave the strong possibility of a carnosaurian trackmaker, as originally proposed.

Regardless of the interpretations proposed by de Lapparent and ourselves, seven of the tracks are preserved as replicas in the Paleontological Museum in Oslo, and so are available for further study by future generations of ichnologists. Natasha Heintz[75] described in detail the successful expedition made back to Festningen to procure plaster casts of the best tracks. Originally the intention had been to obtain latex casts, an objective that in theory should have been easier than making plaster casts on a vertical surface. But the latex compound (Castoflex) that was chosen did not perform well under conditions of low temperature and high humidity, even though it was tested in Norway under similar conditions. As a result, the casts were made by a laborious process of applying thin layers of plaster and plaster-soaked burlap fabric inside a wooden frame. Even this procedure was threatened by the destructive influence of seawater on plaster, and the success of the enterprise was ensured only by carefully washing the rock surface and carefully mixing the plaster only with fresh water obtained from nearby glacial streams.

The extreme northern location of the tracksite, at present-day latitude 78 degrees, was noted by de Lapparent, Heintz, and other members of the excursion. Previously tracks had not been reported north of latitude 56 degrees.[73,76] The various implications were as follows: a uniformly mild climate might have prevailed from the tropics to high latitudes, resulting in widespread Wealden facies. There had to have been a land bridge to Eurasia. The animals must have endured several months of darkness or migrated seasonally to lower latitudes below the Arctic Circle, where there

FIGURE 8.18. Artistic rendition by A. Heintz of the 1960 Spitzbergen expedition. Note the stylized depiction of the dinosaur tracks (redrawn).

was no prolonged winter darkness. Alternatively, polar wandering or continental drift had moved Spitzbergen to higher latitudes since Lower Cretaceous times.

In the early 1960s, not much was known about polar wandering and continental drift, or about the geologic revolution in plate tectonics that was just around the corner. We now know that Spitzbergen has moved, but that during the early Cretaceous it was still at a very high latitude within the Arctic Circle. To get a cartographic picture of high-latitude paleogeography, we can look at the Lower Cretaceous map presented in Dale Russell's 1989 book *The Dinosaurs of North America.*[77] Here Spitzbergen is shown just to the north of Greenland, with land connections to the North American continent, as well as to regions of western Europe, such as England and Spain, discussed earlier in this chapter. Since the 1960s there have been many other discoveries of high-latitude dinosaurs, notably in the Late Cretaceous of Alaska but also in the Southern Hemisphere. Thus, subsequent research has shown the phenomenon of high-latitude migratory dinosaurs, first suggested by the footprints of Spitzbergen, to be quite common.

A second discovery of dinosaur tracks was reported by Marc Edwards and others in 1978[78] following the Continental Shelf Institute's expedition of 1976. These tracks, found 120 km to the southeast of the Festningen discovery, are purported to be those of carnosaurs originating from the Helvetiafjellet Formation. Only two tracks were reported with precise measurements of 28.5 to 29.6 cm long and 22.5 to 23 cm wide. But based on the photographs, it appears that these measurements are transposed. In other words, the tracks are wider than they are long. This fact, coupled with the oval to teardrop shape of the pads, suggests that these tracks are also those of iguanodontids, not theropods.

Thus, we conclude that the so-called *Iguanodon* tracks from Spitzbergen are probably of theropod affinity, and the purported theropod tracks are probably of ornithopod (iguanodontid) affinity. The net result of this reinterpretation remains essentially the same: both theropod and ornithopod tracks appear to be represented in the Early Cretaceous of the Spitzbergen. This implies that both herbivores (prey species) and carnivores (predators or scavengers) regularly migrated to high latitudes, presumably on a seasonal basis.

Dinosaur ichnology has come a long way in less than 40 years. It is remarkable to consider that in 1962 de Lapparent scored only 27 or 28 dinosaur tracksites in the entire world. Even allowing for the fact that he may have overlooked a few, we can confidently assert that there are literally thousands known today. Some of these are being found at new high-latitude locations such as Alaska. Spitzbergen also may reveal more. It is excellent terrain for finding extensive, largely unexplored outcroppings.

BIBLIOGRAPHIC NOTES

1. Delair, J. B. 1963.
2. Prince, N. K. and M. G. Lockley. 1989.

3. Wright, J., D. M. Unwin, M. G. Lockley, and E. C. Rainforth. 1997.
4. Padian, K. and P. Olsen. 1984.
5. Lockley, M. G., M. Huh, S-K. Lim, S-Y. Yang, S. S. Chun, and D. Unwin. 1997.
6. Wright, J. 1996.
7. Lockley, M. G., S-Y. Yang, M. Matsukawa, F. Fleming, and S-K. Lim. 1992.
8. Lockley, M. G. and E. Rainforth. In press.
9. Fuentes Vidarte, C. 1996.
10. Tagart, E. 1846.
11. Jones, T. R. 1862.
12. Sarjeant, W. A. S., J. Delair, and M. G. Lockley. 1998.
13. Casamiquela, R. M. and A. Fasola. 1968.
14. Santos, V. F. dos, M. G. Lockley, J. J. Moratalla, and A. Galopim. 1992.
15. Moratalla, J. J. 1993.
16. Dollo, L. 1905.
17. Norman, D. 1987.
18. Moratalla, J. J., J. L. Sanz, S. Jimenez, and M. G. Lockley. 1992.
19. Moratalla, J. J., J. L. Sanz, and S. Jimenez. 1992.
20. Wright, J. 1999.
21. Paul, G. 1991.
22. Lockley, M. G., V. F. dos Santos, C. A. Meyer, and A. P. Hunt. 1998.
23. Lim, S. K., M. G. Lockley, S-Y. Yang, R. F. Fleming, and K. A. Houck. 1994.
24. Delair, J. B. 1982.
25. Delair, J. B. and A. B. Lander. 1973.
26. Lockley, M. G. 1995.
27. Lockley, M. G. 1994.
28. Beckles, S. H. 1851.
29. Beckles, S. H. 1852.
30. Beckles, S. H. 1854.
31. Batory, D. and W. A. S. Sarjeant. 1989.
32. Woodhams, K. and J. Hines. 1989.
33. Woodhams, K. and J. Hines. In press.
34. Parkes, S. 1993.
35. Radley, J. D., M. J. Barker, and I. C. Harding. 1998.
36. Radley, J. D., 1994.
37. Moratalla, J. J. and J. L. Sanz. 1997.
38. Martin, V. and E. Buffetaut. 1992.
39. Pereda-Suberbiola, J. 1993.
40. Lockley, M. G., K. Conrad, and M. Jones. 1988.
41. Lockley, M. G. and J. G. Pittman. 1989.
42. Lockley, M. G. 1991.
43. Lockley, M. G. and A. P. Hunt. 1995.
44. Lockley, M. G., T. J. Logue, J. J. Moratalla, A. P. Hunt, R. J. Schultz, and J. W. Robinson. 1995.
45. Moratalla, J. J., J. L. Sanz, and S. Jimenez. 1992.

46. Lockley, M. G., C. A. Meyer, and J. J. Moratalla. In press.
47. Alonso, A. and J. R. Mas. 1993.
48. Aguirrezabala, L. M. and L. I. Viera. 1980.
49. Moratalla, J. J., J. L. Sanz, I. Melero, and S. Jimenez. 1988.
50. Kuhn, O. 1958.
51. Lockley, M. G. In press(b).
52. Moratalla, J. J., J. Garcia-Mondejar, M. G. Lockley, J. Sanz, and S. Jimenez. 1994.
53. Casanovas, M., A. Fernandez, F. Perez-Lorente, and J. V. Santafe. 1997.
54. Lockley, M. G., J. O. Farlow, and C. A. Meyer. 1994.
55. Lockley, M. G., C. Meyer, A. P. Hunt, and S. G. Lucas. 1994.
56. Hendricks, H. 1981.
57. Pérez-Lorente, F., C. Cuenca-Bescos, M. Aurell, J. I. Canudo, A. I. Soria, and J. I. Ruiz-Omenaca. 1997.
58. Aguirrezabala, L. M., J. A. Torres, and L. I. Viera. 1985.
59. Casanovas Cladellas, M. L., A. Fernandez-Ortega, F. Perez-Lorente, and J-V. Santafe-Lopis. 1991.
60. Viera, L. I. and J. A. Torres. 1992.
61. Martin-Escorza, C. 1992.
62. Lockley, M. G., C. A. Meyer, and V. F. dos Santos. In press.
63. Lockley, M. G. In press(c).
64. Moratalla, J. J., M. G. Lockley, A. D. Buscalioni, M. Fregenal, N. Melendez, B. R. Perez-Moreno, E. Perez, J. L. Sanz, R. Schultz. 1995.
65. Haubold, H. 1971.
66. Sternberg, C. M. 1932.
67. Carpenter, K., 1984.
68. Bachofen-Echt, A. 1925.
69. Nopsca, F. von. 1923.
70. Dalla Vecchia, F. M., A. Tarlao, and G. Tunis. 1993.
71. Dalla Vecchia, F. M. 1994.
72. Dalla Vecchia, F. M. In press.
73. Lapparent, A. F. de. 1962.
74. Heintz, N. 1962a.
75. Heintz, N. 1963.
76. Heintz, N. 1962b.
77. Russell, D. 1989.
78. Edwards, M. B., E. Rosalind, and E. H. Colbert. 1978.

Titanosaurid sauropods. Based on fossil evidence from the Late Cretaceous of France and Spain. Courtesy of Jean Le Loeuff.

CHAPTER 9

THE END OF THE DINOSAUR TRAIL: UPPER CRETACEOUS

Vivam las pegadas! [Long live the tracks!]

Mário Soares, President of Portugal, October 2, 1993

THE BATTLE OF CARENQUE

When a large bipedal dinosaur made its solitary way across a coastal mud flat on the shores of the Atlantic, in an area that is now a part of Portugal, it did not know that 95 million years later it would be credited, briefly, with the record for the world's longest dinosaur trackway. Today the trackway can be seen on the floor of an old limestone quarry in Carenque, a suburb of Lisbon. For 5 years it was known as the Carenque[1,2] site, but it is now also known as the Pêgo Longo site.[3] Although it is visually spectacular, the individual tracks consist only of big round potholes that reveal little about the anatomy of the dinosaur that made the trackway. Despite uncertainty about the identity of the trackmaker, the site proved important enough to initiate a battle between environmentalists and the relentless march of technological progress.

The "Battle of Carenque" began when it was realized that freeway construction was about to destroy a 95 million year old (Cenomanian) trackway exposed in the floor of an old quarry previously used as a rubbish dump. (Most quarries are literally holes in the ground and so become landfills by the simple process of filling in through weathering, erosion, deposition, and waste disposal by messy humans.) The freeway through Carenque had already been partially constructed and was rapidly converging on the quarry from both sides. Advocates of the development argued that it was too late to do anything except complete the freeway link as planned. Regrettably, the trackway would have to be sacrificed to the advancing thoroughfare of tarmac. Environmentalists and conservationists, however, pointed out that this was an irreplaceable piece of Portugal's natural heritage.

The preservation of dinosaur tracks is a challenge. Because dinosaurs are no longer with us, it is also obvious that fossil footprints are a nonrenewable resource, and once destroyed they are gone forever. Unlike bones and archeological remains, which can be dug up and preserved in a museum, it is practically impossible to remove an entire tracksite, the size of a football field, and put it elsewhere. Usually the best course of action, therefore, is to leave tracksites in place and take measures to protect them from weathering and vandalism. This can best be done by creating interpretive centers or parks, providing public education resources for the local community and tourist attractions for visitors from further afield. Such interpretive and preservation measures usually arise initially from scientific research activities, which are a necessary prerequisite to establish the importance of the site in the first place.

On arrival at Carenque, one is first impressed by the length of the trackway and the excellent view one has from an elevated vantage point overlooking the quarry. The trackway stands out in splendid isolation. When first exposed, the trackway segment measured 127 m in length (figure 9.1) and was the longest that had actually been measured anywhere in the world,[1] hardly something one would want to destroy without good reason. The fact that the trackmaker is hard to identify is a secondary consideration.

Even though some individuals advocated priority for social agendas ahead of natural history and the environment, most were in favor of preserving Portugal's natural heritage. One of the strongest supporters was President Mario Soares, who, along with the Minister of Science and Technology, was lobbied successfully by Antonio Galopim de Carvalho, head of the Paleontology Department at the National Museum of Natural History. After the Ministry commissioned several reports on the tracks, the President visited the site, took one look at the tracks, and proclaimed "Vivam las pegadas de Carenque" ["Long live the footprints of Carenque"]. After 95 million years there seemed no good reason to destroy them. When parliament voted on the preservation of the site, the vote was unanimously in favor. Galopim, a prolific writer, penned a book, probably the only one ever written entirely on the subject of a single tracksite and the community effort to save it from destruction. His title, of course, was *The Battle of Carenque,*[2] a battle he had played a large part in winning. One of his strategies in this battle was to make it known that the community did not want to witness a second extinction of the dinosaurs.[4]

In a creative solution the dinosaur tracks were saved and the road construction completed by "diverting" the freeway's dual carriageways through two parallel tunnels beneath the tracksite. Each tunnel has a dinosaur head at one end and a tail at the other, so that there are two dinosaurs, facing in opposite directions, essentially holding up the tracksite with their backs. During the construction process the trackway was excavated to an extended length of 141 m, again a world record,

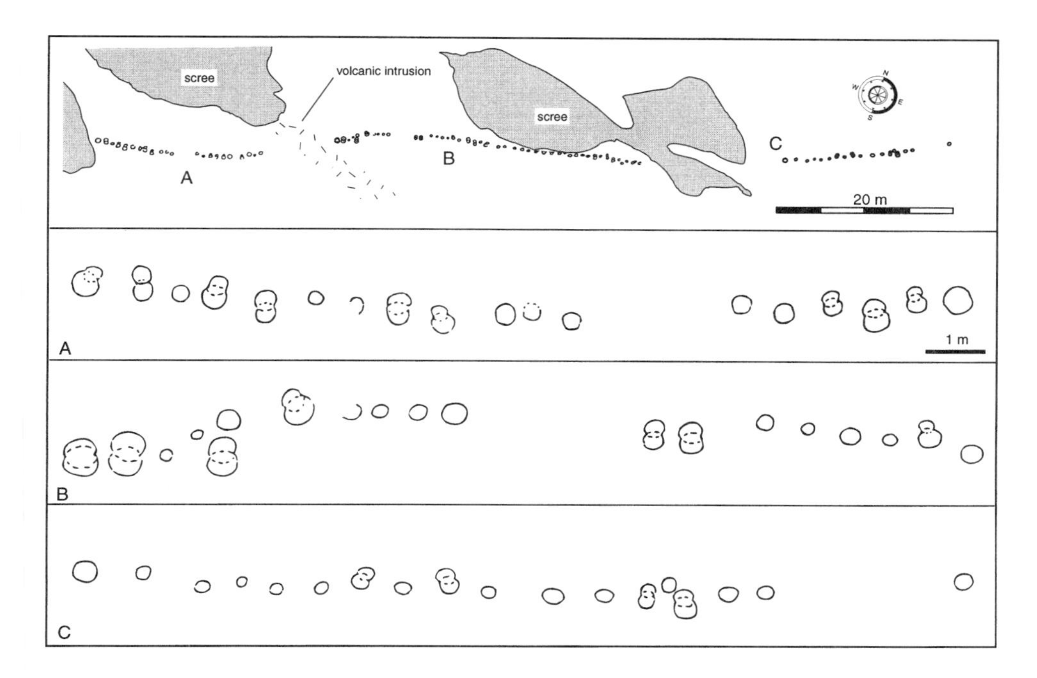

FIGURE 9.1. Previous holder of the record for the world's longest dinosaur trackway, at 141 m from the Carenque site, Upper Cretaceous (Cenomanian), of the Lisbon region, Portugal. Photo (below) shows segment A and part of segment B.

until it was surpassed by the discovery of trackways 142 to 147 m in length at Fatima, also in Portugal (chapter 5).

What can we say about the ichnologic importance of the Carenque site, besides the *Guinness Book of World Records* statistics that it has produced? Although the identity of the trackmaker is a mystery, the trackway is interesting for at least two reasons. The first is that around many of the footprints there appears to be a high concentration of small shells of molluscs and other vertebrates. It appears that these are not distributed in equal abundance elsewhere on the surface, and so may be preserved preferentially as a result of the compaction of the sediment by the dinosaur.[1] The impact of dinosaurs and other large vertebrates on bivalves, plants, and other inhabitants of the substrate has been reported elsewhere.[5] The site therefore provides an interesting opportunity for sedimentologists to study compaction of sediments around footprints and to determine the extent to which dinosaurs were contributing to the fossilization process.

Another interesting aspect of the trackway is that in places there appear to be double prints. In other words, there is a single sequence of alternating left and right footprints in some segments of the trackway, and then a few footprints in the sequence appear to be double (figure 9.1). One possibility is that there were two dinosaurs walking along, one following the other's footprints precisely, most of the time, and that occasionally the follower got out of step. This explanation, however, is rather farfetched and is not supported by any other known examples of well-synchronized trackways. A second possible explanation is that an earthquake may have disturbed the sediment containing these relatively deep tracks, perhaps after they were already buried by an unknown thickness of sediment. In some places the sediment may have been too compact to move, but in others there may have been a shearing so that the circumference of the subcircular track wall was shifted relative to the position of the track wall at another level. Again, such speculative hypotheses need to be tested by a careful study of the sediments themselves. The Carenque site provides many such future opportunities.

MORE DALMATIAN DINOSAURS AND DWARFS

We have already seen that the coastal exposures of Istria have yielded a number of interesting Lower Cretaceous dinosaur tracksites, ranging in age from Berriasian to Albian (see chapter 7) and indicating the presence of both sauropods and theropods. The youngest tracks in this area, however, are Cenomanian in age, and consist of several sauropod and theropod trackways. The best-known sauropod trackway consists of 34 consecutive footprints (figure 9.2) found on the islet of Felonega on the extreme southern tip of the Istrian peninsula.[6–8] The pes tracks average about 40 cm long and 30 cm wide, with very few manus tracks preserved, owing to overprinting and suboptimal preservation. Although the manus area is recorded as being about half of the pes area, this cannot be regarded as a precise measure of het-

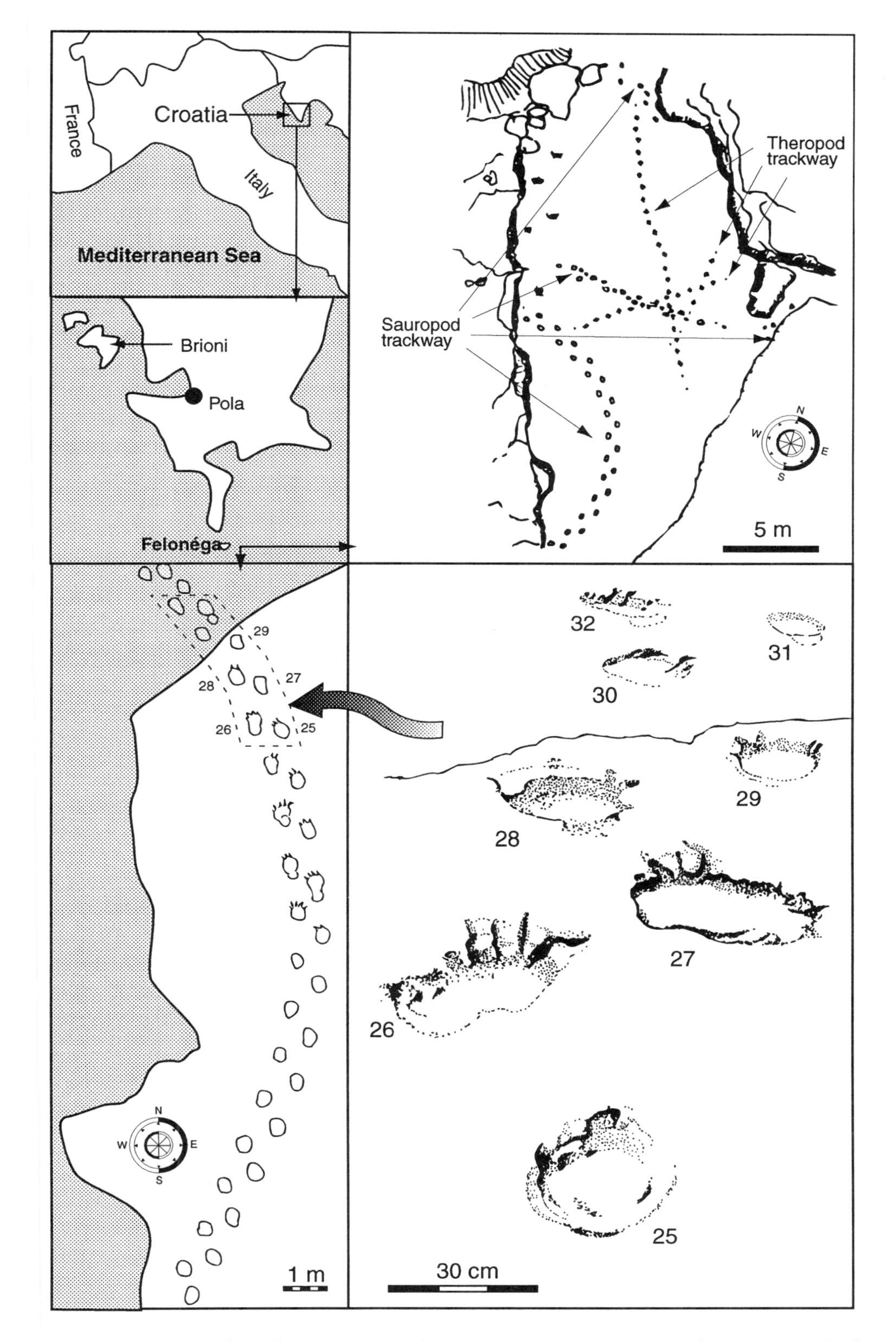

FIGURE 9.2. The Felonega sauropod tracksite, Upper Cretaceous (Cenomanian), of the Istrian peninsula.

eropody. The trackway width is 90 cm, with about 20 cm separating the inside margin of the pes tracks.[9] Hence, it is a wide-gauge trackway of the *Brontopodus* type. Pace angulation (95 degrees), pace (80 cm), and stride (about 120 cm) are all about normal for a walking sauropod.

The total area of track-bearing surface at the Felonega site is approximately 25 by 10 m, with seven trackways recorded,[7,8] of which four are probably those of sauropods and three those of theropods. Again we see the association of sauropod and theropod tracks in platform carbonates (chapters 7 and 8). Because the tracks are exposed in the intertidal zone, their size may be slightly enlarged by erosion. Given this situation, the tracks are relatively small, and perhaps owing to the lack of other sauropod trackways in the Cretaceous of Europe, these are probably the smallest on record. One must go back in time to the Jurassic (see chapter 7) or move geographically to the Cretaceous of Korea[10] to find smaller sauropod tracks. What, if anything, can we say about the small size of these tracks? Probably nothing with certainty. However, an intriguing possibility exists with regard to faunas from this general area. During the Cretaceous, much of this region was an island archipelago, and on the basis of dinosaurs discovered in the region of Rumania, not far to the north, it has been suggested that the faunas may have been dwarfed[11] owing to their island habitats. It is well known that dwarf elephants evolved on the Mediterranean islands of Cyprus, Malta, and Sicily during the Pleistocene.[12]

THE LAST OF THE BRONTOSAURS: TRACKING TITANOSAURS IN THE HIGH PYRENEES

Although dinosaur tracksites are common in Early Cretaceous deposits in Spain, they are much less well known in the Late Cretaceous. This may be partially the result of a lack of systematic search for footprints. Certainly there is no shortage of information on dinosaurs from this epoch, especially in the Pyrenean region: both their bones[13] and their eggs[14,15] have been the subject of several studies. As is the case in many regions, the first reports of important tracksites are only just beginning to emerge.

The first reports of dinosaur tracks from the Pyrenean region were published by Carme Lompart in 1979[16] in reference to possible ornithopod footprints from the Ager Valley, Province of Lleida, not far from the border between Spain and France. In 1984, she and her colleagues described two types of purported ornithopod tracks from another nearby locality, in the vicinity of Orcau, a village north of Lleida.[17] The first type, designated "form A," was named *Ornithopodichnites magna* (meaning "large ornithopod track") and described as having a total length of between 43 and 67 cm and a width of 37 cm. Such an elongate track (i.e., one that appears to be significantly longer than wide) is not typical of large ornithopod footprints, and for this reason it is possible that the tracks are of theropod origin. The second track type or form, labeled *Orcauichnites garumeniensis* (named after the village of Orcau and the Garumnian red beds of the region), is also a three-toed track of uncertain affinity. It is also somewhat longer than wide and could be of theropod affinity. Neither of these tracks is preserved as part of a trackway sequence, and the quality of preservation is poor. In addition, the formal ichnotaxonomy is not properly presented because diagnoses are given, but without descriptions. These various factors, but particularly the poor quality of the material, conspire to make these names of little ichnologic significance (*nomina dubia*). It is to be hoped that better material will come to light that will allow a better understanding of the types of three-toed dinosaur tracks in this region.

Lompart and colleagues also noted the presence of large subcircular tracks, about 50 cm in diameter, that were interpreted as possibly being of sauropod affinity. This interpretation may be correct, not just because titanosaurid sauropods are known to be well represented in the region but because subsequent studies have revealed the presence of abundant sauropod trackways. Anne Schulp and Wouter Brokx,[18] two geologists from the University of Amsterdam, have documented extensive sauropod trackways on a large limestone surface at an old lignite mine in the region of Fumanya, Province of Barcelona. This picturesque location is situated just 1500 m (about 5000 feet) above sea level in terrain built of folds and thrust sheets, again close to the frontier between Spain and France. The tracks occur on the upper surface of the Maastrictian Bona Formation in contact with the lower part of the Tremp Formation, which is in turn overlain by the Garumnian red bed facies, referred to above. The track-bearing layers are rich in plant remains, including large tree trunks, and contain dinosaur egg shell and various molluscs (snails and clams), indicative of brackish water conditions.

The track-bearing surface is very extensive and can be traced for several hundred meters along a steeply dipping bedding plane inclined at about 60 degrees. In order to accurately map and document dozens of trackways on the surface without climbing over a vast and dangerous area, Schulp, Brokx, and their assistants used ropes to measure off a grid before photographing the surface. They then digitized the photographs and, using a computer program, corrected them to remove any distortion. The result was a spectacular map, covering an area of more than 10,000 m^2, in which more than 30 long sauropod trackway segments were recorded (figure 9.3). Although the directions are more or less random, there are a few parallel trackways, suggestive of herding activity, and the lack of any other trackway types suggests that the site represents a monospecific assemblage, presumably indicative of some measure of congregation or gregarious behavior in a relatively small area.

After correcting their maps for photographic distortion, Schulp and Brokx also noted that the trackways had suffered tectonic distortion of the type previously recorded in Devonian trackways from Ireland (see chapter 2). This distortion was also removed by plotting the degree of deformation for each trackway and establishing the direction and amount of maximum compression and maximum extension. Through this exercise it appears that the tracks were compressed and ex-

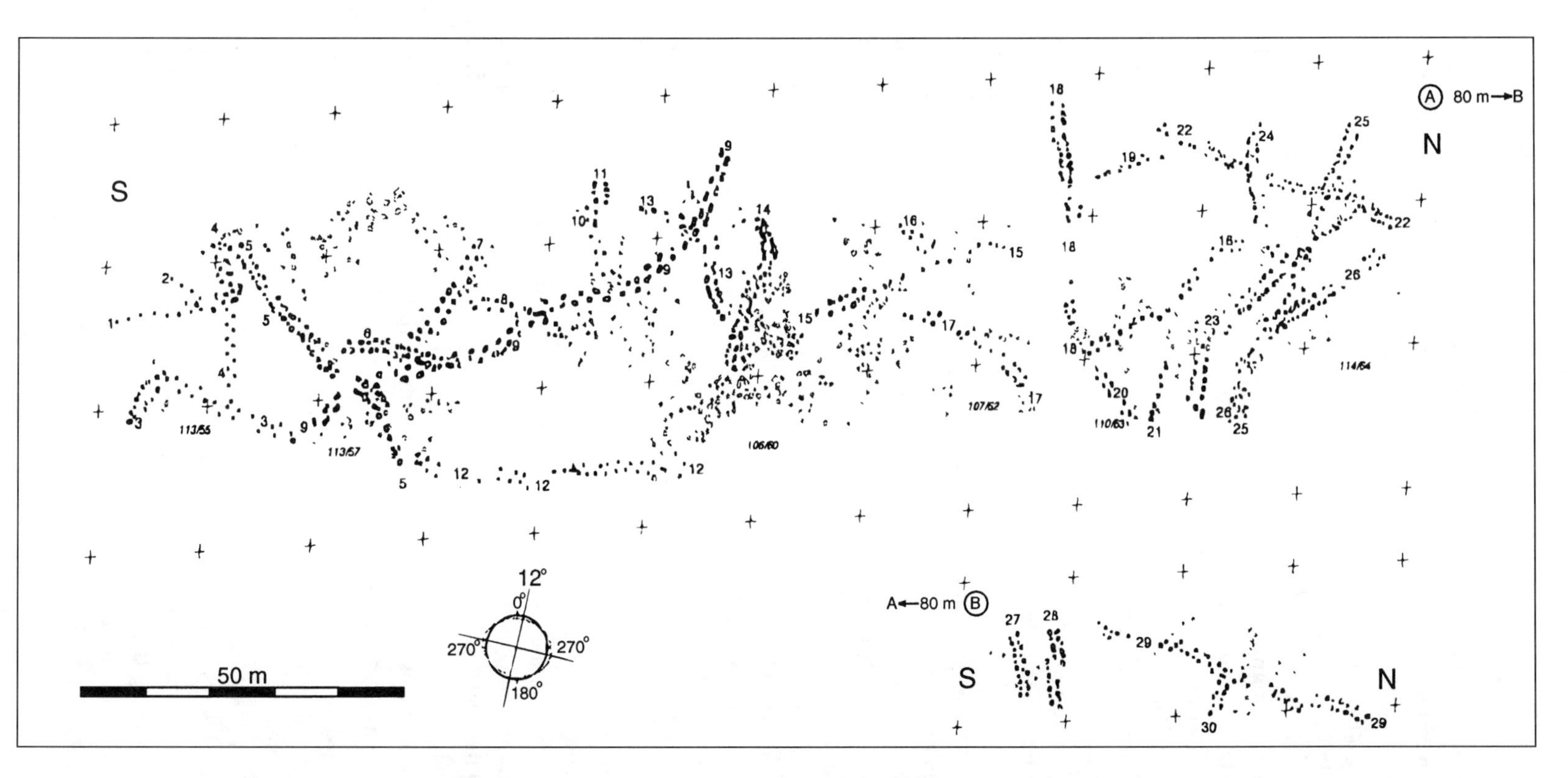

FIGURE 9.3. Map of the Fumanya titanosaurid sauropod tracksite, Upper Cretaceous (Maastrichtian), Province of Barcelona, Spain. Note that the distortion caused by tectonic strain has been removed. *Courtesy of Anne Schulp and Wouter Brokx.*

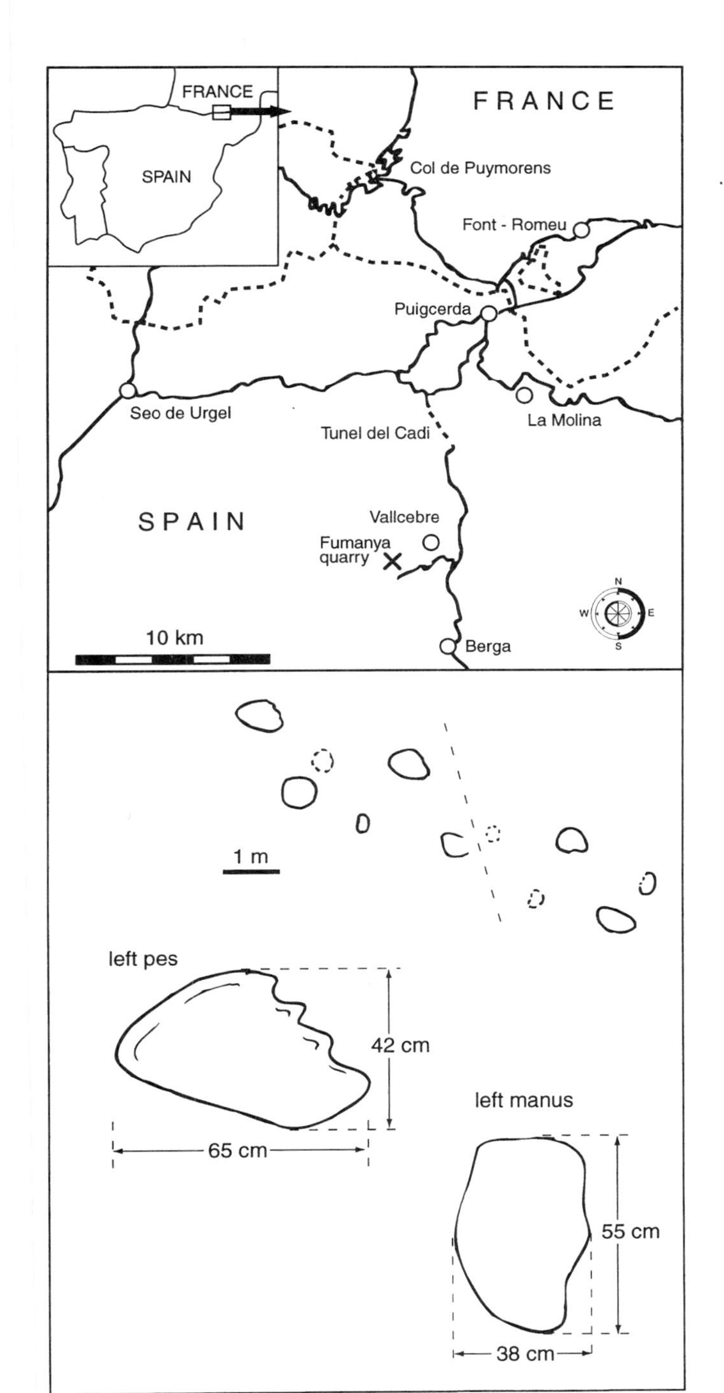

FIGURE 9.4.
Detail of a titanosaurid sauropod trackway, Fumanya site, Upper Cretaceous (Maastrichtian) Spain.

tended about 12 degrees, thereby slightly reducing the potential to accurately reconstruct the outline of footprints as they appeared prior to distortion. As noted by these researchers, such distortion might make normal, symmetrical trackways appear as if they had been made by a limping dinosaur.

Despite the minor tectonic distortion, there are at least 30 long trackway segments, some up to 80 m in length, available for study. These reveal relatively wide-gauge trackways of the *Brontopodus* type, with estimated hip to shoulder (gleno-acetabular) lengths of between 2.4 and 3.2 m. Hind footprints are longer than wide (65 by 40 cm in the best-preserved example), whereas manus tracks are wider than long (55 by 40 cm). As shown in figure 9.4, these configurations and measurements provide some indication of the anatomy of the trackmaker's foot and the degree of heteropody associated with this particular morphology. Prior to this study, well-preserved sauropod tracks were virtually unknown from the Maastrictian. Given that the only Maastrictian sauropods presently known in this region or anywhere in Europe are the titanosaurids, the Fumanya quarry is currently the world's premier titanosaurid tracksite, thus allowing us to match this track morphology with a particular family of sauropods. Equally spectacular titanosaurid tracks have been discovered in Bolivia, but they have yet to be described in detail.

In recent years it has become evident that not all sauropod tracks are the same. Not only can we distinguish wide- and narrow-gauge varieties, but, as noted in chapter 5, we have also established that there is considerable variation in the degree of heteropody found within the group. The Fumanya site acts as a window into the foot morphology and gait of titanosaurids. Like most Cretaceous sauropods, they had wide-gauge trackways and progressed only at a walk. Their manus tracks were quite wide and transverse relative to pes width, but the overall estimated heteropody value (1:3 to 1:4) is in the middle of the known range (1:2 to 1:5).[18] The Schulp and Brokx study, although an excellent preliminary report on the Fumanya site, leaves room for further detailed study of the individual trackways. Such study together with the possibility of documentation of other sites in the region, where undistorted tracks may be preserved, offers the potential to further document the finer details of titanosaurid foot morphology and locomotion. Finally, we note that the main tracksite occurs in a limestone, which represents a marginal marine environment. In this sense the Fumanya site fits the concept of the *Brontopodus* ichnofacies (see chapters 6 and 7). But because it is only one site, although the Orcau site makes two, it is not a very extensive example as yet.

THE LAST EUROPEAN DINOSAURS

One of the most popular topics of discussion in all of science is what killed off the dinosaurs. As a result of this debate, which has been lively and controversial for almost two decades, a number of subsidiary questions have arisen. One of these concerns whether the dinosaurs died off suddenly or not. In order to establish, beyond reasonable doubt, that the dinosaurs were killed off instantaneously by a sudden impact event, it would help to have evidence of high mortality right at the level where the impact is recorded, i.e., at the Cretaceous-Tertiary boundary (also known as the "K/T boundary"). For a long time such evidence was not available, and some paleontologists proposed the counterargument that the lack of skeletal remains for at least 3 m below the boundary implied that the dinosaurs might have died off prior to the K/T event. The same argument was used to explain the absence of marine invertebrates such as ammonites for several meters below the boundary. But there is an obvious counterargument to these counterarguments which suggests that lack of fossils (negative evidence) is not a compelling argument in favor of a pre-K/T demise of dinosaurs and other animals.

A significant step toward resolving this argument came in 1989, when dinosaur tracks were reported only 37 cm below the K/T boundary in Colorado.[5,19,20] This effectively reduced the so-called 3-m gap by an order of magnitude (from 3.0 to 0.37 m). It now appears that tracks are known very close to the K/T boundary in Europe.[21] The tracks in question are found in the Tremp Formation (see previous section) of the Ager basin region of northeastern Spain, in the southern Pyrenees. According to this recent report, the tracks are those of sauropods and ornithopods. Their presence very close to the K/T boundary indicates that dinosaurs were alive and well until just before the catastrophic K/T event. We are somewhat skeptical of this report because little information is given on the track morphology. One level is reported as being dinoturbated by a herd of sauropods, and the locality is referred to as a megatracksite 900 m^2 in extent. A site of this size is not what we consider a megatracksite (see chapter 7), which is usually measured in hundreds of square kilometers.[5]

The presence of dinosaur tracks just below the K/T boundary in Europe, as well as in North America, indicates that dinosaurs were alive and well until the "eleventh hour." At present we do not know how close to the boundary the European tracks were recorded. If this information is forthcoming from the most recent study[21] or from subsequent research, it may eventually be possible to estimate the approximate duration that intervened between the making of the last European dinosaur tracks that have been preserved and the actual impact event. Tracks are reliable evidence of living animals, and it is always possible that dinosaur tracks will be found above the K/T boundary, indicating the survival of a few individuals for days, weeks, months, or even years after the K/T event. Although the K/T boundary has been extensively explored in many regions by geochemists and paleobotanists, there has been little coordinated effort to look for tracks in many areas. Although we do not suggest a high probability of finding dinosaur tracks above the K/T boundary, it is fair to say that a systematic search for vertebrate tracks in boundary sections has not been undertaken in most areas.

The Late Cretaceous dinosaur track record is rather sparse in Europe. Moreover, many of the trackways, such as those from Carenque, cannot readily be attributed to a particular dinosaur group. Part of the reason that tracksites are uncommon is that parts of the upper Cretaceous section are represented only by marine rocks. Even so, in other regions, notably North and South America, Upper Cretaceous terrestrial tracksites are quite common. In North America, they reveal evidence of the classic fauna of duck-billed dinosaurs (hadrosaurs), horned dinosaurs (ceratopsians), and tyrannosaurs, also well known from their skeletal remains, not to mention the tracks of pterosaurs, birds, and many other groups. The fauna in Europe was different, especially with respect to the abundance of titanosaurs known from skeletons and now from tracks. We may hope that in addition to confirmed titanosaurid footprints, the upper Cretaceous of Europe will soon yield additional tracks that give insight into the vertebrate world of large dinosaurs and their many smaller contemporaries.

BIBLIOGRAPHIC NOTES

1. Santos, V. F. dos, M. G. Lockley, J. J. Moratalla, and A. Galopim. 1992.
2. Galopim, A. C. 1994.
3. Santos, V. F. dos. 1998.
4. Galopim, A. C. 1992.
5. Lockley, M. G. 1991.
6. Gogola, M. 1975.
7. Gogola, M. and R. Pavlovec. 1992.
8. Dalla Vecchia, F. M. 1994.
9. Leghissa, S. and G. Leonardi. 1990.
10. Lim, S. K., M. G. Lockley, S-Y. Yang, R. F. Fleming, and K. Houck. 1994.
11. Dalla Vecchia, F. M. In press.
12. Haynes, G. 1991.
13. Le Loeuff, J. 1993
14. Cousin, R., G. Breton, R. Fournier, and J-P. Watte. 1994.
15. Moratalla, J. J. 1993.
16. Lompart, C. 1979.
17. Lompart, C. 1984.
18. Schulp, A, S. and A. B. Brokx. In press.
19. Lockley, M. G. and R. F. Fleming. 1991.
20. Lockley, M. G. and A. P. Hunt. 1995.
21. Lopez-Martinez, N., L. Ardevol, M. E. Arribas, J. Civis, and A. Gonzalez-Delgado. 1998.

A Miocence cat from Spain leaves the oldest known tracks of this well-known group of carnivores. Courtesy of Mauricio Anton.

CHAPTER 10

NEW HORIZONS

THE AGE OF MAMMALS AND BIRDS

Traditionally the Tertiary is known as the "age of mammals." However, if we were to try to estimate the diversity of major tetrapod groups, we would find, as we do at the present time, that birds outnumber mammals by as much as 2 to 1. For this reason, paleontologists sometimes also refer to the Tertiary as the "age of birds." The track record also gives us a similar perspective, for we often find that bird tracks outnumber those of mammals. As the following review indicates, this pattern is typical of the European track record.

If we look at the entire post-Cretaceous track record, including the Quaternary as well as the Tertiary period, we find that it can be broadly subdivided into three phases. The first, corresponding to the Early Tertiary or Paleogene (Paleocene, Eocene, and Oligocene epochs), is characterized by faunas in which perissodactyls (odd-toed ungulates such as rhinos and tapirs) are often well represented and artiodactyls (even-toed ungulates) are rare, at least prior to the Oligocene. This fauna and corresponding tracks represent what had in the past been referred to as an "archaic" mammal fauna. Late Tertiary or Neogene (Miocene and Pliocene epochs) faunas are, by contrast, characteristically "modern," with abundant artiodactyl and bird tracks as well as, at some sites, tracks of distinctly modern-looking carnivores belonging to various cats, dogs, and bears.

For convenience, we can also separate the Quaternary track faunas of the great Pleistocene Ice Age, and those of the more recent Holocene epoch, from those of the Tertiary and note that although they continue to exhibit these modern characteristics, they are also characterized by tracks of giant representatives of the Pleistocene megafauna (e.g., mammoth) and most notably by the tracks of hominids, mainly preserved in caves.

THE QUIET DAWN: PALEOCENE-EOCENE

The vertebrate track record in the Paleocene is very impoverished the world over, as is the fauna itself. The mammal paleontologist Bjorn Kurten characterized the Paleocene as an almost empty stage,[1] a time of eerie emptiness in the aftermath of the great K/T extinctions. Although there are a few reports of tracks from the Paleocene of North America, we know of none from the Paleocene of Europe.

The Eocene of Europe also has a sparse track record. In 1980, however, Paul Ellenberger reported the tracks of birds and artiodactyls from the upper Eocene of Garrigues, Ste. Eulalie Province of Gard.[2] Known for his love of elaborate and lengthy ichnospecies names, he dubbed the bird tracks *Ludicharadripodiscus,* in typical Ellenbergerian style. Ellenberger identified eight track types (in five ichnogenera) which he attributed to large mammals. He attributed *Anoplotheripus* and *Diplartiopus* to even-toed forms, and he attributed *Palaeotheripus* and *Lophiopus* to odd-toed forms. *Hyaenodontipus,* as the name suggests, he attributed to *Hyaenodon,* a carnivore. He also identified a very small purported mammal track, which he named *Ucetipodisus,* and attributed it to a small arboreal primate-like species.

As shown by his map of the site (figure 10.1), tracks cover an area of about 80 m^2, in which the larger tracks, such as *Anoplotheripus, Palaeotheripus,* and *Hyaenodontipus,* are most clearly visible. Ellenberger attributed *Anoplotheripus* to the genus *Anoplotherium,* as the name suggests. These animals were long-tailed ruminants that flourished in Europe in the Eocene and Oligocene. Ellenberger assigned three different sizes of tracks (12 to 18 cm long) to three different ichnospecies. Elongate artiodactyl tracks he assigned to a different ichnogenus, *Diplartipus.* Large *Perissodactyl* tracks (13 cm long) were assigned to *Paleotheripus,* indicating that the trackmaker was *Paleotherium,* a horse-like form that grew, in some species, to the size of a cart horse. Smaller forms (two ichnospecies of *Lophiopus*) ranging from 7 to 11 cm were attributed to the semi-tapir *Lophiodon* or some similar form. Lastly, among the large forms are the *Hyaenodontipus* tracks, attributed to some of the earliest carnivores, known as "creodonts."

The genus *Hyaenodon* to which Ellenberger attributed these tracks is common in the Eocene through Miocene of both the New and Old Worlds. These forms are not members of the modern Hyaena tribe, which is related to cats and civets.

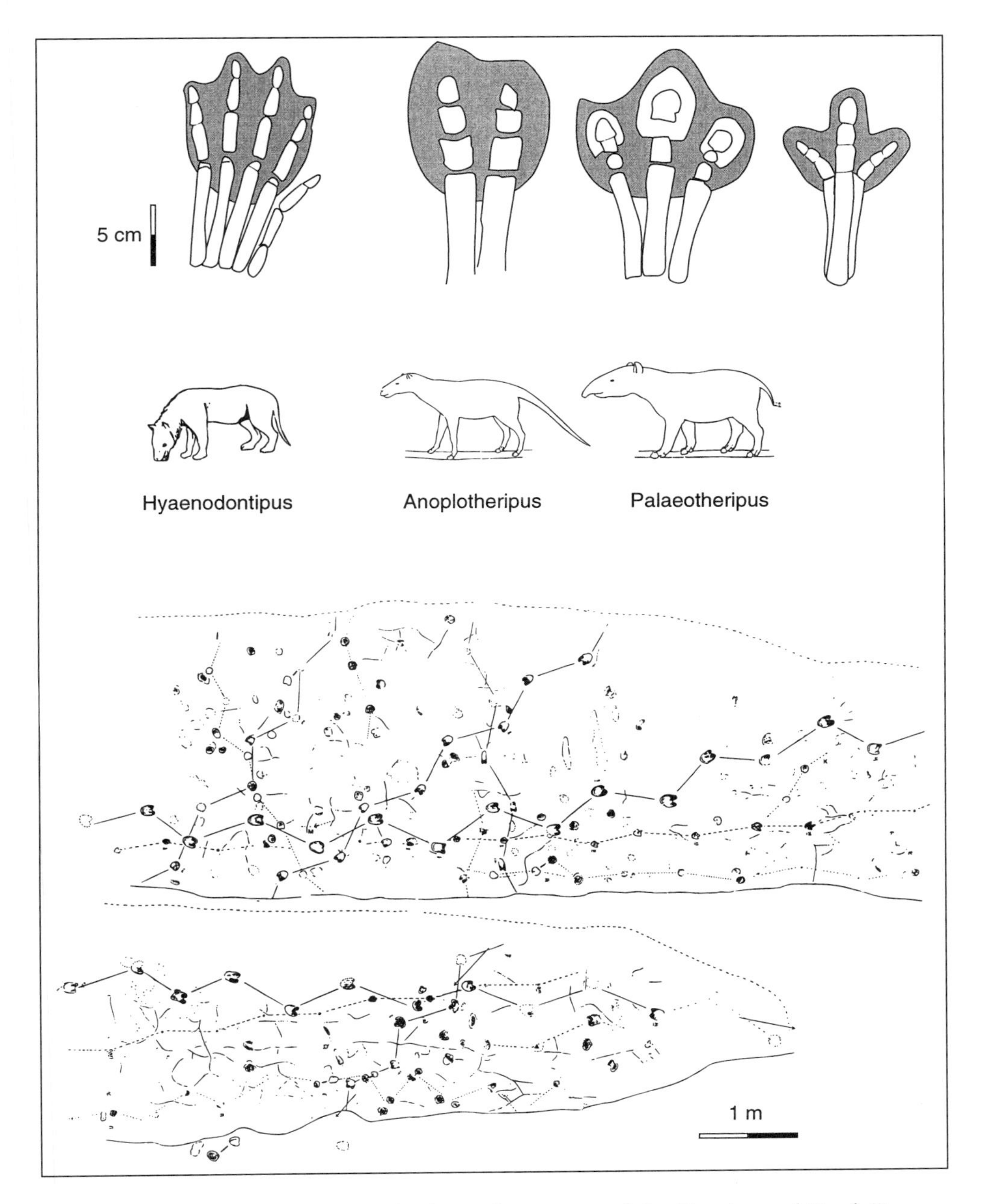

FIGURE 10.1. Mammal tracks from the Eocene of the Province of Gard, France, with foot bones of probable trackmakers and map of the site. *After Ellenberger.*[2]

OLIGOCENE ACT I: TRACKING *RONZOTHERIUM*, AN EARLY RHINO

Possibly the most interesting and best-studied Tertiary footprint locality in Europe is an Oligocene tracksite in the Apt Basin (type locality for the Aptian) at a site named Saignon, near Valcluse in southeastern France. The tracksite was described in 1984 by George Demathieu and colleagues[3] and is not far from another site, at Veins, described in 1969,[4] which yielded a similar track assemblage. The site represents a lower Oligocene pond or lake in which calcareous sediment was deposited in a relatively dry or seasonal savanna setting.

The largest and most obvious tracks at the site are those attributed to the rhino, *Ronzotherium,* and given the ichnospecies name *Ronzotherichnus voconcense.* The manus tracks are typically a little broader (about 14 to 18 cm long and 16 to 20 cm wide) than the pes tracks, which are about the same length but slightly narrower. The stride varies from about 1.2 to 1.5 m, depending on the size, and the trackway is relatively narrow, with a pace angulation averaging about 150 degrees (figure 10.2). The size of these trackmakers can be compared with those of relatively modern rhinos (probably genus *Diceros*), whose tracks are preserved at the famous late Pliocene homind tracksite at Laeotoli in Tanzania, where typical manus tracks range from about 25 to 29 cm. These African tracks, incidentally, were studied by Claude Guerin and George Demathieu[5] and named *Dicerotinichnus laetoliensis.*

The other common mammal tracks known from this site are clearly those of an artiodactyl and have been named *Bifidipes velox.* They average about 9 cm long by 7 cm wide with a stride of about 1 m, suggesting a medium-sized animal such as a deer, although no known fossil species from the region provides a convincing match for size. A third smaller mammal track named *Sarcotherichnus enigmaticus* measures about 5 cm long and wide and displays five digit impressions, and was interpreted by Demathieu and colleagues as the possible track of an early carnivore or creodont. Bird tracks from this site (figure 10.2) were named *Pulchravipes magnificus* and average about 7.5 cm wide, and are clearly made by a shorebird of some type.

Another important Oligocene mammal tracksite was reported by R. Santamaria and colleagues[6] from the vicinity of Agramunt in the Leida region of Spain. They described four new forms, including a new ichnospecies of *Bothriodontipus,* the new ichnogenus *Plagiolophustipus,* and two ichnospecies in the new ichnogenus *Creodontipus.* Clearly, *Bothriodontipus* implies that the trackmaker was the pig-like artiodactyl genus *Bothriodon* or a close relative. These early swine-like animals, from the tribe known as the "anthracotheres," grew to be about 5 feet long and left tracks about the size of those of modern pigs. Similarly, *Plagiolophustipus* implies the genus *Plagiolophus,* which belongs to the tribe known as the "paleotheres," a side branch of the horse family tree, related to the tapir-like lophiodonts, whose tracks were reported from the Eocene of France by Ellen-

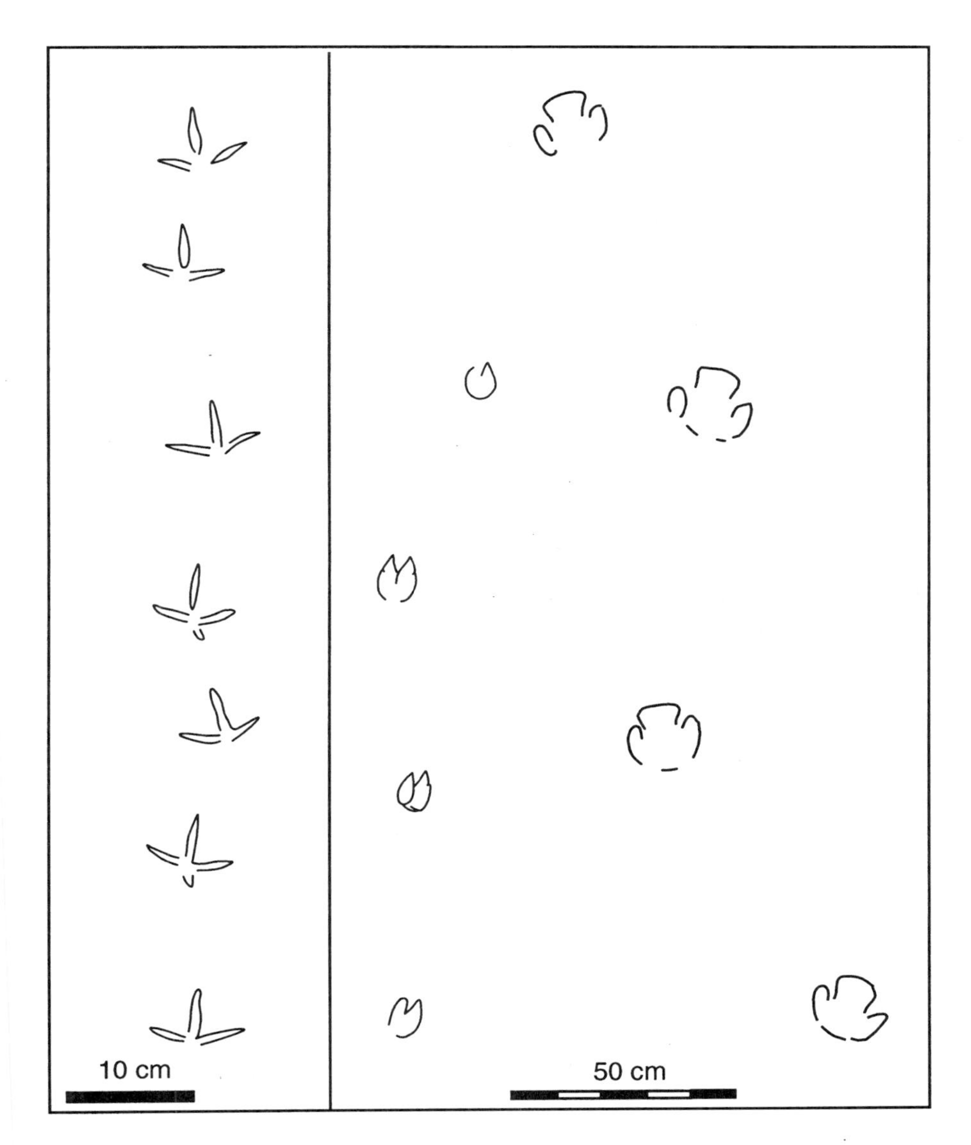

FIGURE 10.2. Rhino tracks (*Ronzotherichnus voconcense*), artiodactyl tracks (*Bifidipes velox*), and bird tracks (*Pulchravipes magnificus*) from the Saignon site, Oligocene, Apt region, France. *Redrawn from Demathieu et al.*[3]

berger.[2] *Creodontipus* implies the track of a creodont, an early group of carnivores, which includes *Hyaenodon* and other genera. In fact, the tracks compare morphologically with Ellenberger's *Hyaenodontipus* (see figure 10.1), although they are somewhat younger. Tracks from the Agramunt locality were made in sediments that represent the distal end of an alluvial fan system where lakes and floodplain habitats are thought to have provided a permanent habitat for the trackmakers.

OLIGOCENE ACT II: AN ABUNDANCE OF WATERFOWL

In 1979 Marc Weidmann and Manfred Reichel[7] published a lengthy paper reviewing multiple reports of bird tracks in the Molasse sediments of the Oligocene and Miocene age in Switzerland. They stated that the presence of birds was noted as early as 1872, thanks to the discovery of feathers, but that in general the body fossil remains of birds (bones and feathers) are rare in the Swiss Molasse. The situation with footprints is quite different. Tracks were first reported in the 1950s and are now known to be abundant and quite widely known in deposits of the same age in other parts of Europe. Although they only briefly mention some of these other sites, the contribution by Weidmann and Reichel is the most illuminating published to date, owing to the abundance of sites and the number and diversity of track types recorded and their descriptions of the paleoenvironmental context of the tracks.

As summarized in an earlier short paper published in 1971 by de Clercq and Holst,[8] on a single Upper Oligocene locality near Lucerne, bird tracks typically occur in fine-grained deposits at the top of thick (about 15 m) cycles that grade up from coarse conglomerates and sandstones to fine track-bearing sands, silts, and muds at the very top of the sequence (figure 10.3). Associated fossil evidence includes plant stems in vertical growth positions as in life, as well as freshwater snails and bivalves, lands snails, and mammals. The tracks at this locality are thought to be those of rails (Rallidae), not waders (Limicolae), because of the long first toe (hallux) and long, pointed claw impressions.

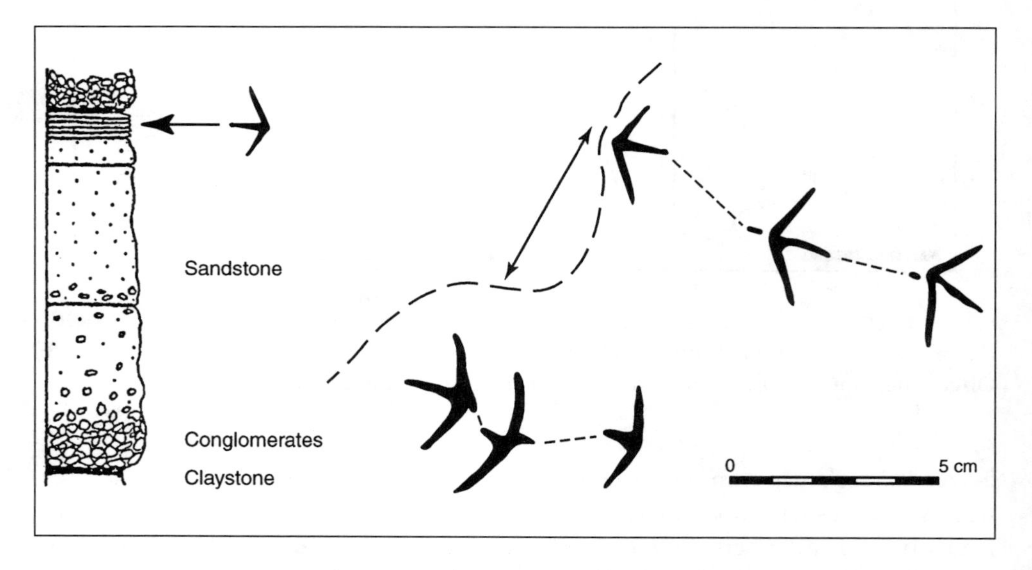

FIGURE 10.3. Typical stratigraphic setting of bird tracks from the Swiss Molasse near Lucerne, with tectonically deformed tracks. *Redrawn from Clercq and Holst[8] and Weidmann and Reichel.[7]*

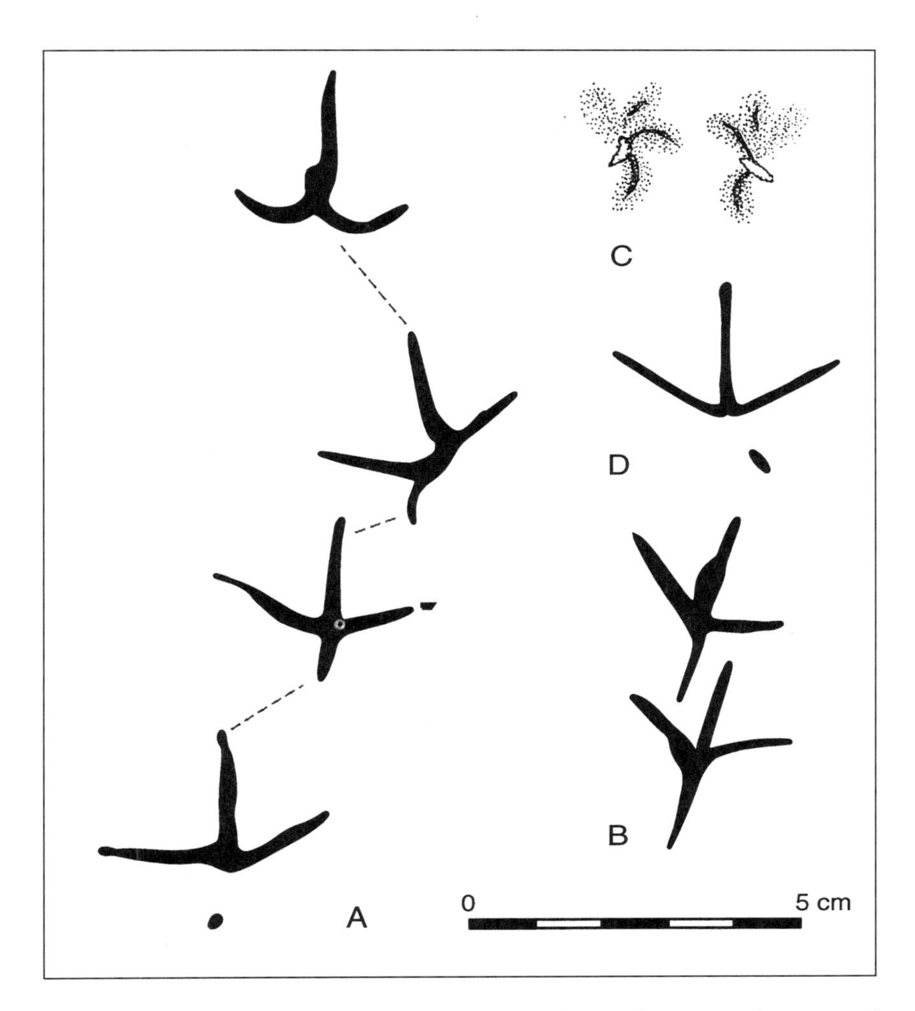

FIGURE 10.4. Bird tracks from the Swiss Molasse show a wide range of morphologies, including web-footed tracks and tracks attributed to arboreal, perching (passerine) birds. *Redrawn from Weidmann and Reichel.*[7]

Returning to the study of Weidmann and Reichel,[7] we find that they went to considerable lengths to differentiate track morphologies and place them in proper stratigraphic context. They also noted that although most of the modern orders of birds were extant in the Oligocene, we do not know how far back families and genera extend, and that most modern species probably arose as recently as the Pleistocene. Despite this uncertainty, they made attempts to categorize the various bird track morphologies. They reviewed the work of Panin and Avram,[9,10] who studied similar track assemblages from the Carpathian region and placed all bird tracks as members of the "ichno order" Avipedia. In this scheme, adapted from the work of the Russian ichnologist Vialov,[11] the waders (Limicolae) are placed in the ichno-

			Tracks indet.	Charadriformes - Solopacidae	Charadriidae	Recurvirostridae	Laridae	ANSERFORMES - Anatidae	GRUIFORMES - Gruidae	Rallidae	PASSERIFORMES - Motacillidae
MIOCENE	Upper Freshwater Molasse	TORTONIAN	↓							↓	
		SERRAVALLIEN									
		LANGHIAN									
	Upper Marine Molasse	BURDIGALIAN	↓	↓	↓	↓	↓				
OLIGOCENE	Lower Freshwater Molasse	AQUITANIAN	↓	↓							
		CHATTIAN	↓	↓				↓	↓	↓	↓
	Lower Marine Molasse	RUPELIAN									

FIGURE 10.5. Stratigraphic distribution of bird tracks in the Oligocene and Miocene of the Swiss Molasse. *Redrawn from Weidmann and Reichel.*[7]

family Charadriipedidae, and the tracks attributed to plovers (*pluviers*), snipe (*becasseaux*), and sandpipers (*chevaliers*) in the ichnogenus *Charadriipeda*. Altogether, Weidmann and Reichel differentiated eight different categories of bird tracks, including Charadriformes (four types), Anseriformes (one type), Gruiformes (two types), and Passeriformes (one type), that is, waders, ducks, herons, and perching birds, respectively (figure 10.4). Such an assemblage is to some extent predictable, being dominated by waterbirds, particularly the Charadriformes, which are described as "omnipresent." As these researchers point out, the tracks of perching, arboreal birds are rare. In fact, it has been remarked that the track record of birds is almost completely dominated by the tracks of shorebirds.[12] Thus, the passerine bird tracks reported here are among the only footprints of this type currently known from the fossil record.

The Weidmann-Reichel study is also noteworthy for the care taken in presenting the stratigraphic context of the footprints. The 11 track assemblages described are placed in four stages: the Chattian and Aquitanian (corresponding to the upper Oligocene) and the Burdigalian and Tortonian (corresponding to the

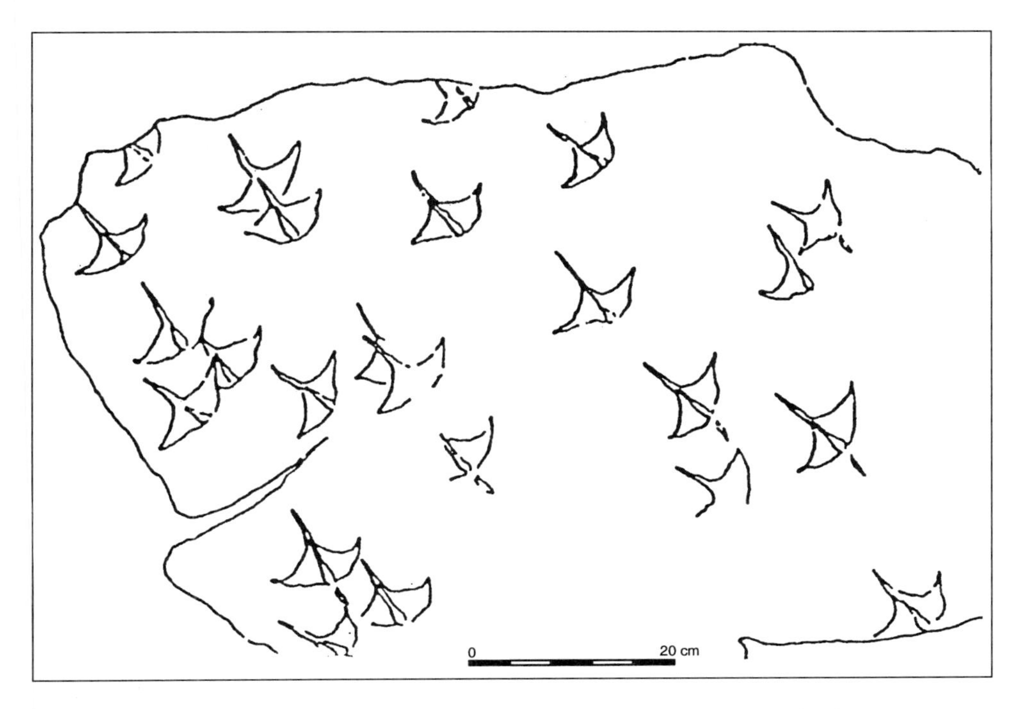

FIGURE 10.6. Bird tracks from the Oligocene of Spain, indicate flocking or gregarious behavior. *Redrawn from de Raaf et al.*[15]

Lower Miocene) (figure 10.5). Other noteworthy reports of bird tracks in similar facies come from the Lutétien of the Carcassonne and Pyrenean region.[13] The presence of bird tracks in fine-grained Molasse-type sediments sandwiched between thick sandstone and conglomerate sequences has given rise to a number of interesting sedimentary structures, and several researchers have commented on the juxtaposition of tracks with flute casts and load casts. It also has been noted that some tracks have been tectonically distorted, as noted in studies of Devonian and Cretaceous tracks (see chapters 2 and 9). Again, ichnologists have to be careful not to base morphologic studies and taxonomic conclusions on tracks that are distorted.

As pointed out by Jean-Claude Plaziat long before many of the aforementioned studies were published, it had become apparent by the mid-1960s that Tertiary bird tracks were much more common in Europe than generally thought.[14] Although the Weidmann-Reichel study is the most comprehensive and commendable to date, there is plenty of room for further research. It is apparent that many distinctive track morphologies can be identified and placed in stratigraphic context. Such distinctive features as webbing (complete and partial or palmate) are observed in some morphotypes. For example, J. F. M. de Raaf and colleagues[15] reported abundant web-footed bird tracks from the Oligocene of northern Spain, noting an extraordinary abundance and high quality of preservation. Also notable at this site is the strong preferred orientation of tracks (figure 10.6), suggesting a flock walking deliberately in the same direction. Such preferred orientations are not common among bird tracks, with the notable exception of the Early Cretaceous *Archaeornithipus* site described in chapter 8 (see figure 8.2). Recent studies of tracks from the Cretaceous and Tertiary of North America[16] reveal the presence of beak marks or feeding traces associated with some waterbird trackways. Similar traces may presumably be found at European bird tracksites, given further study. A study of the comparative ichnology between Tertiary bird tracksites in Europe and elsewhere could prove very illuminating, especially as a bird track taxonomy exists, allowing one to determine if ichnofaunas are globally or regionally distributed or characterized by any special paleoecologic attributes. Other European bird tracksites of Oligocene age include a site at Peralta de la Sal, near Lerida, Spain, described by Hernandez-Pacheco in 1929.[17]

A MIOCENE MENAGERIE

Although outside the geographic region of Western Europe discussed in this book, the Miocene track-bearing deposits of the sub-Carpathian mountains of Romania are worthy of brief mention. They were first described in 1962 by Panin and Avram,[10] who introduced a number of new names for bird and mammal tracks that were used in subsequent descriptions of European tracksites, for example, the aforementioned Oligocene tracksites of Switzerland. The Romanian track assemblages are notable for their diversity and include footprints attributable to proboscidians, carnivores (Felidae and Canidae), artiodactyls, and four types of birds (Ardeidae, Gruidae, Charadriidae, and Anatidae) (figure 10.7). At least one tracksite is notable for the extensive trampling caused by proboscidians over an area of at least 100 m^2. Also of particular interest are the largest of the bird tracks (*Ardeipeda gigantea*), which measure up to 23 cm long and presumably represent large heron- or egret-like species.

We can add to the menagerie of Miocene trackmakers recorded in Romania by taking a quick tour of other European sites that represent this epoch, most notably in Italy and Spain. Trackways of three different types of large mammals have been reported from the Conglomerato di Osoppo, in the Province of Udine in northeastern Italy, by Fabio Dalla Vecchia and Marco Rustioni.[18] The tracks occur on a small exposure of bedding plane (about 100 m^2) in a sequence of coarse conglomerates in which one would not normally expect to find tracks. The track-bearing unit is a thin (20 cm) sequence of ripple-marked and mudcracked sandstones sandwiched between cross-bedded conglomerates and evidently represents a low-energy deposit where fine material settled during or after flooding. All the tracks run perpendicular to the ripple crests, indicating

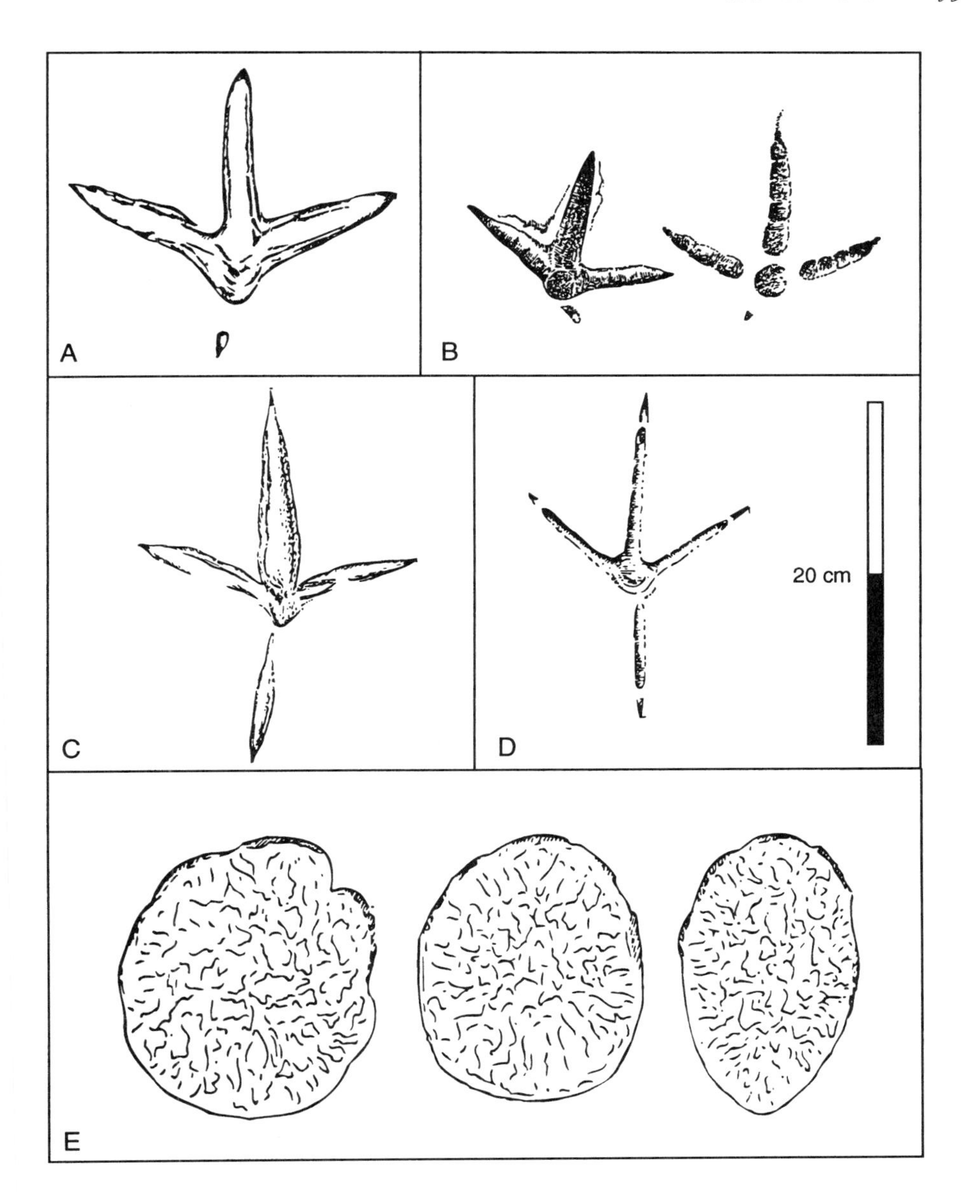

FIGURE 10.7. Miocene tracks from the sub-Carpathian region of Romania include large birds and proboscideans. *After Panin and Avram.*[10]

progression of the animals parallel to the direction of flow of the drainage. This sequence has not produced fossils that are useful in determining the age of these conglomerates. According to these researchers the trackways are probably those of *Hipparion,* a large bovid, and possibly a small rhinoceros (figure 10.8). None of the tracks are particularly well preserved, but the presence of purported *Hip-*

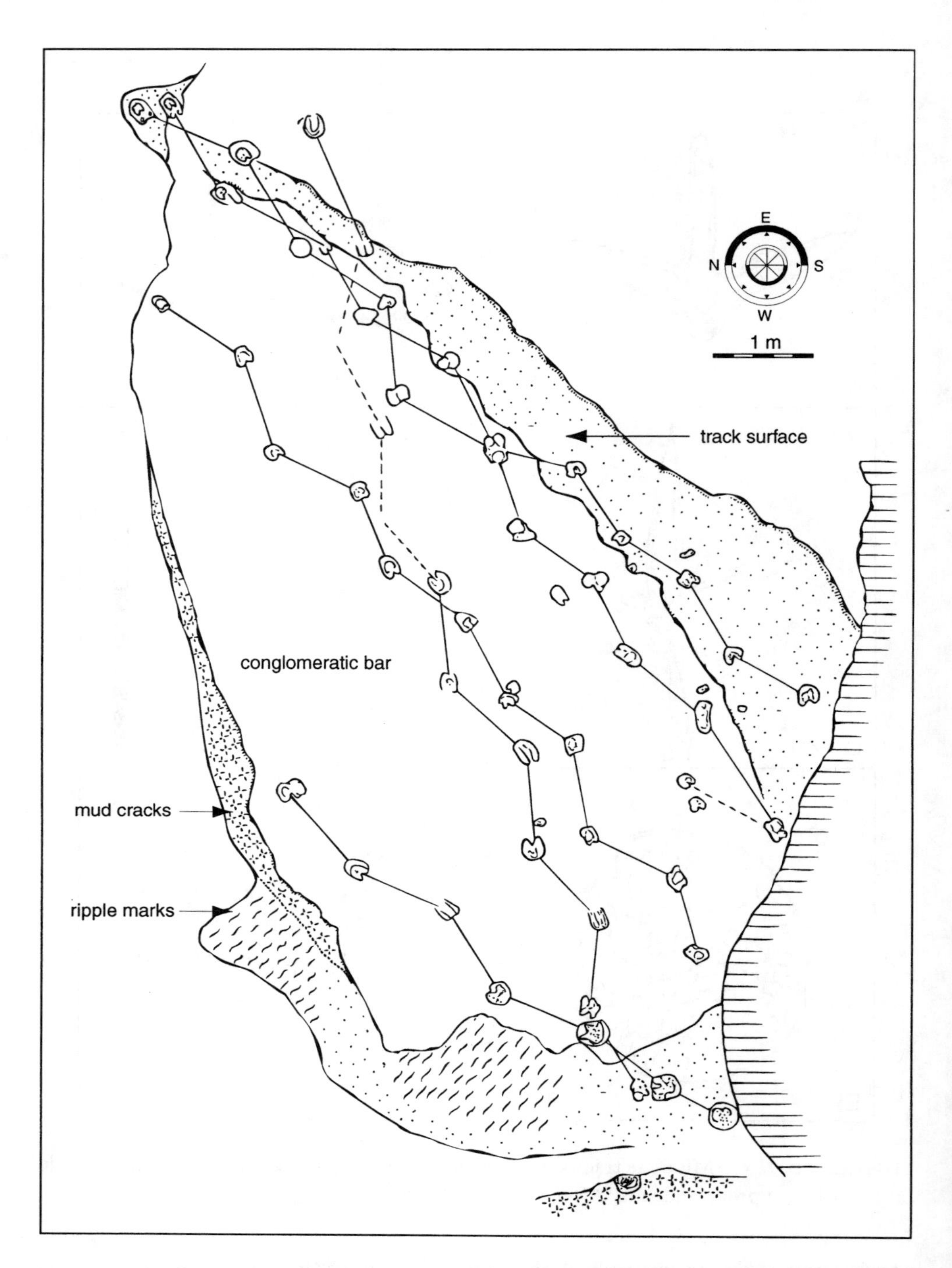

FIGURE 10.8. Upper Miocene mammalian trackways from the Conglomerato di Osoppo, northeastern Italy, indicate the presence of artiodactyls. *After Dalla Vecchia and Rustioni.*[18]

parion trackways, comprising three of the five trackways in the assemblage, can be used to suggest an age of upper Miocene, that is, post-Messinian (Turolian). As noted below, possible *Hipparion* tracks are also reported from the lower Pliocene of Spain[19] and from the middle Pliocene of east Africa.[20] This is an example of the utility of tracks in suggesting an age for a poorly dated sedimentary sequence.

In addition to the eastern European site in Romania we must also mention a classic Lower Miocene site, discovered at Ipolytarnóc in northern Hungary in 1900, where we find more than 1600 footprints of mammals and birds. Mammal tracks were attributed to carnivores (*Bestiopeda, Carnivoripeda,* and *Mustelipeda*), rhinoceras (*Rhinoceripeda*), and "peccaries" (*Megapecoripeda* and *Pecoripeda*). Reports of "mastodon tracks" from this site are dubious. Bird tracks were named *Ornithotarniocia* (after the locality), *Aviadactyla* (meaning "bird-toed"), *Tetraornithopedia* (meaning "four-toed bird footprint"), and *Passeripeda* (meaning "foot of a songbird or passerine"). Again the latter form is very rare in the fossil record. This site has been preserved in a "Conservation Hall" and can be visited by the public.[21]

TRACKING ANCESTORS OF THE CAT: MIOCENE OF SPAIN

Footprints of Tertiary carnivores (cats, dogs, bears, etc.) are generally rare, especially in Europe. Fortunately, however, a spectacular new site has come to light at Salinas de Añana, in Alava Province, Spain. The site is Miocene in age and records what are probably the oldest tracks (figure 10.9) that can be attributed to the cat family (Felidae). In *The Big Cats and Their Fossil Relatives,* Alan Turner and Mauricio Anton[22] suggest that the tracks from this site were made by five individuals of the genus *Pseudaelurus,* an animal about the size of a European wildcat moving with a digitigrade stance at a slow and casual walk. The parallel configuration of these trackways suggests the possibility of gregarious behavior.

The site is remarkable for the high quality of preservation of tracks. Although a detailed scientific report has not yet been published, in 1993 the popular Spanish magazine *Blanco y Negro* published a 10-page color article featuring the site under the title (loosely translated) "Who Went There 20 Million Years Ago?" In this article, we learn that not only is the site rich in cat tracks but that it also reveals the tracks of ancestral hyenas, mongoose (similar to the modern genus *Herpestes*), the superficially rabbit-like ruminant *Cainotherium* (which stood only about 15 cm high at the shoulders), and a larger ruminant of unknown affinity that stood about 48 cm at the shoulders. Tracks of at least one bird species are also recorded at this site.

Although we can only glean so much from a magazine article, the following points emerge. At least six species, including three carnivores, are represented. The paleoenvironment at this time was savanna-like, with local lakes and ponds that acted as a magnet to mammals, birds, and other wildlife. Mauricio Anton, who authored parts of the article, is quoted as suggesting that the tracks, all exquisitely

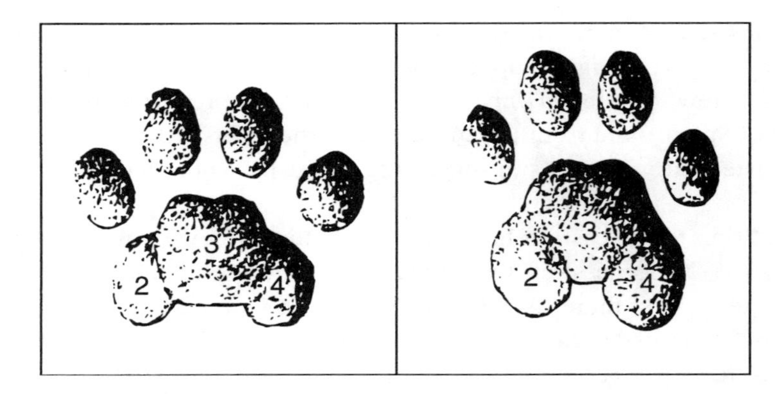

FIGURE 10.9. Europe's oldest cat tracks from Miocene Spain. *After Turner and Anton.*[21]

preserved, were made in a short period of time, estimated at no more than 36 hours, which allowed for the passage of both diurnal and nocturnal species. The article touches on tantalizing glimpses of unusual behavior, such as a pair of trackways, one made by a small carnivore, the other by a small ruminant, that both made a sudden 90 degree turn and sped up. Such activity is obviously suggestive of a predator-prey interaction or the efforts of small animals to get out of the way of larger ones at a crowded watering hole. We shall have to await the publication of a full report on the site to see how such interesting trackways are interpreted.

PLIOCENE INTERLUDE

In reference to possible *Hipparion* tracksites, we have already mentioned one of the most interesting Tertiary track localities reported, in 1992, by C. Lancis and A. Estévez[19] from lower Pliocene deposits at the eastern end of Sierra del Colmenar, south of Alicante, Spain. No more need be said on this topic, except that the probable equid tracks display very strange preservation (figure 10.10), suggesting that the trackmaker sank deep into a soft substrate that then collapsed around the footprints. What makes the site most interesting is the presence of large bear tracks (figure 10.10) up to 30 cm long. These are attributed either to *Ursus ruscinensis,* the oldest representative of the genus in Europe, or to *Agriotherium,* a larger genus represented in lower Pliocene sediments. Based on the size of the footprints (30 cm in length) and the long stride (2.5 m), the size of the bear was estimated at more than 2 m in length when in a quadrupedal posture. As noted in the following sections, bear tracks and traces are common in cave deposits in the Pleistocene. This, however, is the oldest reported occurrence in Europe and represents an outdoor setting described as a paleosol substrate in a palustrine setting.

The Pliocene is the shortest Tertiary epoch, and as such has a sparse track record in some regions, including Europe. This is, in part, simply a function of the

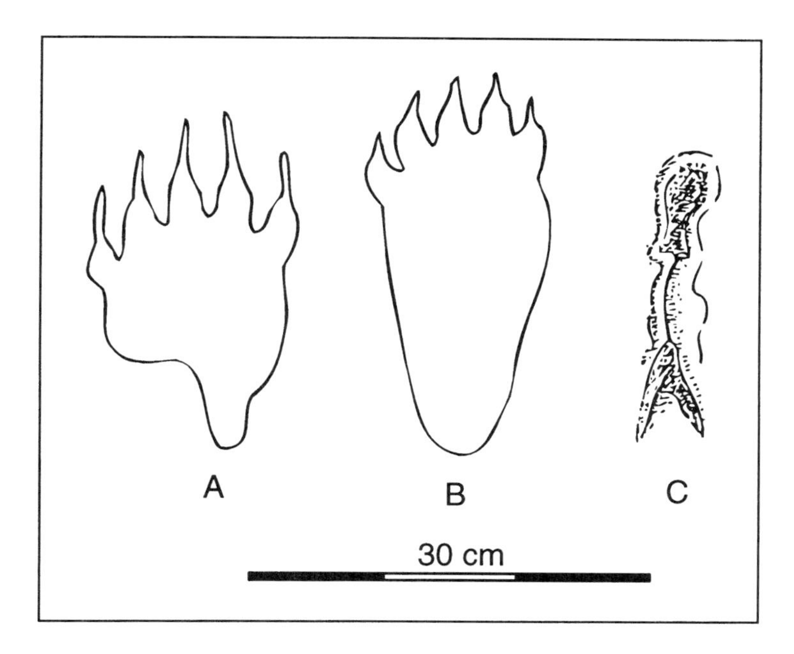

FIGURE 10.10. A and B: Late Pleistocene bear tracks. C: Distorted equid tracks from the Sierra del Colmenar, south of Alicante, Spain. *After Lancis and Estévez.*[19]

short duration of the epoch and the lack of intensive study of Tertiary tracks in general. Pliocene tracksites are known from North America, and the most famous of all Pliocene tracksites, at Laeotoli in Tanzania, has been the subject of intensive study. The most notable feature of this world-famous tracksite is the hominid trackways, which herald the emergence in the track record of footprints of several representatives of the family Hominidae. As we shall see in the sections that follow, such tracks are quite common in Pleistocene sediments of Europe, most notably in cave deposits, where, although not immune to weathering and erosion, they were generally protected from the worst ravages of the elements.

PLEISTOCENE: ICE AGE TRACKMAKERS

A recent study by W. Koenigswald and his colleagues[23] of a newly discovered Ice Age (Pleistocene) tracksite in the middle of a sewage treatment plant on the banks of the Emscher River in Bottrop, northern Germany, has revealed splendid trackways of the cave lion (*Panthera leo spelaea*), the first record of the species in middle Europe, caught in the open, outside its cave habitat. Tracks of wolf, bison, horse, and reindeer also have been reported from this locality (figure 10.11). Reconstruction of the setting indicates that these animals were crossing an open area of alluvium in a subglacial environment approximately 20,000 to 30,000 years ago. The reindeer tell us that the climate of Germany was then more like that of Lapland.

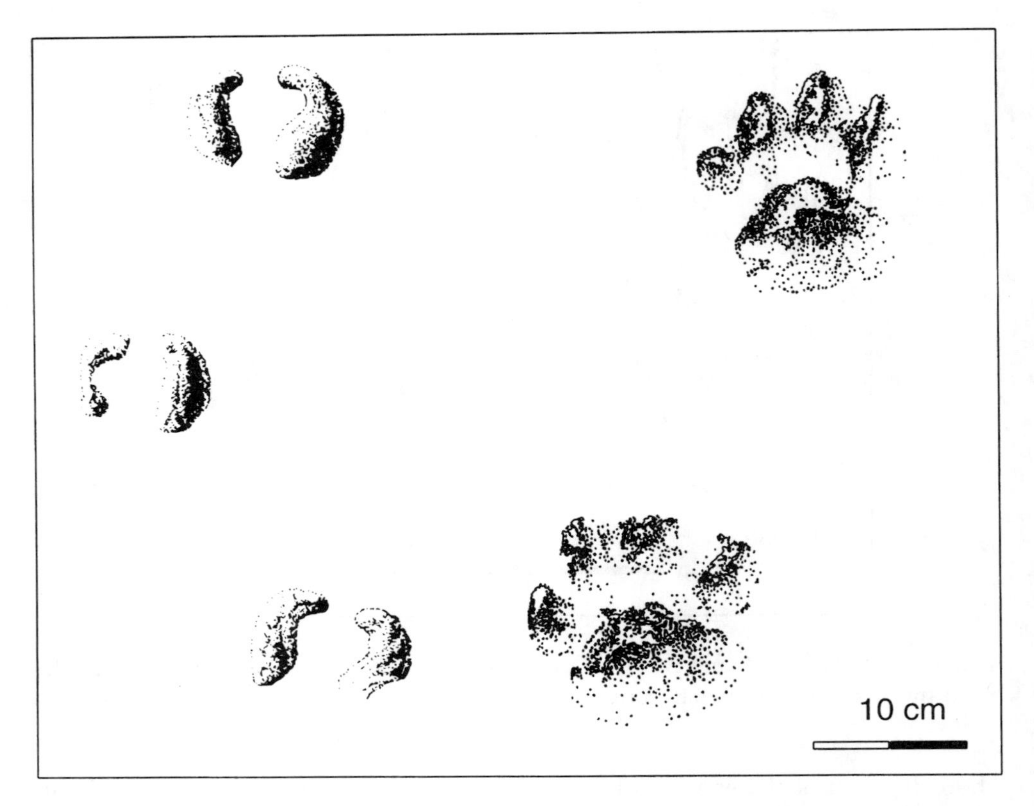

FIGURE 10.11. A Late Pleistocene tracksite on the banks of the Emscher River in Bottrop, northern Germany, provides insight into the fauna that inhabited glacial outwash plains 20,000 to 30,000 years ago. Footprints include those of reindeer (left) and cave lion (right). *After Koenigswald et al.*[23]

We have now moved into an epoch when our immediate ancestors coexisted with the trackmakers described in our ichnologic articles. In fact, it is an intriguing thought that our ancestors, less than a thousand generations removed, may have gazed on these very tracks and the animals that made them long before the Bottrop River had its present name and course. As we shall see in the sections that conclude this chapter, this is not a typical preserved Pleistocene tracksite, even though many such sites must have existed, revealing what is clearly a quite representative fauna. This is because most Pleistocene tracksites are found in caves, where the most common trackmaking species were cave bears and humans.

It is perhaps disappointing not to have a large, well-preserved "outdoor" Pleistocene tracksite where all the classic elements of the Pleistocene megafauna (mammoth, woolly rhino, etc.) are well represented by their footprints. No doubt some such traces exist, as they do in other parts of the world, and there is little doubt that new "outdoor" tracksites will be found in the future. As indicated above, tracks of proboscidians and rhinos are known from pre-Pleistocene (Neogene) sediments, and the lack of footprints in the track record is compensated for by a new type of ichnologic facies, that found painted on the walls of caves and carved on antlers and bones.

SUBTERRANEAN TRACKING: HOMINID ICHNOLOGY

One of the most vivid concepts in all of prehistory is the concept of the cave man or, should we say in this politically correct age, cave person. Although this may conjure up images of humans in animal skins warming themselves by fires, the direct skeletal evidence of our cave-dwelling ancestors is often scrappy and incomplete, and sometimes not from caves at all. The indirect evidence of footprints, but also cave art, is often much more evocative, giving rise to images of ancestors who were imaginative, artistic, and emerging into a new psychological realm of self-awareness. At least symbolically, our Ice Age ancestors form the link between our animal past and our technologically advanced future. The question often posed, however, by thoughtful students of human evolution is, were they fundamentally different in terms of their physical, mental, and spiritual makeup? The answer, in simple physical terms, is usually no. As members of the species *Homo sapiens sapiens* (as distinct from the Neanderthal subspecies), we find little in the way of obvious anatomic differences, at least among representatives from the past 50,000 years or so. Let us turn then to what the track record and cave art have to add. For those who object to the inclusion of cave art under the broad heading of ichnology or trace fossils, let us make two points. First, footprints and cave art coexist at many sites, and second, in the final section, we shall make the conceptual academic argument that cave art can be considered a special category of ichnology.

Although it is usually cave art that provides the most vivid evidence of habitation of caves by our Paleolithic Pleistocene ancestors, there are a large number of caves in which the footprints of humans and other mammals such as bear and hyena have been reported, sometimes in conjunction with cave art but sometimes in caves that lack paintings. Few of these reports find their way into the geologic and paleontologic literature, but are found instead in the annals of archeology and anthropology. It is outside the scope of this book to list all the sites with hominid tracks or to begin to list all known examples of caves in which vertebrate tracks have been reported or documented. We shall therefore restrict ourselves to a few selected examples. But first a few generalities.

Given that there is currently no synthesis available describing all known vertebrate and hominid tracksites in Pleistocene caves, it is to be expected that the available information is somewhat scattered. Even if we were to compile all such occurrences, it would be hard to put them in chronologic order, with reliable dates. This is because dating of cave deposits is notoriously difficult. Even when reliable radiocarbon dates are available, we must be aware of what is being dated. Is it the track-bearing floor, a bone, artifact, or cave painting, and how representative is one date of everything else in the cave? It is clear from caves with multiple floors and large attritional bone accumulations that they were occupied, in some cases, for prolonged periods. For the simple reason that most cave tracksites appear to be upper Pleistocene or late Paleolithic in age, we have confined ourselves to discussion of this interval. Because most paleontologists are most interested in skeletal remains, the information on nonhominid vertebrate tracks is sparse, although there have been casual reports of hyena[24] and fox[25] tracks at the Aldène and Fontanet caves, respectively, and, as noted below, many reports of cave bear footprints, scratch marks, and hibernation dens.

The human track record, interestingly, seems to be dominated by the footprints of children, and in all cases the tracks indicate unshod individuals, with little or no sign of deformity due to injury or frostbite.[25] Cave temperatures would have remained constant at around 14°C. When we arise from the realm of cave floor ichnology to examine the pictorial record on the walls, it is interesting to observe that there are apparently few unequivocal examples of depictions of animal tracks, even though they are quite common as engravings, petroglyphs, and pictographs in other regions, notably Australia[26] and North America. The reasons for this are unknown, though they could be related in some way to both age and sample size. Some animals are drawn in such a way that their feet look like tracks rather than profile views of their hooves.

There are fewer than 300 cave art sites in Europe compared with thousands of well-illuminated outdoor sites in North America and Australia, and on the latter continent they span a longer time interval. European cave art, however, does serve one further ichnologic purpose. It allows us a look at the hands of early humans, completing the partial picture provided by the footprints. Hand prints (strictly, hand paintings, stencils, or outlines) have been the subject of considerable controversy owing to the large number that have been interpreted, probably incorrectly, as evidence of deformity. We leave the reader to ponder the enigma of a sample where the footprint record is biased toward uninjured juveniles and the handprint record toward adults purported to show a high incidence of deformity.

Addressing first the nonhominid track record, we should note that cave bears appear to take first prize as the most common and ubiquitous trace makers in late Pleistocene caves. There are many examples of sites where we either find footprints or other traces. For example, Chauvet cave, possibly the oldest reasonably well dated cave art site,[27] contains the spoor of cave bear in the form of footprints. At many other sites we find reports of tracks, numerous scratch marks and hollows, bears' nests, or hibernation dens, which like parts of the walls are often smoothed and polished by the constant rubbing of many generations of fur. The famous Grotte Aldène in France is a good example of a cave that contains abundant bear spoor, mainly in the form of scratches and bears' nests about 6 to 10 feet wide and 1.5 to 2 feet deep. Grotte Aldène is dated at about 15,000 years before the present (BP) and contains abundant human footprints and charcoal created by human activity that was later crushed underfoot by hyenas.[24] Grotta del Fiume in Italy has also yielded well-preserved cave bear tracks.[28]

One might think that because European cave art depicts so many animals, it would include many examples of tracks. The only reference to tracks that we could find[25] refers to possible horse and bird footprints. These are interesting because they also have been interpreted as representations of female genitalia (vulvae). The author of this theory was the famous "father of prehistory" and originator of the tradition of cave art research, Abbé Henri Breiul, who "should not be considered an expert on this particular motif."[29] It seems probably that despite Breiul's influence in making many archeologists believe that depictions of vulvae were common and that sexuality played a big part in Late Paleolithic spiritual life, most of these motifs are open to other interpretations, such as tracks. Paul Bahn, one of today's leading experts on cave art, has discussed this problem in a paper with the memorable title of "No Sex Please We're Aurignacians."[29] It is outside the scope of this book to discuss this debate, but it is possible that Breiul's interpretations are correct.[30]

One of the most fascinating examples of a cave tracksite is the complex at Niaux, a labyrinth that extends for several kilometers, where footprints were discovered at multiple locations in 1906, 1949, and again in 1970 to 1972. The site was documented by Leon Pales in one of the most comprehensive studies of cave footprints ever published.[31] At one of these sites we see two dozen tracks, made by two children, that form a distinctive, repeated "rectilinear" pattern indicating that the trackmaking youngsters were consciously or deliberately creating a footprint design (figure 10.12). The footprints are only about 20 cm long, compared with those of modern adults, which typically range from 25 to 28 cm.

The configuration of footprints at Niaux is clearly an example of an ichnologic track assemblage, and not cave art, painting, or sculpture in the normal sense. The implications, however, are interesting. It is possible to regard tracks made during

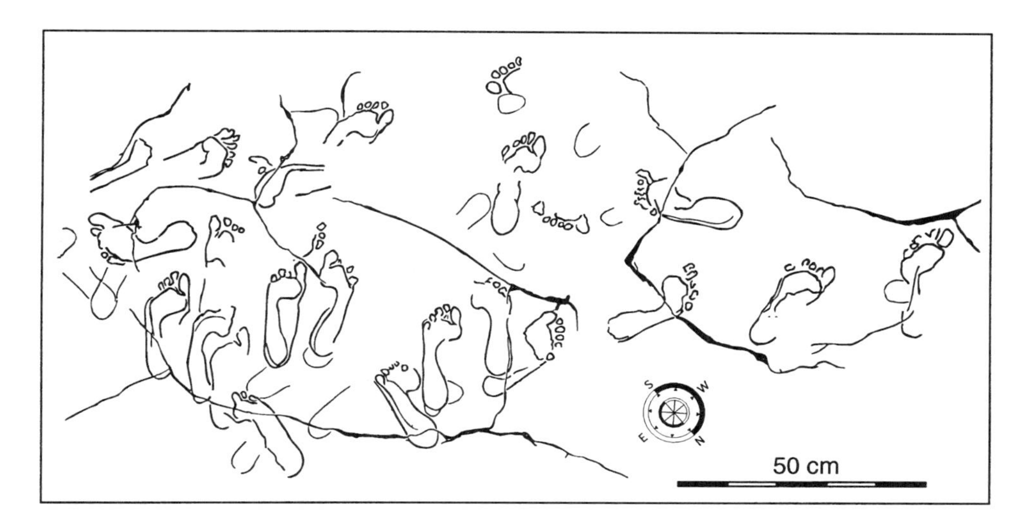

FIGURE 10.12. Kids at play in the Pleistocene. Map of juvenile tracks from Niaux cave. *After Pales.*[31]

normal locomotion as the result of unconscious or instinctive activity. Although the decision to move from point A to point B can be considered conscious, at least for humans, where one places one's feet when moving across unobstructed ground is not really a conscious process. Not so with the Niaux tracks. They were made by individuals who had taken it into their heads to deliberately make patterns in the mud. In a strict ichnologic sense, this could be interpreted as evidence for a new type of behavior—play. (This is not to say that other young mammals do not play, simply that they do not make deliberate patterns in the mud with their feet.) To our knowledge, it is the only subfossil example of "foot play" and so seems to suggest that children, like the creators of wall paintings, who were presumably adults, also engaged in artistic creations. In a sense, such footprint playground designs are the juvenile equivalent of the more sophisticated or adult artwork preserved in wall paintings.

Similar deliberate designs in cave floor mud are recorded at Tuc d'Audoubert.[32,33] Here we find rows and groups of small circular impressions made by the fingertips of individuals of unknown age. These have no obvious functional significance, and so may be considered a type of doodling that could have been made by playful children or idle adults. Such interpretations, however, are speculative, and it could be that the "fingerprints" served an important symbolic function, such as counting game animals, numbers in the tribe or kinship group, days of duration of a particular activity, and so on.

In the Grotte de Cabrerets (also known as Peche Merle), Comte Begouen[34] reported the trackway of an adult accompanied at regular, stride-length intervals on the right-hand side by a rounded trace interpreted very cleverly by Abbé Leymozi, discoverer of the site, as the impressions made by the end of a walking stick. ("Chaque couple de pas de ce sujet est accompagné, du côté droite, d'une dépression arrondie qui a été très judicieusement interprétée par l'abbé Leymozi, découvreur de la grotte, comme une empreinte de baton").

Although this section is devoted mainly to footprints found in caves, a new type of outdoor hominid tracksite has been reported recently from the British Isles. These sites are younger, ranging from about 6000 to 4000 BP, but are nevertheless prehistoric, representing Mesolithic and Neolithic cultures. What are probably the older sites (at least 5720 to 6250 BP) at two locations on the northern banks of the Severn River estuary, near Uskmouth in Wales, have yielded a total of four hominid trackways (figure 10.13) that have been studied in considerable detail.[35] Using the global average ratio of foot length/height of 15%, the trackways indicate two adults 1.73 to 1.8 m tall and a juvenile 1.4 m tall, from the main Uskmouth site, and a much larger individual almost 2.00 m tall from the second site. The latter individual would have had a size 12 shoe (British), which is size 13 in America, and by Mesolithic standards was perhaps a giant. Tracks of deer, aurochs (our ancestral cattle), and birds also have been found.

A similar site has been found on the English coast near Formby, near Liverpool. Published dates suggest a minimum age of 3500 years, but new dates suggest 5500 BP. Here more than 150 human trackways have been documented in association with

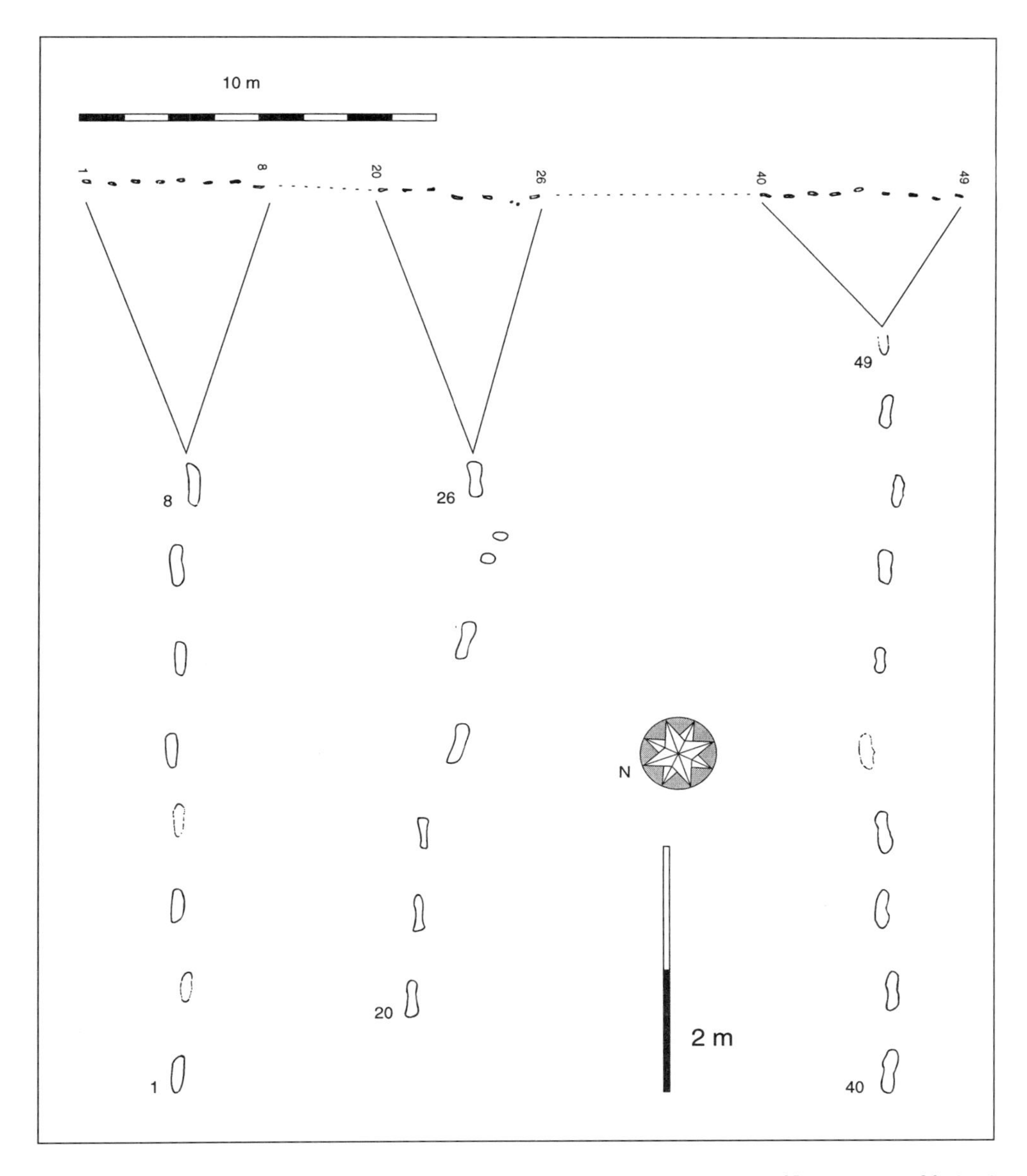

FIGURE 10.13. Holocene footprints from the coast of England[35] and Wales[36] shed light on the activities of our Mesolithic and Neolithic ancestors from about 4000 to 6000 years ago. These footprints are associated with the tracks of aurochs, deer, unshod horses, and birds.

the footprints of aurochs, red deer, roe deer, unshod horses, cranes, and oyster catchers.[36] The assemblage is therefore similar to that found in the Severn estuary. The main difference is that the human tracks are so much more abundant, allowing men (mean height 1.66 m), women (mean height 1.45), and children to be differentiated. It appears also that the men were moving faster, possibly following or herding red deer, whereas the women and children were moving slowly, possibly carrying or gathering seafood. "Another important feature of some human footprints is the presence of abnormalities (e.g., toes missing, fusion of two toes, congenital bursitis and arthritis) which can be appreciated in the prints, giving unusual detail about the medical condition of the people who made them."[36]

If we move from the temperate shores of Great Britain to the balmy shores of the Mediterranean Balearic isles, we follow a trail that leads from men to mice. Here a very different Quaternary tracksite was reported in 1993 by J. Quintana[37] in coastal sand dunes on the island of Menorca. These footprints have been attributed to the recently extinct Balearis islands cave goat (genus *Myotragus*), whose remains were abundant until Neolithic times, and *Hypnomys,* which belongs to the doormouse family. The latter are surely among the smallest prehistoric trackmakers described in this entire book, with footprints only 2 cm in length and width. It appears that the mouse tracks were preserved where large blocks of older Miocene sediment fell onto the dunes, creating very localized areas protected from erosion. The goat tracks, by contrast, are more widespread because their larger size leads to much deeper tracks and a greater potential for preservation. The distinguished invertebrate ichnologist Richard Bromley tells us he has also found goat tracks on the island of Mallorca (personal communication 1997).

CAVE ART TO FORENSICS: THE SIGNATURE OF MODERN HUMANITY

A strong argument can be put forward that cave art is a type of trace fossil. In a literal sense the word "trace" originates from the old French *trac,* meaning "to draw." In a geologic or paleontologic sense, trace fossils are marks made by organisms on substrates. These substrates are not necessarily soft sedimentary substrates, for it is widely accepted that invertebrates such as sponges and bivalves that bore or burrow into wood and rock are making trace fossils. It is also a widely held opinion that trace fossils are manifestations of behavior. For precisely this reason, no one argues against the view that cave art is an expression of a new type of behavior that appears to distinguish modern humans, possibly only our species, from more distant hominid ancestors. Whether archeologists accept the definition of cave art as a type of trace fossil is largely a matter of personal preference and convention. There can be no doubt that, in a broad conceptual sense, cave art represents an interesting and very distinctive trace fossil category.

The problem with this wider definition of trace fossils is that it has far-reaching implications because anything that is made or constructed by humans as a manifestation of behavior then becomes a trace fossil. Thus, any technological creation or artifact is in one sense a trace fossil because it was constructed by taking a natural object or substrate and modifying it, in most cases by hand or by means of tools constructed by the human hand. If we follow the history of tool making, it essentially began as deliberate modification of rock and wood substrates to facilitate human activity, particularly as it pertained to hunting and gathering food. Again, many trace fossils are directly related to feeding, whether they be worm burrows that systematically harvest the substrate (fodichnia, or literally "food traces") or teeth marks of scavengers on dinosaur bone. So if these are trace fossils, surely a trace made by a stone blade on a butchered bone is also a trace fossil, as some researchers assert.[38] The stone tool is just a substitute tooth. In fact, teeth of animals may substitute as spear or arrow points in many artifacts. If we continue to explore the link between simple tools and feeding and survival activities, we will readily find that archeologists agree that the main subjects depicted in cave art are hunting and its product: food. Using a less masculine perspective, William Irwin Thompson has suggested that Late Paleolithic cultures were matriarchal.[30] Note the abundance of so-called Venus figurines at this time. In his fascinating and in-depth thesis, he argues that engraved bones in some instances probably represent lunar tally sticks for counting the month of the year and months of pregnancy. In this thesis women may have been the first to count and write, for these tally sticks are a form of symbolic writing (or trace fossil). In a behavioral sense, they would be deliberate, unlike the random cuts on bone that were the result of butchering animals.

It seems that in this decade of ichnologic progress there are many important new Holocene sites coming to light. The presence of tracks of hominids that reveal abnormalities may give us new insights into the day-to-day condition of our ancestors. Note the lack of abnormalities in tracks from caves that are dominated by the footprints of youngsters. The presence of tracks of aurochs and unshod horses may suggest that we can look at the development of domestication through an ichnologic lens. For example, a Bronze Age tracksite in Sweden offers insights into the domestication and use of horses earlier than previously believed.[39] It is also clear that in the future we are likely to have to consider two types of hominid track-bearing assemblages: cave or subterranean and outdoor sites. It already appears that the outdoor sites have the potential to be large, and although often in danger of marine erosion, they offer researchers the benefit of the clear light of day. We predict that this will be a growing field. Some archeologists have already spoken with unbridled enthusiasm of the insights offered by ichnology. For example, in reference to the Swedish site, we read:

> Techniques were adopted from the palaeontological discipline of ichnology—the study of fossil vertebrate footprints. The global discovery of several billion preserved dinosaur footprints in the last thirty years has brought about the rapid development of advanced techniques for the recording analysis and interpretation of trace fossils of this kind—methods which have a direct application to the study of archaeologically-investigated tracks and prints.

Although this commentary is a little overenthusiastic, and debatable in some of its assertions, it is symptomatic of the enthusiasm felt by archeologists discovering the interdisciplinary link between their field and the revitalized subdiscipline of ichnology as practiced by paleontologists in the 1990s.

If we accept that footprints, handprints, and fingerprints found in caves are trace fossils that are frequently as young as or younger than contemporary cave art, what do we make of the representations of human hands that form part of several murals? Are these more akin to trace fossils than the paintings of animals, and does such a distinction have any great significance beyond adjusting our definitions of trace fossils? Certainly they tell us something specific about the morphology of the humans responsible for the paintings, or at least on site, at the time some of the art was produced. Again, this is largely a matter of preference. It is beyond the scope of this book to enter into a deeper discussion of the scope of ichnology in embracing certain types of cave art. We therefore close with three examples of convergence between cave art and symbolic representation (new hominid ichnology) and the world of paleontology and trace fossils as traditionally defined.

The first example involves the mighty mammoth. Although well known as the classic Ice Age mammal, its tracks are not well known or documented in Europe, even though footprints are known from sites in North America and its bones are plentiful. When first discovered more than 200 years ago, fossilized mammoths were an enigma, which gave rise to legends about giants. Many were still frozen in the permafrost and so appeared to be creatures that had just recently perished. Indeed, they provided a larder of "fresh" meat for wolves and perhaps even for humans, who also used them as a major source of ivory before the African elephant was exploited. From a scientific perspective, however, before the advent of modern, radiocarbon dating, it was hard to tell whether these creatures had ever coexisted with our ancestors. Even when stone tools were found in conjunction with their skeletal remains, it was possible to interpret such occurrences as evidence that our ancestors had chopped up the frozen carcasses.

It was only when cave paintings that depicted the mammoth were discovered that we had direct evidence that they had coexisted with humans. Moreover, the paintings were important in demonstrating the proper angle at which the curved spiral tusks were carried in life and in showing the presence of a large, fleshy hump over its shoulders. In this regard cave paintings are like trace fossils because they record evidence of the living animal long after it has disappeared. Having said this, an intriguing but unlikely possibility exists. In theory one could find skeletal remains, and tracks, of individual mammoths that have been depicted in cave paintings, and even find remains of the artists who painted them. But even if such remains exist to match certain paintings, it seems impossible that we could ever prove the correspondence.

Before leaving this topic, we should add that despite the many individual animals—including mammoth-scale megafauna painted on cave walls and the presumed tracking skills of Paleolithic hunters—there are virtually no examples of tracks associated with European cave art. The possible exception is the depiction of animal hooves, as if they have been turned to show the track-making surface,[25] a type of dual or "twisted" perspective that has stimulated discussion among archeologists when also applied to the orientation of horns and other anatomic features. There are also no known examples of cave paintings of human footprints despite the abundance of hand stencils. There are, however, at least two sites in Great Britain where isolated human footprints have been depicted on rock surfaces. This lack of tracks may simply be a result of the relatively small size of the cave art sample and the ephemeral nature of paintings, especially those depicting relatively small objects.

Our second example comes from consideration of stencils and paintings or prints of human hands. In one sense these may be nothing more than the signatures that early artists attached to their cave art, although in another sense they may be a celebration of the power of the hand. Did the improved self-awareness that we associate with *Homo sapiens*'s expanded artistic and aesthetic faculties also lead to a conscious awareness of the advantages of manual dexterity? This is not such a farfetched proposition when we consider that modern archeologists have been obsessed with explaining that the hand has been all important in the development of tool-using technologies, in fact predating the phenomenal expansion of the brain by a million years or more. Some sites such as Gargas in the high Pyrenees have revealed hundreds of hand stencils.[40] Was this a hand cult or simply a mass signing of artwork? We shall probably never know the full story.

Those who have studied the site in detail suggest that the stencils are predominantly of the left hand, implying right-handed artists stenciling their left hands and sometimes parts of their lower forearms. (There is the possibility, if the hands were reversed, i.e., palm out, that we are dealing with a much higher proportion of right-hand stencils.) The recently discovered, and partially submerged, Cosquer cave on the Mediterranean coast near Marseilles also has produced more than 50 hand stencils, and fine examples also come from Fuente del Salin (Santander), Spain (figure 10.14).[41] Many of the handprints at Gargas, and some at other sites, show what may at first appear to be one or more missing or shortened fingers. The proportion of apparently deformed or mutilated hands at Gargas is particularly high. However, the theory that the prints represent some form of widespread ritual mutilation, or serious disease or debilitation,[40] such as leprosy or frostbite, is hotly contested by those who rightly question the probability of such a high incidence of deformity when other, simpler explanations exist.[25] It is possible that the fingers were bent palmward with the palm upward, an interpretation much more in keeping with the manual dexterity evident in the accompanying artwork. It seems rather ridiculous to propose that humans were honing new levels of artistic virtuosity and manual dexterity while simultaneously lopping off their fingers in large numbers. The many hands that wave to us from cave walls appear to be the expression of a positive development in human adaptation, not the expression of a macabre self-mutilating cult.

Our third and final example is taken from the world of fingerprints. As we shift progressively from the mainly subconsciously registered footprints of our ancestors to traces of deliberate foot games and handprints associated with cave paintings, let us consider how we move inexorably toward the contemporary world of

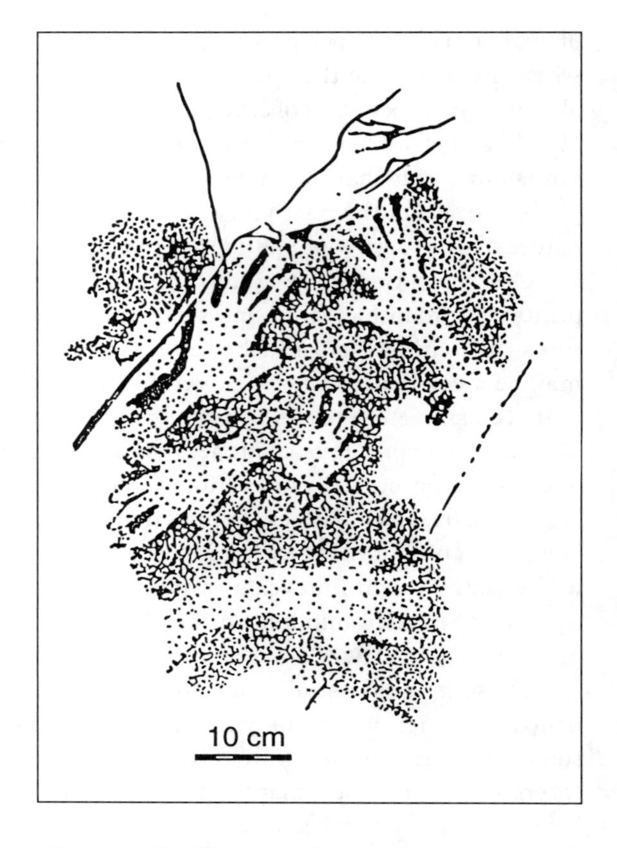

FIGURE 10.14. Cave art is a new type of hominid ichnology sometimes depicting the size and shape of the hands and other parts of the anatomy of our ancestors and sometimes depicting animal tracks. *After Bahn and Vertut.*[25]

signatures, writing, and fingerprints that we take for granted in modern society. Although we may not think of it as such, the billions of human fingerprints on record with various government agencies are the largest global database in existence in all of vertebrate ichnology. The early proddings of fingertips into cave floor clay at sites such as Tuc d'Audoubert foreshadowed a world in which we record our unique physical identities with such tiny traces (figure 10.15). Symbolically this delicate and diminutive interface between organism and environment is pivotal in our awareness of human manipulation of the world at our fingertips and may be more significant than we generally recognize in evolutionary debates about feedback between organism and environment. But it is not just hands and minds that are molded by the work they do and the objects they touch in life. Octogenarian Phylis Jackson, now making a name for herself in prehistoric studies after a lifetime of hands-on experience as a chiropodist, has demonstrated that Anglo-Saxon and Celtic populations have quite different feet,[42,43] not just at present but in samples back to the Neolithic (figure 10.15). Our ethnic and cultural signatures may indeed be in our feet and footprints as well as in our handwriting.

The French paleontologist-philosopher-priest Teilhard de Chardin, whose special interest and expertise focused on fossil mammals and human ancestors, noted among his voluminous writings that humans had probably developed the same in-

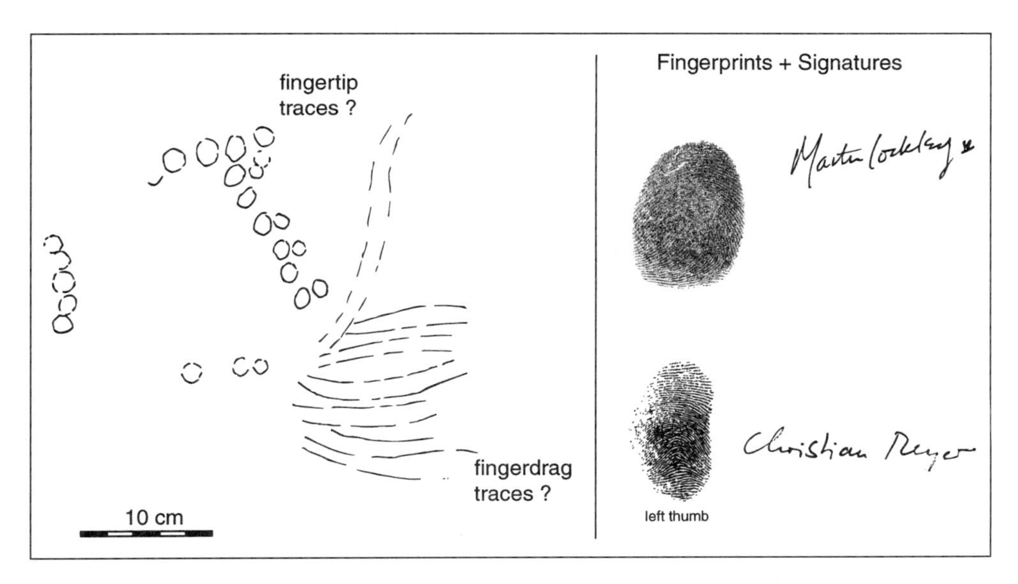

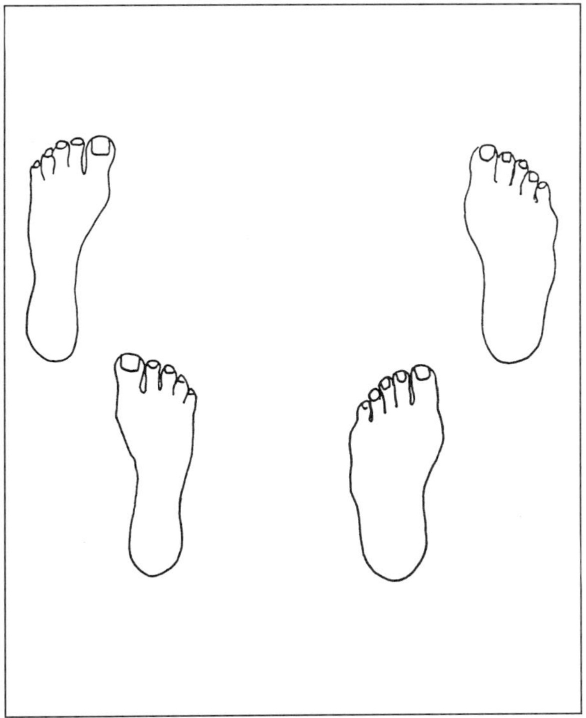

FIGURE 10.15. Fingerprints, the biggest vertebrate ichnology database in existence, are known as late Pleistocene trace fossils in cave floor mud[29,31] and serve to identify all individuals of the human species. The authors' fingerprints are given as a signature of authenticity on this work. Recent work suggests that we may be able to distinguish different regional or endemic groups using their footprints.[42] Narrow feet on the left are Celtic; broad feet on the right are Anglo-Saxon.

telligence and aesthetic senses that we have today by the time they began painting on cave walls and leaving footprints of their unshod feet on late Pleistocene cave floors. Regardless of whether or not we accept the label of trace fossils for these pictorial messages, most students of human evolution are happy to accept the psychological implications of this remarkable innovation in hominid evolution. With the birth and maturation of our species, *Homo sapiens,* came a new level of self-awareness and awareness of other creatures, mainly our closest vertebrate relatives, large mammals. By depicting them graphically, we were embarking on the path to ever more sophisticated forms of symbolic communication, which eventually culminated in language and writing as we know it today. We have already seen that, like fossil footprints, the cave paintings of mammoths and other extinct animals provide behavioral and biological clues to the ecology of the times.

Viewed from this broad perspective, all forms of modern pictorial representation and the skill we call writing evolved progressively from the symbolic language of pictures and pictographs created alongside our ancestors' footprints and handprints in the Pleistocene twilight only 30,000 years ago. It is these skills in symbolic written communication and the sharing of pictures of tracks that have allowed the hundreds of ichnologists and paleontologists cited in this volume to contribute the many details of Europe's track record that have made this book possible. In our sharing and exchanging of information on the movements, size, numbers; and habits of vertebrate animals, are we so different from our Paleolithic ancestors who drew on their cave walls to convey such similar messages?

BIBLIOGRAPHIC NOTES

1. Kurten, B. J. 1971.
2. Ellenberger, P. 1980.
3. Demathieu, G., L. Ginsburg, C. Guerin, and G. Truc. 1984.
4. Bessonat, C., H. R. Dughi, and F. Sirugue. 1969.
5. Guerin, C. and G. Demathieu. 1993.
6. Santamaria, R., G. Lopez, and M. L. Casanovas-Cladellas. 1989.
7. Weidmann, M. and M. Reichel. 1979.
8. Clercq, S. W. G. and H. R. H. Holst. 1971.
9. Panin, N. 1965.
10. Panin, N. and E. Avram. 1962.
11. Vialov, O. 1972.
12. Lockley, M. G., S-Y. Yang, M. Matsukawa, F. Fleming, and S-K. Lim. 1992.
13. Mangin, P. 1962.
14. Plaziat, J. C. 1964.
15. Raaf, J. F. M. de, C. Beets, and G. Kortenbout van der Sluijs. 1965.
16. Yang, S-Y., M. G. Lockley, R. Greben, B. R. Erikson, and S-Y. Lim. 1994.
17. Hernandez-Pacheco, F. 1929.

18. Dalla Vecchia, F. and M. Rustioni. 1996.
19. Lancis, C. and A. Estévez. 1992.
20. Renders, E. 1984.
21. Kordos, L. 1983.
22. Turner, A. and M. Anton. 1997.
23. Koenigswald, W. v., M. Walders, and M. Sander. 1995.
24. Casteret, N. 1948.
25. Bahn, P. G. and J. Vertut. 1988.
26. Chaloupka. G. 1993.
27. Chauvet, J-M., E. B. Deschamps, and C. Hillaire. 1996.
28. Bocchini, A. and M. Coltorti. 1978.
29. Bahn, P. G. 1986.
30. Thompson, W. I. 1981.
31. Pales, L. 1976.
32. Vallois, H. V. 1927.
33. Vallois, H. V. 1931.
34. Begouen, H. 1927.
35. Aldhouse-Green, S. H. R., A. W. R. Whittle, J. R. L. Allen, A. E. Caseldine, S. J. Culver, M. H. Day, J. Lundquist, and D. Upton. 1995.
36. Roberts, G., S. Gonzalez, and D. Huddart. 1996.
37. Quintana, J. 1993.
38. Noe-Nygard, N. 1995.
39. Price, N. 1995.
40. Barriere, C and M. Sueres. 1993.
41. Clottes, J. 1998.
42. Jackson, P. 1995.
43. Jackson, P. 1997.

APPENDIX

WHERE TO SEE AND FIND DINOSAUR TRACKS AND OTHER FOSSIL FOOTPRINTS IN EUROPE

As indicated in the preceding chapters, there are many dinosaur tracksites in Europe that may be visited by the serious tracker or enthusiastic paleontologist. The locations of these are often indicated in the literature cited in this book. Some are situated on private land; others are found in parks or other lands accessible to the public. Most if not all documented sites have been studied by professional paleontologists and geologists with the permission of land owners or local and regional administrative authorities.

It is important to stress that anyone wishing to visit fossil footprint sites should first establish who owns the land and what constraints and regulations apply to obtaining permission to visit such locations. We provide this appendix on where to visit tracks only for the purpose of encouraging students and other enthusiasts to learn more about tracks. For this reason we avoid providing information on sites that are on private land or in little-known locations where they are not well protected. We stress that collecting of tracks should be avoided at all costs, except during authorized scientific investigations. There are many reports of unauthorized fossil collecting that have resulted in considerable controversy and bad feeling. Anyone wishing to obtain information on tracks is encouraged to approach local museums and universities or the authors of scientific papers in order to find out what additional information might be available. In some cases replicas may be available through museum gift shops or catalogs.

For the purposes of brevity and clarity, the following list of sites is restricted mainly to well-known, publicly accessible indoor and outdoor sites, where the track enthusiast is likely to find useful interpretive information such as guidebooks, interpretive panels and displays, museum exhibits, and so forth. We have not given precise locations of any sites for the reasons cited above. Besides, these locations are given in the many papers cited in this book, and the serious tracker is encouraged to seek them out as part of the tracking adventure. It is also important, we believe, to try to appreciate dinosaur tracksites in their geologic and environmental context. This is particularly easy to appreciate in the case of sites such as Cabo Espichel in Portugal and Lavini di Marco in Italy, which are both very accessible, and Piz dal Diavel in Switzerland, which is not normally accessible. Such sites are found in spectacular scenic settings. In some cases we have deliberately described these sites in more detail. Cabo Espichel, for example, offers the chance for a great walking adventure.

Although we strongly advocate that visitors to any site display the utmost respect for the local resources by leaving them well alone, it is possible that the observant visitor may find something new. This is particularly true in the case of remote and frequently eroded sea coasts, mountainous regions, river beds, and areas where fresh rock is being exposed by road construction and other such activities. In such cases it is desirable to report such finds to local museums or universities.

The amateur enthusiast may help the discovery and documentation process by taking photographs and recording the precise location of finds for professionals and other local authorities. In such circumstances, if the discovery really is new, it is good to keep a record of communications (letters, E-mail, etc.) and it is reasonable to expect or even request that the professionals acknowledge the contribution of the discoverer in any reports or articles that are published. If the site is of minor significance or another example of a well-known track type, it may not be a high priority for study by experts, and the discoverer should not expect great acclaim. If, on the other hand, the site is highly significant, it may be appropriate to contact several experts. In such cases it is helpful to encourage constructive dialog between all potentially interested parties. Amateurs may even get involved in the scientific documentation process if circumstances permit their participation.

The following is a list of sites, by geographic region (countries) and by paleontologic age, where the enthusiastic tracker can see fossil footprints and find useful interpretive information and, in some cases, resident experts on vertebrate paleontology and ichnology. We are well aware of the fact that there are many museums all over Europe where small collections are housed, and it would be impossible to begin to list them all. We have therefore selected those sites, museums, and collections that we consider most important to the science of vertebrate ichnology, most relevant to the material outlined in this book, and most representative of the local geologic and paleontologic landscapes of the regions concerned. For each country we have attempted to summarize the extent of predinosaurian, dinosaurian, and postdinosaurian track assemblages, which, where represented, tend to fall naturally into certain geologic and geographic regions.

ENGLAND

Fossil footprints of Permian and Triassic age were discovered abundantly in various sandstone quarries in the English Midlands during the nineteenth century. Most of these quarries are now overgrown or filled in, and the chances of finding tracks are very poor. However, there are many museums, for example, those at Liverpool University (see chapter 3) and Birmingham University (see chapter 2) where specimens are housed. A complete listing of museums with Triassic footprints is given by Tresise and Sarjeant.[1] The Oxford University Museum houses some of the historically famous Permian tracks from Scotland (see chapter 2) and other famous paleontologic specimens. There are also significant collections housed in the British Museum of Natural History, including a trackway on display outside the museum (figure A.1). Tracks are also held in the Royal Scottish Museum, Edinburgh, and the Hunterian Museum at Glasgow University. Care should be taken in establishing whether these tracks are of iguanodontid affinity (British Museum and Royal Scottish Museum specimens) or of theropod affinity, as in the case of the recently reinterpreted Hunterian Museum specimen (see chapter 8).

Jurassic dinosaur tracks have been discovered periodically along the Yorkshire coast in the region of Whitby, especially where rocks have fallen after severe weather. This is a dangerous region of coastline, and casual searching for dinosaur tracks is not advised without a good knowledge of the coastline, tidal cycles, and potential dangers involved.

Dinosaur tracks have been reported and collected from a number of coastal rock outcroppings of latest Jurassic and Cretaceous age along the south coast of England from the Portland Bill-Purbeck area, near Dorchester, in the west to the Isle of Wight and the Sussex coast, and near Hastings in the east. New track discoveries may be made along the coast, especially after severe storms. Tracks of pterosaurs and various dinosaurs (iguanodontids, theropods, and ankylosaurs) are on display at the Dorchester Museum and the Isle of Wight Museum at Sandown. Some tracks from this region have found their way to the British Museum of Natural History.

FRANCE

Permian tracks are abundant at a number of regions (e.g., Lodève) in southern France. Georges Demathieu and Paul Ellenberger have been instrumental in col-

FIGURE A.1. Iguanodontid trackway from the Isle of Purbeck, on display outside the British Museum of Natural History in London (see chapter 8).

lecting and documenting these sites. These gentlemen have amassed collections of tracks and track replicas at the University of Dijon and the University of Montpellier, respectively. These are essentially research collections and are not normally accessible to the public.

The coast of the Vendée region produced some of the first Early Jurassic dinosaur tracks to be described in detail. A large slab was reconstructed for display at the Sables d'Olonne Museum, and a slab is also on display at the Aathal Museum in Switzerland (see chapter 5).

The lithographic limestones of Cerin, near Lyon, are famous for their fossils and now infamous for their tracks, claimed to be those of turtles and hopping dinosaurs (see chapter 7). These tracks are described in a guidebook,[2] and examples can be seen at the Natural History Museum in Lyons.

A replica of the important stegosaur-like trackway from Corgnac sur l'Isle, Province of Dordogne, is on display at the Dinosaur Museum in Esperaza, south of Carcassone, and the original tracksite also is open to the public at selected times (see chapter 5). The Esperaza museum is open to the public and worth a visit for the many dinosaur specimens on display. It is in the heart of some of the richest Upper Cretaceous dinosaur beds in Europe, yielding abundant titanosaurid dinosaurs and eggs. The site is not far from the Spanish frontier and the Fumanya titanosaurid dinosaur tracksite (see section on Spain).

The Saignon tracksite near Valcluse is one of the few Tertiary tracksites that is open to the public. The tracks include those of the Oligocene rhinoceros *Ronzotherium* (figure A.2), as well as various artiodactyls, carnivores, and birds (see chapter 10). The site is geologically situated in the Apt basin, which is the type section for the Aptian stage of the Lower Cretaceous.

GERMANY

As outlined in chapter 2, the red beds of the Rotliegende have yielded many Permian track assemblages. Among the best known are those from the Cornberg Sandstone and the Tambach Sandstone. Examples of the former can be seen in the converted monastery, part of which which is now the Cornberg Sandstone Museum, in Cornberg, which also includes exhibits of sandstone carvings and masonry. Cornberg tracks are also on display in the nearby Kreisheimatmuseum at Rotenburg-Fulda. The Natural History Museum at Gotha has many fine large slabs of track-bearing mud-cracked sandstone on display where the visitor may examine excellent examples of long trackways assigned to such ichnogenera as *Ichnotherium* and *Dimetropus* (see chapter 2). Both museums are open to the public and include reconstructions of Permian trackmakers and dioramas depicting the faunas and ancient environments they inhabited.

The Geiselthal Museum at Martin Luther University in Halle is also home to an extensive research collection of Permian fossil footprints, molds, and replicas

FIGURE A.2. Tracks from the Saignon tracksite near Valcluse, one of the few Tertiary tracksites that is open to the public. The tracks include those of an Oligocene rhinoceros (see chapter 10).

collected by Hartmut Haubold of that institution. Much of the track collection is not on display, but the museum contains diverse paleontologic exhibits and is open to the public.

Barkhausen, near Bad Essen, near Osnabrück, is an outdoor interpretive site where the visitor may examine brontosaur and theropod trackways on a steeply dipping bedding surface of Late Jurassic age. Replicas of the two track types are also on display at the Osnabrück Natural History Museum. Existing guidebooks are somewhat misleading in describing the sauropod trackway configurations and directions. See chapter 7 for our interpretations. The large megalosaurid tracks are easy to identify (figure A.3), and a guidebook is available locally.[3]

Münchehagen, near Hannover, is undoubtedly one of the best-developed dinosaur tracksites in Europe. It is essentially a paleontologic theme park, which attracts around 150,000 visitors every year. The site features a walk through time, illustrated with large models of vertebrates, beginning in the Paleozoic with large Devonian fish and extending through the Mesozoic "Age of Dinosaurs" to the "Age of Mammals." The tracks (see chapter 8), which are Early Cretaceous in age, are covered to protect them from weathering and are incorporated into the theme park exhibit as the central point of the walk through time trail (figure A.4).

FIGURE A.3. The Barkhausen tracksite reveals Late Jurassic tracks of sauropods and theropods (megalosaurs) (see chapter 7). Scale=2 m.

FIGURE A.4. The Münchehagen tracksite, built around Early Cretaceous sauropod and ornithopod tracks, is one of the largest dinosaur attractions in Europe (see chapter 8). *Photograph courtesy of Detlev Thies.*

ITALY

Triassic dinosaur tracks can be located in the Parco delle Dolomite Friulane (Pordonone). At this writing a walking route to tracksite destinations, interpretive signs, and an exhibit within the park at Claut are under construction. The Dolomites are of course a classic geologic destination. Triassic tracks can also be seen in the school at Dogna (Udine).

One of the most beautiful and historically interesting dinosaur tracksites in Italy is the Lavini di Marco site, near Rovereto, near Trento in northern Italy. The tracks at this location are Lower Jurassic in age and include the oldest sauropod trackways (figure A.5) in Europe, as well as various theropod tracks. The site is situated in a region designated as an alpine floral reserve, and is also famous as a World War II battlefield. The museum at nearby Rovereto includes outdoor dinosaur sculptures pertaining to the Lavini de Marco site and a state-of-the-art computer collections management system where information on the site can easily be accessed. Dinosaur sculptures are also on display at the railway station. Information on the Lavini di Marco site can also be obtained from the Natural Science Museum in Trento.

Cretaceous tracks also can be seen in Italy, including a block on display at the Natural Science Museum in Faenza (Ravenna). This site was not mentioned in chapters 8 and 9. However, one may see an important collection of representative casts from the Croatian sites (Istria) in the Museo Paleontologico Cittidino in Monfalcone (Gorizia). These sites are hard to visit, as the experience of Fabio Dalla Vecchia has indicated (see chapter 9), so at the present time we cannot in good conscience recommend visits to this otherwise beautiful area.

PORTUGAL

There are five dinosaur tracksites in Portugal that have been designated as national monuments. The oldest of these, the Middle Jurassic Galinha quarry site, near Fatima (see chapter 6), although reported only in 1994, is already famous for revealing the world's longest sauropod trackways (at least 147 m). It is also the largest and most accessible Middle Jurassic tracksite in Europe, indeed in the entire world. The site has a small visitor's center and a walking trail furnished with interpretive signs. The site is not far from Fatima, a popular destination for religious pilgrimages.

Cabo Espichel is undoubtedly one of the most beautiful dinosaur tracksite regions in Europe. There are several tracksites on the Cabo Espichel peninsula, near Sesimbra, south of Lisbon. The easiest to find are on the cape (cabo) itself within walking distance of the monastery. The oldest tracksites can be seen looking south from the peninsula. These represent sauropod tracks from the lower part of the Upper Jurassic sequence. Large theropod tracks can be found beside the path leading

FIGURE A.5. The Lavini di Marco tracksite is a famous historical locality (see chapter 5).

down to the bay (Praia de Cavalo) south of the peninsula. These show alternating long and short steps indicative of a "limping dinosaur"[4] (see chapter 7). There is a path down to this bay, but it is only for the adventurous and experienced hiker. In this entire region, winds and falling rock debris are always a potential hazard.

To the north of the monastery in Lagosterios Bay one can find more sauropod tracks, situated on the cliffs below the small shrine, famous for commemorating the local legend of the mule tracks (Pegadas de Mula; see chapter 7). These cannot be seen from above and must be approached by taking the old road to the north and walking down to the bay. On the way down, one sees large bedding planes of coarse sandstone, which lie close to the Jurassic-Cretaceous boundary. These exhibit extensive trampling or dinoturbation (figure A.6), which have yet to be described in detail. Turning south to the southern end of the beach, one reaches the Upper Jurassic limestone sequences that contain many layers with sauropod tracks. The site is best approached at low tide, and it is difficult and dangerous to climb on these surfaces if one is inexperienced. However, it is easy to see the large trackways closest to the beach.

These same trackways also can be seen from the clifftops north of Lagosterios Bay (figure A.7), also reached by a branch of the same road that skirts the east side of the bay. Here the strata are Early Cretaceous in age and slightly less steeply inclined. One may find the circular tracks of a large biped, named *Neosauripus*,[5] an

FIGURE A.6. Trampled beds at Lagosterios Bay (Cabo Espichel) from near the Jurassic-Cretaceous boundary reveal theropod tracks.

FIGURE A.7. View of the Cabo Espichel tracksite from the north side of Lagosterios Bay reveals multiple track-bearing layers in Upper Jurassic limestone (see chapter 7).

ichnotaxon that is no longer considered valid. Despite the dubious taxonomic status of this trackway, it is very photogenic (figure A.8).

At Cabo Mondego, near Figueria da Foz, the coastal outcroppings are considered the global type section for part of the Middle Jurassic and contain an unparalleled sequence of ammonite-rich marine deposits of Aalenian to Callovian (Middle Jurassic) age. Ammonites are easy to find and photograph but should not be collected. Of interest to the dinosaur tracker is the overlying Upper Jurassic sequence from which the first so-called megalosaurid tracks were reported (see chapter 7). Several surfaces can be seen that still exhibit large theropod tracks. The region is historically interesting as one of the first in the world from which Late Jurassic tracks were reported.

The Carenque site, in the greater Lisbon area (see chapter 9), is a fine example of a site to which enormous public resources have been committed. At this writing the first phase in establishing an outdoor park had been completed. During this process the tracks were covered so as to protect them while two tunnels were excavated below. Visible manifestations of this work are seen in the tunnels themselves, which have entrances and exits embellished with stylistic dinosaur heads and tails. The second phase of creating the park will involve uncovering the tracks and creating a visitor's center and interpretive trails. The visitor should not expect

FIGURE A.8. The trackway of a large bipedal dinosaur of probable theropod affinity from the Cretaceous of the north side of Lagosterios Bay.

to see the main trackway until this phase of the project is completed, although the surrounding geology of the old quarry is visible.

SPAIN

The Province of La Rioja situated some 200 km north-northeast of Madrid is one of the richest dinosaur tracksite regions in Europe. The largest concentration of accessible sites is in the eastern part of the province near Arnedo. A glossy, well-designed guidebook[6] lists the most accessible sites as Prejano, Munilla, Enciso, El Villar-Poyales, Navalsaz, Ambas Aguas, Muro de Aguas, Cornago, Igea, Gravelos, Treviljano, and Cabezón de Caméros. This guidebook is highly recommended because many of the sites have two or more names. For example, possibly the best known of these localities, and the most accessible, is the Cornago site, also known as Los Cayos, where a part of the extensive track-bearing outcropping is covered by a permanent shelter (figure A.9). The tracks at this site are mainly those of large Lower Cretaceous theropods.

The visitor to this region will be greeted by prominent road signs labeled "Ruta de icnitas," clearly indicating arrival in an ichnologically rich area. Most of the dinosaur tracks in this region are either Late Jurassic or Early Cretaceous in age. In

FIGURE A.9. Cretaceous theropod dinosaur tracks from the Cornago site, also known as Los Cayos, La Rioja Province, Spain (see chapter 8).

addition to sites with abundant theropod tracks such as Cornago (Los Cayos) and Enciso (Valdecevillo), distinctive large ornithopod (iguanodontid) tracks are found at Prejano (Valdete and Magdalena outcroppings). As mentioned in the text, sauropod, pterosaur, and turtle tracks are also known from this region. Another important site mentioned in the text (see chapter 8) is at Cabezón de Caméros.

Like the trackway from the Magdalena outcropping, which represents a quadrupedal ornithopod, a trackway from the Sierra de Regumel site, near Burgos, in neighboring Castilla y Leon Province, just to the west of La Rioja, also reveals one of the few trackways of quadrupedal ornithopods known from Europe. Despite the abundance of dinosaur tracks in these two provinces, which, geologically, represent the Cameros basin, the region is not a megatracksite in the strict sense. In fact, it is hard to determine the exact age of all the different tracksites and know where they fit in a chronologic scheme (see chapter 8).

SWITZERLAND

Tracks are on display at the Aathal Dinosaur Museum, which was the first museum in Europe to host the Tracking Dinosaurs Exhibit (see preface). These tracks include the fine specimen from Vendée (see figure 5.3) and many other track repli-

FIGURE A.10. The Lommiswil dinosaur tracksite, near Solothurn, Switzerland, was the first discovered in the Upper Jurassic. It is also one of the most accessible, educational, and visually most spectacular.

FIGURE A.11. The Piz dal Diavel site, revealing Triassic prosauropod and theropod tracks, is one of the most spectacular high-altitude locations in Europe. Photograph courtesy of H. Turner.

cas from around the world. The museum is worth visiting for the many dinosaurs and other fossil vertebrates on display.

The Lommiswil Jurassic dinosaur tracksite, near Solothurn (see chapter 7), is the largest, most accessible, and most clearly visible dinosaur tracksite in Switzerland (figure A.10). It is easy to pick out large dinosaur tracks and trackways on a steep surface in the old quarry. The site is equipped with interpretive signs, and casts of representative tracks are on display in the nearby Solothurn Naturmuseum.

The Moutier Jurassic dinosaur tracksite, or "dinosaur disco" (see chapter 7), is close to the town of Moutier in the Jura mountains, an area that is popular for hiking and rock climbing. Work on this site has been featured in dinosaur documentaries, and shows that close examination of the tracks requires some basic mountaineering experience and the use of ropes and safety equipment. Tracks are not easy to see from some angles.

The Piz dal Diavel site is one of the most spectacular high-altitude locations in all of Europe. Unfortunately, the site is not accessible to the general public and is hard to reach without mountaineering experience. A photograph (figure A.11) is included here to show the site's fine geologic and scenic qualities. The large tracks are thought to be those of prosauropods (see chapter 4).

WALES

Dinosaur tracks have been collected from at least two locations in Wales and placed in the collections of the National Museum (Geology Department) in Cardiff. Tracks from the nearby coastal exposures at Barry include abundant examples of Late Triassic *Grallator,* which were excavated and placed on exhibit as part of the permanent display.

OTHER AREAS

As noted elsewhere in this book, it was not possible to include detailed accounts of tracksites outside Western Europe, though sites in Poland, Hungary, and Romania were mentioned briefly. For paleontonlogists and track enthusiasts who may wish to explore in these regions, the Ipolytarnóc site[7] in northern Hungary mentioned in chapter 10 is open to the public.

BIBLIOGRAPHIC NOTES

1. Tresise, G. and W. A. S. Sarjeant. 1997.
2. Bernier, P. 1985.
3. Freise, H. 1988.
4. Lockley, M. G., A. P. Hunt, J. J. Moratalla, and M. Matsukawa. 1994.
5. Santos V. F. dos, M. G. Lockley, J. J. Moratalla, and A. Galopim. 1992.
6. Moratalla, J. J., J. L. Sanz, I. Melero, and S. Jimenez. 1988.
7. Kordos, L. 1983.

REFERENCES

Abel, O. 1935. *Vorzeitliche Lebenspuren.* Jena: Gustav Fischer.

Aguirrezabala, L. M., J. A. Torres, and L. I. Viera. 1985. El Weald de Igea (Cameros-La Rioja). Sedimentología, bioestratigrafia y paleoichnología de grandes reptiles (dinosaurios). *Munibe. Sociedad de Ciencias Naturales Aranzadi, San Sebastian* 37: 111.

Aguirrezabala, L. M. and L. I. Viera. 1980. Icnitas de dinosaurios en Bretun (Soria). *Munibe. Sociedad de Ciencias Naturales Aranzadi, San Sebastian* 32: 257–79.

Ahlberg, A. and M. Siverson. 1991. Lower Jurassic dinosaur footprints in Helsinborg, southern Sweden. *Geologiska Föreningens I Stockholm Förhandlingar* 113: 339–40.

Aldhouse-Green, S. H. R., A. W. R. Whittle, J. R. L. Allen, A. E. Caseldine, S. J. Culver, M. H. Day, J. Lundquist, and D. Upton. 1995. Prehistoric human footprints from the Severn Estuary at Uskmouth and Magor Pill, Gwent, Wales. *Archaeologia Cambrensis* 141: 14–55.

Alexander, R. McN. 1976. Estimates of speeds of dinosaurs. *Nature* 261: 129–30.

Alonso, A. and J. R. Mas. 1993. Control tectónico e influence del eustatismo en la sedimentación del Cretácio inferior de la cuenca de Los Cameros. *Cuadernos de Geología Ibérica* 17: 285–310.

Anderton, S. 1997. *Footprints in the Sands of Time. France.* Stow on the Wold, England: Centralhaven.

Andrews, J. E. and J. D. Hudson. 1984. First Jurassic dinosaur footprint from Scotland. *Scottish Journal of Geology* 20: 129–34.

Andrieux, C. 1974. Premiers résultats sur l'étude du climat de salle des peintures de la Galérie Clastres (Niaux, Ariège). *Annales de Spéléologie* 29: 3–25.

Antunes, M. T. 1976. Dinosaurios eocretacios de Lagosterios. *Ciéncias da Terra* 1: 1–35.

Arkell, W. J. 1933. *The Jurassic System in Great Britain.* Oxford, England: Oxford University Press.

Avanzini, M. 1995. Impronte di sauropodi nel Giurassico Inferiore del Becco di Filadonna (Piattaforma di Trento—Italia settentrionale) Studi Trentini di Scienze naturali. *Acta Geologica* 72: 117–22.

Avanzini, M., K. van den Dreissche, and E. Keppens. 1997. A dinosaur tracksite in an early Liassic tidal flat in northern Italy: paleoenvironmental reconstruction from the sedimentology and geochemistry. *Palaois* 12: 538–51.

Avanzini, M. and G. Leonardi, G. 1993. I dinosauri dei Lavini di Marco ed I grand vertebrati fossili del Trentino Alto Adige. *Natura Alpina* 44: 1–16.

Bachofen-Echt, A. 1925. Die Entdeckung von Iguanodon-fährten im Neokom der Insel Brioni. *Sitzungsberichte Akadamie der Wissenschaften Mathematik Naturwissenschaften Kl* February 12, 1925.

Bahn, P. G. 1986. No sex, please, we're Aurignacians. *Rock Art Research* 3: 99–120.

Bahn, P. G. and J. Vertut. 1988. *Images of the Ice Age.* New York: Facts on File.

Bakker, R. T. 1986. *The Dinosaur Heresies.* New York: William Morrow.

Bakker, R. T. 1996. The real Jurassic Park: dinosaurs and habitats at Como Bluff, Wyoming. In: M. Morales, ed. *The Continental Jurassic (Symposium Volume),* pp. 35–49. Flagstaff, AZ: Museum of Northern Arizona.

Barriere, C. and M. Sueres. 1993. Les mains de Gargas. *Les Dossiers d'Archéologie* 178: 46–55.

Batory, D. and W. A. S. Sarjeant. 1989. Sussex iguanodon footprints and the writing of *The Lost World.* In: D. D. Gillette and M. G. Lockley, eds. *Dinosaur Tracks and Traces,* pp. 13–18. New York: Cambridge University Press.

Beaumont, G. and G. Demathieu. 1980. Remarques sur les extremités antérieures des sauropodes (reptiles, saurischiens). *Compte Rendu des Séances de la Société de Physique et d'Histoire Naturelle de Genève* 15: 191–8.

Beckles, S. H. 1851. On supposed casts of footprints in the Wealden. *Quarterly Journal of the Geological Society of London* 7: 117.

Beckles, S. H. 1852. On the Ornithoidichnites of the Wealden. *Quarterly Journal of the Geological Society of London* 8: 396–7.

Beckles, S. H. 1854. On the Ornithoidichnites of the Wealden. *Quarterly Journal of the Geological Society of London* 10: 456–64.

Begouen, C. M. 1927. Les empreintes de pieds préhistoriques. Institut International d'Anthropologie, III session, Amsterdam, pp. 323–7.

Bennett, S. C. 1993. Reinterpretation of problematic tracks at Clayton Lake State Park, New Mexico: not one pterosaur, but several crocodiles. *Ichnos* 2: 37–42.

Bennett, S. C. 1997. Terrestrial locomotion of pterosaurs: a reconstruction based on *Pteraichnus* trackways. *Journal of Vertebrate Paleontology* 17: 104–13.

Benton, M. J., D. M. Martill, and M. A. Taylor. 1995. The first Lower Jurassic dinosaur from Scotland: a ceratosaur theropod from Skye. *Scottish Journal of Geology* 31: 177–82.

Bernier, P. 1985. *Une lagune tropicale au temps des dinosaures.* Lyon, France: Centre National de la Recherche Scientifique, Musée de Lyon.

Bernier, P., G. Barale, J-P. Bourseau, E. Buffetaut, G. Demathieu, C. Gaillard, and J-C. Gall. 1982. Trace nouvelle de locomotion de chélonien et figures d'émersion associées dans les calcaires lithographiques de Cerin (Kimmeridgien superieur, Ain, France). *Geobios* 15: 447--67.

Bernier, P., G. Barale, J-P. Bourseau, E. Buffetaut, G. Demathieu, C. Gaillard, J-C. Gall, and S. Wenz. 1984. Découverte des pistes de dinosaures sauteurs dans les calcaires lithographiques de Cerin (Kimmeridgien superieur, Ain, France)—implications paléoecologiques. *Géobios Mémoire Spécial* 8: 177–85.

Bessonat, C., H. R. Dughi, and F. Sirugue. 1969. Un important gisement d'empreintes de pas dans le Paléogène du Bassin d'Apt-Forcalquier. *Comptes Rendus Hebdomadaire, Académie des Sciences, Paris.*

Bird, R. T. 1944. Did Brontosaurus ever walk on land? *Natural History* 53: 61.

Bocchini, A. and M. Coltorti. 1978. Unghiate ed impronte di Ursus spelaeus nella Grotta del Fiume nella Gola di Frasassi (Ancona). *Rassegna Speleologia Italiana* Memoria XII: 138–41.

Briggs, D. E. G., W. D. I. Rolfe, and J. Brannan. 1979. A giant myriapod trail from the Namurian of Arran, Scotland. *Palaeontology* 22: 273–91.

Broderick, H. 1909. Note on footprint casts from the Inferior Oolite, near Whitby, Yorkshire. *Proceedings of the Liverpool Geological Society of London* 10: 327–35.

Buckland, W. 1824. Notice on the Megalosaurus, or great fossil lizard of Stonesfield. *Transactions of the Geological Society of London* 1: 390–6.

Buckland, W. 1858. *The Bridgewater Treatises on the Power, Wisdom and Goodness of God, as Manifest in the Creation. Treatises VI, Geology and Mineralogy Considered with Reference to Natural Theology,* 3rd ed. London: George Routledge & Co.

Calkin, B. J. 1968. *Ancient Purbeck: An Account of the Geology of the Isle of Purbeck and Its Early Inhabitants.* Dorchester, England: Friaray Press.

Carpenter, K. 1984. Skeletal reconstructions and life restorations of *Sauropelta* and *Paleoscincus* (Ankylosauria: Nodosauridae) from the Cretaceous of North America. *Canadian Journal of Earth Sciences* 21: 1491–8.

Casamiquela, R. M. and A. Fasola. 1968. Sobre pisadas de dinosaurios del Cretacico Inferior de Colchagua (Chile). *Publicaciones Departmento de Geología, Chile Universidad* 30: 1–24.

Casanovas, M., A. Fernandez, F. Perez-Lorente, and J. V. Santafe-Llopis. 1997. Sauropod trackways from site El Sobaquillo (Munilla, La Rioja, Spain) indicate amble walking. *Ichnos* 5: 101–7.

Casanovas Cladellas, M. L., A. Fernandez-Ortega, F. Perez-Lorente, and J-V. Santafe-Llopis. 1991. Ichnitas de dinosaurios en Valdevajes (La Rioja). *Revista Española Paleontología* 7: 97–9.

Casanovas Cladellas, M. L. and J-V. Santafe Llopis. 1971. Icnitas de reptiles mesozoicos en la provincia de Logrono. *Acta Geologica Hispania* 6: 139–42.

Casanovas Cladellas, M. L. and J-V. Santafe Llopis. 1974. Dos nuevos yacimentos de ichnitas de dinosaurios. *Acta Geologica Hispania* 9: 88--91.

Caster, K. E. 1938. A restudy of the tracks of *Paramphibius. Journal of Paleontology* 12: 3–60.

Casteret, N. 1948. The footprints of prehistoric man: vivid new evidence of our ancestors of fifteen thousand years ago. *Illustrated London News,* pp. 110, 408–9.

Chaloupka, G. 1993. *Journey in Time: The World's Longest Continuing Art Tradition: The 50,000-Year-Old Story of the Australia Aboriginal Rock Art of Arnhem Land.* Chatswood, Australia: Reed/William Heinemann.

Charig, A. and B. Newman. 1962. Footprints in the Purbeck. *New Scientist* 14: 234–5.

Chauvet, J-M., E. B. Deschamps, and C. Hillaire. 1996. *Dawn of Art: The Chauvet Cave: The Oldest Known Paintings in the World.* New York: Harry N. Abrams Publishers.

Chure, D. and J. McIntosh. 1989. *A Bibliography of the Dinosauria: Exclusive of the Aves, 1670–1986.* Grand Junction, CO: Museum of Western Colorado.

Clack, J. A. 1997. Devonian tetrapod trackways and trackmakers: a review of the fossils and footprints. *Palaeogeography Paleoclimatology Paleoecology* 130: 227–50.

Clark, N. D. L., J. D. Boyd, R. J. Dixon, and D. A. Ross. 1995. The first Middle Jurassic dinosaur from Scotland: a cetiosaurid? (Sauropoda) from the Bathonian of the Isle of Skye. *Scottish Journal of Geology* 31: 171–6.

Clercq, S. W. G. and H. R. H. Holst. 1971. Footprints of birds and sedimentary structures from the subalpine Molasse near Fluhli (Canton of Luzern). *Eclogae Geologicae Helvetiae* 64: 63–9.

Clottes, J. 1998. The "three Cs": fresh avenues towards European palaeolithic art. In: C. Chippendale and P. S. C. Tacun, eds. *The Archeology of Rock Art.* New York: Cambridge University Press.

Coates, M. I. and J. A. Clack. 1995. Romer's gap: tetrapod origins and terrestriality. *Bulletin du Musée National d'Histoire Naturelle, Paris* (séries 4) 17: 373–88.

Colbert, E. H. and D. Merrilees. 1967. Cretaceous dinosaur footprints from Western Australia. *Journal of the Royal Society of Western Australia* 50: 21–5.

Conti, M. A., G. Leonardi, N. Mariotti, and U. Nicosia. 1997. Reevaluation of *Pachypes dolomiticus Leonardi et alii,* 1975, a Late Permian Pareiosaur footprint. In: H. Haubold, ed. *Workshop on Ichnofacies and Ichnotaxonomy of the Terrestrial Permian,* p. 13. Halle, Germany: Martin Luther University.

Courel, L. and G. Demathieu. 1995. Tentative stratigraphic correlation using ichnological data from continental sandstone series and marine faunas in the Middle Triassic of Europe. *Albertiana* 15: 83–91.

Cousin, R., G. Breton, R. Fournier, and J-P. Watte. 1994. Dinosaur egg laying and nesting in France. In: K. Carpenter, K. Hirsch, and J. Horner, eds. *Dinosaur Eggs and Babies,* pp. 56–74. New York: Cambridge University Press.

Dalla Vecchia, F. M. 1994. Jurassic and Cretaceous sauropod evidence in the Mesozoic carbonate platforms of the southern Alps and Dinards. *Gaia: Revista de Geociencias,* Museu Nacional de Historia Natural, Lisbon, Portugal 10: 65–73.

Dalla Vecchia, F. M. 1996. Archosaurian trackways in the Upper Carnian of Dogna Valley (Udine, Fruili, NE Italy). *Natura Nascosta* 12: 5–17.

Dalla Vecchia, F. M. In press. Theropod tracks in the Cretaceous Adriatic-Dinaric carbonate platform (Italy and Croatia). *Gaia: Revista de Geociencias,* Museu Nacional de Historia Natural, Lisbon, Portugal.

Dalla Vecchia, F. M. and M. Rustioni. 1996. Mammalian trackways in the Conglomerato di Osoppo (Udine, NE Italy) and their contribution to its age determination. *Memorie Science Geologica (Padova)* 48: 221--32.

Dalla Vecchia, F. M., A. Tarlao, and G. Tunis. 1993. Theropod (Reptilia, Dinosauria) footprints in the Albian (Lower Cretaceous) of the Quieto/Mirna river mouth (NW Istria, Croatia) and dinosaur population of Istrian region during the Cretaceous. *Memorie Science Geologica (Padova)* 45: 139–48.

Day, M. H. 1991. Origine(s) de la bipédie chez les hominides. *Cahier de Paléoanthropologie,* pp. 199–213. Paris: Editions du CRNS.

Delair, J. B. 1963. Notes on Purbeck fossils with descriptions of two hitherto unknown forms from Dorset. *Proceedings of the Dorset Natural History and Archaeological Society* 84: 92–100.

Delair, J. B. 1982. Multiple dinosaur trackways from the Isle of Purbeck. *Proceedings of the Dorset Natural History and Archaeological Society (for 1980)* 102: 65–7.

Delair, J. B. 1989. A history of dinosaur footprint discoveries in the British Wealden. In: D. D. Gillette and M. G. Lockley, eds. *Dinosaur Tracks and Traces,* pp. 19–25. New York: Cambridge University Press.

Delair, J. B. and P. A. Brown. 1975. Worbarrow Bay footprints. *Proceedings of the Dorset Natural History and Archaeological Society (for 1974)* 96: 14–6.

Delair, J. B. and A. B. Lander. 1973. A short history of the discovery of reptilian footprints in the Purbeck Beds of Dorset, with notes on their stratigraphical distribution. *Proceedings of the Dorset Natural History and Archaeological Society (for 1972)* 94: 17–20.

Delair, J. B. and W. A. S. Sarjeant. 1985. History and bibliography of the study of fossil vertebrate footprints in the British Isles: supplement 1973–1983. *Palaeogeography Palaeoclimatology Palaeoecology* 49: 123–60.

Demathieu, G. 1970. Contribution de l'ichnologie à la connaisance de l'évolution des reptiles pendant la période triasique. *Comptes Rendus Académie Sommaire Société Géologique France* fascicle 4: 122–3.

Demathieu, G. 1989. Appearance of the first dinosaur tracks in the French Middle Triassic, and their probable significance. In: D. D. Gillette and M. G. Lockley, eds. *Dinosaur Tracks and Traces,* pp. 201–7. New York: Cambridge University Press.

Demathieu, G. 1990. Problems in the discrimination of tridactyl dinosaur footprints, exemplified by the Hettangian trackways, the Causses, France. *Ichnos* 1: 97–110.

Demathieu, G. 1993. Empreintes de pas de dinosaures dans le Causses (France). *Zubia* 5: 229–52.

Demathieu, G., L. Ginsburg, C. Guerin, and G. Truc. 1984. Etude paléontologique, ichnologique et paléoecologique du gisement Oligocène de Saignon (Bassin d'Apt, Valcluse). *Bulletin du Musée National d'Histoire Naturelle* 6: 153–83.

Demathieu, G. and H. Haubold. 1978. Du problème de l'origine des dinosauriens d'après les données de l'ichnologie du Trias. *Geobios* 11: 409–12.

Demathieu, G. and H. Haubold. 1982. Reptilfährten aus dem Mittleren Buntsandstein von Hessen (BRD). *Hallesches Jahrbuch Geowissenschaften* 7: 97–110.

Demathieu, G. and H. W. Oosterink. 1983. Die Wirbeltier-Ichnofauna aus dem unteren Muschelkalk von Winterswijk (Die Reptilienfährten aus der Mitteltrias der Niederlande). *Staringia (Nederlandse Geologische Vereniging)* 7: 52.

Demathieu, G. and J. Sciau. 1995. L'ichnofaune Hettangienne d'archosauriens de Sauclières, Aveyron, France. *Bulletin de la Société d'Histoire Naturelle Autun* 151: 5–46.

Demathieu, G. and M. Weidmann. 1982. Les empreintes des pas des reptiles dans le Trias du Vieux Emosson (Finhaut, Valais, Suisse). *Eclogae Geologicae Helvetiae* 75: 721–57.

Desnoyers, J. 1859. Note sur les empreintes de pas d'animaux dans le gypse des environs de Paris, particulièrement de la vallée de Monmorency. *Comptes Rendus Hebdomadaire Séance Académie des Sciences Paris* 16: 936–44.

Dodson, P. 1990. Counting dinosaurs: how many kinds were there? *Proceedings of the National Academy of Sciences* 87: 7608–12.

Dollo, L. 1905. Les allures des Iguanodon, d'après les empreintes des pieds et de la queue. *Bulletin Scientifique de la France et de la Belgique* (series 50) 40: 1–12.

Duncan, H. 1831. An account of the tracks and footprints of animals found impressed on sandstone in the quarry of Corncockle Muir. *Transactions of the Royal Society of Edinburgh* 11: 194–209.

Edwards, M. B., E. Rosalind, and E. H. Colbert. 1978. Carnosaur footprints in the Lower Cretaceous of eastern Spitzbergen. *Journal of Paleontology* 52: 940–1.

Ellenberger, P. 1972. Contribution à la classification des pistes de vertèbres du Trias: les types du Stormberg d'Afrique du Sud (part 1).In: *Palaeovertebrata, Mémoire Extraordinaire,* pp. 1–152. Montpellier, France: Laboratoire de Paléontologie des Vertèbres.

Ellenberger, P. 1974. Contribution à la classification des pistes de vertèbres du Trias: les types du Stormberg d'Afrique du Sud (part 2). In: *Palaeovertebrata, Mémoire Extraordinaire,* pp. 1–155. Montpellier, France: Laboratoire de Paléontologie des Vertèbres.

Ellenberger, P. 1976. Une piste avec traces de soies epaisses dans le Trias inférieur à moyen de Lodève (Herault, France): *Cynodontipus polythrix nov. gen., nov. sp.* les Cynodontes en France. *Geobios* 9: 769–87.

Ellenberger, P. 1980. Sur les empreintes de pas de gros mammifères de l'Eocène supérieur de Garrigues Ste Eulalie (Gard). *Paleovertebrata, Montpellier, Mémoire Jubilaire R. Lavocat* 37–77.

Ellenberger, V. 1953. *La Fin tragique des bushmen.* Paris: Amiot Dumont.

Ensom, P. 1987. Dinosaur tracks in Dorset. *Geology Today* 3: 182–3.

Ensom, P. 1988. Excavations at Sunnydown Farm. Langton Matravers, Dorset: amphibians discovered in the Purbeck Limestone Formation. *Proceedings of the Dorset Natural History and Archaeological Society* 109: 148–50.

Feist, M., R. D. Lake, C. Wood. 1995. Charophyte biostratigraphy of the Purbeck and Wealden of southern England. *Palaeontology* 38: 407––42.

Foster, J. R. and M. G. Lockley. 1997. Probable crocodilian tracks and traces from the Morrison Formation (Upper Jurassic) of eastern Utah. *Ichnos* 5: 121–9.

Foster, J. R., M. G. Lockley, and J. Brockett. 1999. Problematic vertebrate (turtle) tracks from the Morrison Formation of eastern Utah. In: D. D. Gillette, ed. *Utah Geological Survey,* pp. 185–191. Salt Lake City, UT.

Friese, H. 1988. Die Dinosaurierfährten von Barkhausen im Wiehengebirge. *Veröffentlichungen des landkreises Osnabrück Heft* 1. 36p.

Fuentes Vidarte, C. 1996. Primeras huellas de aves en el Weald de Soria (España). Nuevo ichnogenera, *Archaornithipes* y nueva ichnoespecie *A. meiiidei. Estudios Geológicos* 52: 63–75.

Fuglewicz, R., T. Ptaszynski, and K. Rdzanek. 1990. Lower Triassic footprints from the Swietokrzyskie (Holy Cross) Mountains, Poland. *Acta Paleontol Polonica* 35: 109–64.

Furrer, H. 1993. *Entdeckung und Untersuchung der Dinosaurierfährten im Nationalpark.* Zernez, Switzerland: Cratschla, Ediziuns Specials, 1.

Gabouniya, L. K. 1951. Dinosaur footprints from the Lower Cretaceous of Georgia. *Comptes Rendu Académie des Sciences de l'URSS* 81: 917--9.

Gabouniya, L. K. and V. Kurbatov. 1982. Jurassic dinosaur tracks in the south of Central Asia. In: T. N. Bogdanova, L. I. Khozatsky, and A. A. Istchenko, eds. 1988. *Fossil Traces of Vital Activity and Dynamics of the Environment in Ancient Biotopes. Transactions of the XXX Session of All Union Paleontological Society and VII Session of Ukranian Pleontological Society,* pp. 44–57. Kiev: Naukova Dumka.

Galopim, A. C. 1992. La segunda extinction de los dinosaurios? O el caso de la pista de ichnitas fossiles de Carenque (Sintra Portugal). *Gaia: Revista de Geociencias,* Museu Nacional de Historia Natural, Lisbon, Portugal 4: 43–5.

Galopim, A. C. 1994. *Dinossaurios e a Batalha de Carenque.* Lisbon: Editorial Noticias.

Gand, G. 1976a. *Coelurosaurichnus palissyi. Bulletin de la Société d'Histoire Naturelle Autun* 79: 11–4.

Gand, G. 1976b. *Coelurosaurichnus sabniensis. Bulletin de la Société d'Histoire Naturelle Autun* 79: 16–22.

Garcia Ramos, J. C. and M. Valenzuela. 1977a. Huellas de pisada de vertebrados (dinosaurios y otros) en el Jurassic Superior de Asturias. *Estudios Geológicos* 33: 207–14.

Garcia Ramos, J. C. and M. Valenzuela. 1977b. Estudio e interpretación de la ichnofauna (vertebrados e invertebrados) en el Jurassic de la costa Asturiana. *Cuadernos de Geología* 10: 13–22 (for 1979, published in 1981).

Gatesy, S. M. and K. M. Middleton. 1998. Reconstructing theropod foot function using 3-D computer-animated track simulation. *Journal of Vertebrate Paleontology* 18: 45A.

Gierlinski, G. 1991. New dinosaur ichnotaxa from the Early Jurassic of the Holy Cross Mountains, Poland. *Palaeogeography Paleoclimatology Paleoecology* 85: 137–48.

Gierlinski, G. 1995. Thyreophoran affinity of *Otozoum* tracks. *Przeglad Geologiczny* 43: 123–5.

Gierlinski, G. 1997. Sauropod tracks in the Early Jurassic of Poland. *Acta Palaeontol Polonica* 42: 533–8.

Gierlinski, G. 1999. Tracks of a large thyreophoran dinosaur from the Early Jurassic of Poland. *Acta Paleontol Polonica* 44:231–234.

Gierlinski, G. and A. Ahlberg. 1994. Late Triassic and Early Jurassic dinosaur footprints in the Hogänäs Formation of southern Sweden. *Ichnos* 3: 99–105.

Gillette, D. D. and M. G. Lockley, eds. 1989. *Dinosaur Tracks and Traces.* New York: Cambridge University Press.

Gogola, M. 1975. Sledi iz davivne na Jugu Istre. *Proteus* 37: 229–32.

Gogola, M. and R. Pavlovec. 1978. Se enkrat o sledovih dinozavrov. *Proteus (Ljubljana)* 40: 192–6.

Gomes, J. P. 1915–16. Descoberta de rastos de saurios gigantescos no Jurassico do cabo Mondego. *Comun da Comis Serv Port* 11: 132–4 (em Manuscritos de Jacinto Pedro Gomes). Publicaçao postuma.

Gordon, E. 1892. *The Life and Correspondence of William Buckland, D.D., F.R.S.* London: Murray.

Guerin, C. and G. Demathieu. 1993. Empreintes et piste de Rhinocerotidae (Mammalia, Perissodactyla) du gisement Pliocène terminal de Laeotoli (Tanzanie). *Geobios* 26: 497–513.

Haderer, F. O. 1988. Ein dinosauroider Fährtenrest aus dem Unteren Stubensandstein (Obere Trias, Nor, km_4) des Strambergs (Württemberg). *Stuttgarter Beiträge zur Naturkunde B* 138: 1–12.

Haderer, F. O. 1992. Ein weitere grallatorider Fährtenrest aus dem Stubensandstein des Strambergs (Nordwürttemberg). *Jahrbuch Gesellschaft Naturkunde Württemberg* 147: 5–10.

Hamilton, A. 1952. The case of the mysterious Hand Animal. *Natural History* 61: 296–301, 336.

Haubold, H. 1971. *Ichnia Amphiborum et Reptiliorum Fossilum. Handbuch der Palaeoherpetologie Teil 18.* Stuttgart: Gustav Fischer Verlag.

Haubold, H. 1984. *Saurierfährten. A. Ziemsen-Verlag.* Wittenberg: Lutherstadt.

Haubold, H. 1986. Archosaur footprints at the Terrestrial Triassic-Jurassic transition. In: K. Padian, ed. *The Beginning of the Age of Dinosaurs,* pp. 189–201. New York: Cambridge University Press.

Haubold, H. 1996. Ichnotaxonomy and classification of tetrapod footprints of Permian age. *Hallesches Jahrbuch Geowissenschaften B* 18: 23–88.

Haubold, H., ed. 1997. Ichnofacies and ichnotaxonomy of the terrestrial Permian. Workshop. Martin Luther University, Halle-Wittenberg Institute of Geoscience and Geiselthalmuseum.

Haubold, H. and G. Katzung. 1978. Paleoecology and paleoenvironments of tetrapod footprints from the Rotliegend (Lower Permian) of Central Europe. *Palaeogeography Paleoclimatology Paleoecology* 23: 307–23.

Haubold, H. and W. A. S. Sarjeant. 1974. Vertebrate footprints and the stratigraphical correlation of the Keele and Enville beds of the Birmingham Region. *Proceedings of the Birmingham Natural History Society* 22: 257–68.

Haynes, G. 1991. *Mammoths, Mastodons, and Elephants: Biology, Behavior and the Fossil Record.* New York: Cambridge University Press.

Heintz, N. 1962a. Geological excursion to Svalbard in connection with the XXI International Geological Congress in Norden 1960. *Arbok Norsk Polarinstitutt* 1960: 98–106.

Heintz, N. 1962b. Dinosaur footprints and polar wandering. *Arbok Norsk Polaristitutt* 1960: 34–43.

Heintz, N. 1963. Casting dinosaur footprints at Spitzbergen. *Curator* 6: 217–25.

Hendricks, H. 1981. Die Saurierfährte von Münchenhagen bei Rehburg-Loccum (NW Deutschland). *Abhandlungen Landesmuseum Naturkunde Münster in Westfalen* 43: 1–22.

Hernandez-Pacheco, F. 1929. Pistes de aves fossiles en el Oligocene de Peralta de la Sal (Lerida). *Memorias de la Sociedad España Historía Natural Madrid* 379–82.

Hitchcock, E. 1848. An attempt to discriminate and describe the animals that made the fossil footmarks of the United States, and especially of New England. *Transactions of the American Academy of Arts and Science* 3: 129–256 (first description of Anomoepus).

Hitchcock, E. 1858. *Ichnology of New England. A Report on the Sandstone of the Connecticut Valley Especially Its Fossil Footmarks.* Boston: W. White [reprinted by Arno Press in the Natural Sciences in America Series].

Horner, J. R. and J. Gorman. 1990. *Digging Dinosaurs.* New York: Workman Publishing.

Huene, F. von. 1941. Die Tetrapoden-fährten im toskanischen Verrucano und ihre Bedeutung. *Neues Jahrbuch Mineralogie Geologie Paläontologie Abteilung B* 86: 1–34.

Huxley, T. H. 1877. The crocodilian remains found in the Elgin Sandstone, with remarks on the ichnites of Cummingstone. *Memoir Geological Survey Monographs* 3: 1–52.

Ishigaki, S. 1986. Dinosaur footprints of the Atlas Mountains. *Nature Study (Japan)* 32: 6–9.

Ivens, C. R. and G. Watson. 1995. *Records of Dinosaur Footprints on the Northeast Yorkshire Coast, 1895–1993.* Middlesborough, England: Roseberry Publications.

Jackson, P. 1995. Footloose in archaeology. *Current Archaeology* 144: 466–70.

Jackson, P. 1997. Footloose in archaeology. *Current Archaeology* 156: 456–7.

Jardine, W. Kt. 1850. Note to Mr Harkness's paper on the position of the impressions of footsteps in the Bunter sandstone of Dumfries-shire. *Annals of the Magazine of Natural History* (series 2) 6: 208–9.

Jenkins, F. A., N. H. Shubin, W. W. Amaral, S. M. Gatesy, C. R. Schaff, L. B. Clemmensen, W. R. Downs, A. R. Davidson, N. Bonde, and F. Osbaeck. 1994. Late Triassic continental vertebrates and depositional environments of the Fleming Fjord Formation, Jameson Land, East Greenland. *Meddelelser om Gronland Geoscience* 32: 3–25.

Johnson, E. W., D. E. G. Briggs, R. J. Suthren, J. L. Wright, and S. P. Tunnicliff. 1994. Non-marine arthropod traces from the subaerial Ordovician Borrowdale Volcanic Group, English Lake District. *Geology Magazine* 131: 395–406.

Jones, T. R. 1862. Tracks trails and surface markings. *Geologist* 5: 128–39.

Kaever, M. and A. F. de Lapparent. 1974. Dinosaur footprints of Jurassic age from Barkhausen. *Bulletin de la Société Géologique de France* 16: 516–25.

Kaup, J. J. 1835. Thier Fährten von Hildburghausen: Chirotherium oder Chirosaurus. *Neues Jarbuch für Mineralogie, Geognosie, Geologie und Petrefaktenkunde* 327–8.

King, M. J. 1998. Triassic vertebrate footprints of the British Isles [PhD thesis]. University of Bristol, England.

King, M. J. and M. J. Benton. 1996. Dinosaurs in the Early and Middle Triassic? The footprint evidence from Britain. *Palaeogeography Palaeoclimatology Palaeoecology* 122: 213–25.

Koenigswald, W. v., M. Walders, and M. Sander. 1995. Jungpleistozäne Tierfährten aus der Emscher-Niederterrasse von Bottrop-Welheim. *Münchner Geowissenschaftliche Abhandlungen* 27: 5–50.

Kordos, L. 1983. Footprints in Lower Miocene Sandstone at Ipolytarnóc, N. Hungary. Geologica Hungarica ser. Palaeontologica fasc. 44–46. p. 263–415 (with 17 plates).

Kuhn, O. 1958. *Die Fährten der vorzeitlichen Amphibien und Reptilien.* Bamberg: Verlagshaus Meisenbach.

Kurten, B. J. 1971. *The Age of Mammals.* New York: Columbia University Press.

Kurten, B. J. 1976. *The Cave Bear Story: Life and Death of a Vanished Animal.* New York: Columbia University Press.

Lancis, C. and A. Estévez. 1992. Las icnitas de mammíferos del sur de Alicante (España). *Geogaceta* 12: 60–94.

Lange-Badré, B., M. Dutrieux, J. Feyt, and G. Maury. 1996. Découverte d'empreintes de pas de dinosaures dans le Jurassique supérieur des causses du Quercy (Lot, France) *Comptes Rendu Académie des Sciences Paris* 323: 89–96.

Lanzinger, M. and G. Leonardi. 1991. Piste di Dinosauri del Giurassico Inferiore ai Lavini di Marco (Trento). In: G. Muscio, ed. *Dinosaurs: Il Mondo dei Dinosauri,* pp. 89–94. Trento: Kaleidos.

Lapparent, A. F. de. 1962. Footprints of Dinosaurs in the Lower Cretaceous of Vestspitsbergen-Svalbard. Arbok: Norsk Polarinstitutt, 1960: 14–21.

Lapparent, A. F. de and C. Montenat. 1967. Les empreintes des pas de reptiles de l'infralias du Veillon (Vendée). Société Géologique de France, nouvelle série. *Mémoire* 107: 1–41.

Lapparent, A. F. de and G. Zbyszewsky. 1957. Les dinosauriens du Portugal. *Memorias dos Servicio Geologico de Portugal* 2: 1–63.

Lapparent, A. F. de, G. Zbyszewski, F. Moitinho de Almeida, and O. Viega Ferreira. 1951. Empreintes de pas de dinosauriens dans le Jurassique du Cap Mondego (-Portugal). *Comptes Rendu Sommaire de la Société Géologique de France* 14: 251–2.

Leakey, M. D. and R. L. Hay. 1979. Pliocene footprints in the Laetoli beds at Laetoli, northern Tanzania. *Nature* 278: 317–23.

Lee, M. 1994. The turtle's long-lost relatives. *Natural History* 103: 63–5.

Leghissa, S. and G. Leonardi. 1990. *Una Pista di Sauropode Scoptera nei Calcari Cenomanian dell'Istria.* Genoa: Edito del Centro Cultura Giuliana Dalmata.

Le Loeuff J. 1993. European titanosaurids. *Revue de Paléobiologie* 7: 105–17.

Le Loeuff, J. L., M. G. Lockley, C. Meyer, and J-P. Petit. 1999. Discovery of a thyreophoran trackway in the Hettangian of Central France. *Comptes Rendu de l'Académie des Sciences Paris.* 328: 215–219.

Leonardi, G. 1994. *Annotated Atlas of South America Tetrapod Footprints (Devonian to Holocene).* Brasilia: Publication of the Companhia de Pesquisa de Recursos Minerais, Brasilia.

Leonardi, G. and M. Avanzini. 1994. Dinosauri in Italia. In: G. Ligabue, ed. *Il Tempo dei Dinosauri,* pp. 69–81. Milan: Le Scienze Quaderno 76.

Leonardi, G. and M. G. Lockley. 1995. A proposal to abandon the ichnogenus *Coelurosaurichnus* Huene, 1941. A junior synonym of *Grallator* E. Hitchcock 1858. *Journal of Vertebrate Paleontology* 15: 40A.

Leonardi, P., M. A. Conti, G. Leonardi, N. Mariotti, and U. Nicosia. 1975. *Pachypes dolomiticus* n. gen n. sp.; Pareiasaur footprint from the Val Gardena Sandstone (Middle Permian) in the western Dolomites (N. Italy). *Atti Accademia Nazionale Lincei Rendi Scienzia Fisica Mathematica Naturale* 57: 221–3.

Lessertisseur, J. 1955. Traces fossiles d'activité animale et leur signification paléobiologique. *Mémoire de la Société Géologique de France* 74: 1–150.

Lim, S. K., M. G. Lockley, S-Y. Yang, R. F. Fleming, and K. A. Houck. 1994. Preliminary report on sauropod tracksites from the Cretaceous of Korea. *Gaia: Revista de Geociencias,* Museu Nacional de Historia Natural, Lisbon, Portugal 10: 109–17.

Lockley, M. G. 1991a. The dinosaur footprint renaissance. *Modern Geology* 16: 139–60.

Lockley, M. G. 1991b. *Tracking Dinosaurs: A New Look at an Ancient World.* New York: Cambridge University Press.

Lockley, M. G. 1993. Ichnotopia: a review of the Paleontological Society short course on trace fossils. *Ichnos* 2: 337–42.

Lockley, M. G. 1994. Dinosaur ontogeny and population structure: interpretations and speculations based on footprints. In: K. Carpenter, K. Hirsch, and J. Horner, eds., *Dinosaur Eggs and Babies.* New York: Cambridge University Press.

Lockley, M. G. 1995. Track records. *Natural History* 104: 46–51.

Lockley, M. G. In press(a). Permian perambulations become understandable. A report on a Workshop on the Ichnofacies and Ichnotaxonomy of the Terrestrial Permian convened by Hartmut Haubold in Niersten, Rotenburg, Cornberg, Tambach, Gotha and Halle, Germany, March 8–11, 1997.

Lockley, M. G. In press(b). Philosophical perspectives on theropod track morphology: blending qualities and quantities in the science of ichnology. *Gaia: Revista de Geociencias,* Museu Nacional de Historia Natural, Lisbon, Portugal.

Lockley, M. G. In press(c). Pterosaur and bird tracks from a new locality in the Late Cretaceous of Utah. In: D. D. Gillette, ed. *Utah Geological Survey.*

Lockley, M. G., K. Conrad, and M. Jones. 1988. Regional scale vertebrate bioturbation: new tools for sedimentologists and stratigraphers. *Geological Society of America, Abstracts with Program* 20: 316.

Lockley, M. G., J. O. Farlow, and C. A. Meyer. 1994. *Brontopodus* and *Parabrontopodus* Ichnogen. nov. and the significance of wide- and narrow-gauge sauropod trackways. *Gaia: Revista de Geociencias,* Museu Nacional de Historia Natural, Lisbon, Portugal 10: 135–46.

Lockley, M. G. and R. F. Fleming. 1991. Latest Cretaceous dinosaur track assemblages: implications for biostratigraphy, paleoecology and the K/T debate. Geological Society of America, Abstracts with Program.

Lockley, M. G. and A. P. Hunt. 1994. A track of the giant theropod dinosaur *Tyrannosaurus* from close to Cretaceous/Tertiary boundary, northern New Mexico. *Ichnos* 3: 213–8.

Lockley, M. G. and A. P. Hunt. 1995. *Dinosaur Tracks and Other Fossil Footprints of the Western United States.* New York: Columbia University Press.

Lockley, M. G. and A. P. Hunt. 1998. A probable Stegosaur track from the Morrison Formation of Utah. In: K. Carpenter, D. Chure, and J. Kirkland, eds. *The Upper Jurassic Morrison Formation: An Interdisciplinary Study. Modern Geology* 23: 331–42.

Lockley, M. G., A. P. Hunt, and C. Meyer. 1994. Vertebrate tracks and the ichnofacies concept: implications for paleoecology and palichnostratigraphy. In: S. Donovan, ed. *The Paleobiology of Trace Fossils,* pp. 241–68. New York: Wiley and Sons.

Lockley, M. G., A. P. Hunt, J. J. Moratalla, and M. Matsukawa. 1994. Limping dinosaurs? Trackway evidence for abnormal gaits. *Ichnos* 3: 193–202.

Lockley, M. G., A. P. Hunt, M. Paquette, S-A. Bilbey, and A. Hamblin. 1998. Dinosaur tracks from the Carmel Formation, northeastern Utah: implications for Middle Jurassic paleoecology. *Ichnos* 5: 255–67.

Lockley, M. G., M. J. King, S. Howe, and T. Sharp. 1996. Dinosaur tracks and other archosaur footprints from the Triassic of South Wales. *Ichnos* 5: 23–41.

Lockley, M. G., M. Lim, S-K. Huh, S-Y. Yang, S. S. Chun, and D. Unwin. 1997. First report of pterosaur tracks from Asia, Chollanam Province Korea. *Journal of the Paleontology Society of Korea,* Special Publication No. 2: 17–32.

Lockley, M. G., T. J. Logue, J. J. Moratalla, A. P. Hunt, R. J. Schultz, and J. W. Robinson. 1995. The fossil trackway *Pteraichnus* is pterosaurian, not crocodilian: implications for the global distribution of pterosaur tracks. *Ichnos* 4: 7–20.

Lockley, M. G., C. A. Meyer, A. P. Hunt, and S. G. Lucas. 1994. The distribution of sauropod tracks and trackmakers. *Gaia: Revista de Geociencias,* Museu Nacional de Historia Natural, Lisbon, Portugal 10: 233–48.

Lockley, M. G., C. A. Meyer, and J. J. Moratalla. In press. *Therangospodus:* trackway evidence for the widespread distribution of a Late Jurassic theropod with well-padded feet. *Gaia: Revista de Geociencias,* Museu Nacional de Historia Natural, Lisbon, Portugal.

Lockley, M. G., C. A. Meyer, and V. F. dos Santos. 1994. Trackway evidence for a herd of juvenile sauropods from the Late Jurassic of Portugal. *Gaia: Revista de Geociencias,* Museu Nacional de Historia Natural, Lisbon, Portugal 10: 27–36.

Lockley, M. G., C. A. Meyer, and V. F. dos Santos. 1996. *Megalosauripus, Megalosauropus* and the concept of Megalosaur footprints. In: M. Morales, ed. *The Continental Jurassic (Symposium Volume),* Bulletin 60, pp. 113–8. Flagstaff, AZ: Museum of Northern Arizona.

Lockley, M. G., C. A. Meyer, and V. F. dos Santos. In press. *Megalosauripus,* and the problematic concept of Megalosaur footprints. *Gaia: Revista de Geociencias,* Museu Nacional de Historia Natural, Lisbon, Portugal.

Lockley, M. G., C. A. Meyer, R. Schultz-Pittman, and G. Forney. 1996. Late Jurassic dinosaur tracksites from Central Asia: a preliminary report on the world's longest trackways. In: M. Morales, ed. *Continental Jurassic* (Symposium Volume), Bulletin 60, pp. 271–3. Flagstaff, AZ: Museum of Northern Arizona.

Lockley, M. G., V. Novikov, V. F. dos Santos, L. A. Nessov, and G. Forney. 1994. Pegadas de Mula: an explanation for the occurrence of Mesozoic traces that resemble mule tracks. *Ichnos* 3: 125–33.

Lockley, M. G. and J. G. Pittman. 1989. The megatracksite phenomenon: implications for paleoecology, evolution and stratigraphy. *Journal of Vertebrate Paleontology* 9: 30A.

Lockley, M. G. and E. Rainforth. In press. The tracks record of Mesozoic birds and pterosaurs: an ichnological and paleoecological perspective. In: L. Chiappe and L. M. Witmer, eds. *Mesozoic Birds Above the Heads of Dinosaurs.* Berkeley: University of California Press.

Lockley, M. G. and V. F. dos Santos. 1993. A preliminary report on sauropod trackways from the Avelino site, Sesimbra Region, Upper Jurassic, Portugal. *Gaia: Revista de Geociencias,* Museu Nacional de Historia Natural, Lisbon, Portugal 6: 38–42.

Lockley, M. G., V. F. dos Santos, and A. P. Hunt. 1993. A new Late Triassic tracksite in the Sheep Pen Sandstone, Sloan Canyon, New Mexico. *New Mexico Museum Natural History Science Bulletin* 3: 285–8.

Lockley, M. G., V. F. dos Santos, C. Meyer, and A. P. Hunt. 1998. A new dinosaur tracksite in the Morrison Formation, Boundary Butte, southeastern Utah. In: K. Carpenter, D. Chure, and J. Kirkland, eds. The Upper Jurassic Morrison Formation: an interdisciplinary study. *Modern Geology* 10: 317–30.

Lockley, M. G., V. F. dos Santos, M. M. Ramalho, and A. Galopim. 1993. Descoberta de novas jazidas de pegagdas de dinosaurios no Jurassico Superior de Sesimbra. *Gaia: Revista de Geociencias,* Museu Nacional de Historia Natural, Lisbon, Portugal 15: 40–3.

Lockley, M. G., S-Y. Yang, M. Matsukawa, F. Fleming, and S. K. Lim. 1992. The track record of Mesozoic birds: evidence and implications. *Philosophical Transactions of the Royal Society of London* 336: 113--34.

Lompart, C. 1979. Yacimento de huellas de pisadas de reptil en el Cretacico Superior prepirenacio. *Acta Geologica Hispania* 14: 333–6.

Lompart, C. 1984. Un nuevo yacimiento de ichnitas de dinosaurios en las facies Garumnienses de la Conca de Tremp (Leida España). *Acta Geologica Hispania* 19: 143–7.

Lopez-Martinez, N. Ardevol, L. Arribas, M. E. Civis, and A. Gonzalez-Delgado. 1998. *The Geological Record in Non-marine Environments around the K.T Boundary (Tremp Formation, Spain).* Paris: Société Géologique de France.

Lull, R. S. 1918. Fossil footprints from the Grand Canyon of the Colorado. *American Journal of Science* 45: 337–46.

Lull, R. S. 1953. Triassic life of the Connecticut Valley. *Bulletin of the Connecticut State Geology Natural History Survey* 181: 1–331.

Mangin. P. 1962. Trace des pattes d'oiseaux et Flute castes associés dans un facies flysch du Tertiaire, pyrénéen. *Sedimentology* 1: 163–6.

Martin, V. and E. Buffetaut. 1992. Les Iguanodons (Ornithischia-Ornithopoda) du Crétace Inférieur de la région de Saint Dizier (Haute-Marne). *Revue de Paléobiologie* 11: 67–96.

Martin-Escorza, C. 1992. Gregarismo y dinosaurios. *Revista Española Paleontología* 7: 97.

Mazin, J. M., P. Hantzpergue, J-P. Bassollet, G. Lafaurie, and P. Vignaud. 1997. The Crayssac site (Lower Tithonian, Quercy, Lot, France): discovery of dinosaur trackways in situ and first ichnological results. *Comptes Rendu de l'Académie des Sciences Paris* 325: 733–9.

Mazin, J. M., P. Hantzpergue, G. Lafaurie, and P. Vignaud. 1995. Des pistes de ptérosaures dans le Tithonien de Crayssac (Quercy, France). *Comptes Rendu de l'Académie des Sciences Paris* Série IIa: 417–24.

McKeever, P. J. and H. Haubold. 1996. Reclassification of vertebrate trackways from the Permian of Scotland and related forms from Arizona and Germany. *Journal of Paleontology* 70: 1011–22.

Mensink, H. and D. Mertmann. 1984. Dinosaurier-Fährten (*Gigantosauripus asturiensis* n.g.n.sp.; *Hispanosauripus hauboldi* n.g.n.sp.) im Jura Asturiens bei La Griega und Ribadesella (Spanien). *Neues Jahrbuch für Geologie und Paläontologie Monatshefte* 1984: 405–15.

Meyer, C. A. 1993. A sauropod dinosaur megatracksite from the Late Jurassic of northern Switzerland. *Ichnos* 3: 29–38.

Meyer, C. A. 1997. High resolution three dimensional mapping and conventional methods compared—a case study of fossil footprints. In: H. Haubold, ed. *Ichnofacies and Ichnotaxonomy of the Terrestrial Permian; Workshop,* p. 61. Halle-Wittenberg: Martin Luther University, Halle-Wittenberg Institute of Geoscience and Geiselthalmuseum.

Montenat, C. 1967. Empreintes de pas de reptiles dans le Trias moyen du plateau du Daus près d'Aubenas (Ardèche). *Bulletin Scientifique Bourgogne* 25: 369–89.

Moratalla, J. J. 1993. Restos indirectos de dinosaurious del registro espanol: Paleoichnologica de la Cuenca de Cameros (Jurasico superior-Cretacico inferior) y Palecología del Cretacico superior [doctoral thesis]. Faculdad de Ciencias, Universidad Autonoma, Madrid.

Moratalla, J. J., J. Garcia-Mondejar, M. G. Lockley, J. Sanz, and S. Jimenez. 1994. Sauropod trackways from the Lower Cretaceous of Spain. *Gaia: Revista de Geociencias,* Museu Nacional de Historia Natural, Lisbon, Portugal 10: 75–83.

Moratalla, J. J., M. G. Lockley, A. D. Buscalioni, M. Fregenal, N. Melendez, B. R. Perez-Moreno, E. Perez, J. L. Sanz, and R. Schultz. 1995. A preliminary note on the first tetrapod trackways from the lithographic limestones of Las Hoyas (Lower Cretaceous, Cuenca, Spain). *Geobios* 28: 777–82.

Moratalla, J. J. and J. L. Sanz. 1997. Cameros Basin megatracksite. In: P. J. Currie and K. Padian, eds. *Encyclopedia of Dinosaurs,* pp. 87–90. San Diego: Academic Press.

Moratalla, J. J., J. L. Sanz, and S. Jimenez. 1992. Hallazgo de nuevos tipos de Huellas en la Rioja. *Estrato Revista Riojana de Archeología,* pp. 63–6.

Moratalla, J. J., J. L. Sanz, and S. Jimenez. 1993. Dinosaur tracks from the Lower Cretaceous of Regumiel de la Sierra (province of Burgos, Spain): inferences on a new quadrupedal ornithopod trackway. *Ichnos* 2: 1–9.

Moratalla, J. J., J. L. Sanz, S. Jimenez, and M. G. Lockley. 1992. A quadrupedal ornithopod trackway from the Early Cretaceous of La Rioja (Spain): inferences on gait and hand structure. *Journal of Vertebrate Paleontology* 12: 150–7.

Moratalla, J. J., J. L. Sanz, I. Melero, and S. Jimenez. 1988. *Yacimientos Paleoicnológicos de la Rioja (Huellas de Dinosaurios).* Iberdrola y Gobierno de la Rioja.

Nieves, J. 1993. Quien paso por aqui hace 20 millones de años. *Blanco y Negro Magazine* (Madrid) Año CII, 3871: 56–65.

Noe-Nygard, N. 1995. Ecological, sedimentary and geochemical evolution of the late-glacial to post-glacial Amose lacustrine basin, Denmark. *Fossils and Strata* 37: 1–436.

Nopsca, F. von. 1923. Die fossilen Reptilien. *Fortschritte Geologie Palaeontologie* 2: 1–210.

Norman, D. 1987. Wealden dinosaur biostratigraphy. In: P. J. Currie and E. H. Koster, eds. *Fourth Symposium on Mesozoic Terrestrial Ecosystems, Short Papers,* pp. 165–70. Drumheller, Canada: Tyrrell Museum.

Olsen, P. E. 1980. Fossil Great Lakes of the Newark Supergroup in New Jersey. In: W. Manspeizer, ed. *Field Studies of New Jersey Geology and Guide to Field Trips,* pp. 352–398. New York State Geological Association 52nd Annual Meeting. New Brunswick, NJ: Rutgers University Press.

Olsen, P. E. and P. Galton. 1984. A review of the reptile and amphibian assemblages from the Stromberg of southern Africa, with special emphasis on the footprints and the age of the Stromberg. *Paleontologica Africana* 25: 87–110.

Oppel, A. 1862. Über Fährten im lithographischen Schiefer. *Paläontologische Mitteilungen Museum Bayrische Staatssammlung* 1: 121–5.

Padian, K. and P. E. Olsen. 1984. The fossil trackway *Pteraichnus:* not pterosaurian, but crocodilian. *Journal of Paleontology* 58: 178–84.

Pales, L. 1976. Les empreintes de pieds humains dans les cavernes. *Archéologie Institut Paléontologie Humaine* 36: 1–166.

Pales, L. and M. T. de St. Pereuse. 1976. *Les Gravures de la Marche. II. Les Humains.* Paris: Ophyrys.

Panin, N. 1965. Coexistence de traces des vertèbres et de mécanoglyphes dans la molasse Miocène des Carpathes orientales. *Revue Roumaine Géologie Géophysique Géographique* 9: 141–63.

Panin, N. and E. Avram. 1962. Noi urme de vertebrate in Miocenul Subcarpatilor Romanesti. *Studie Cercet Geologia Geofizica Geografia* 3: 455–84.

Parkes, A. S. 1993. Dinosaur footprints in the Wealden at Fairlight, East Sussex. *Proceedings of the Geological Association* 104: 15–21.

Parrish, M. J. 1989. Phylogenetic patterns in the manus and pes of Early Mesozoic Archosauromorpha. In: D. D. Gillette and M. G. Lockley, eds. *Dinosaur Tracks and Traces,* pp. 249–58. New York: Cambridge University Press.

Paul, G. 1988. *Predatory Dinosaurs of the World.* New York: Simon & Schuster.

Paul, G. 1991. The many myths, some old, some new of dinosaurology. *Modern Geology* 16: 69–99.

Paul, G. 1998. *Predatory Dinosaurs of the World.* New York: Simon & Schuster.

Peabody, F. 1948. Reptile and amphibian trackways from the Lower Triassic Moenkopi Formation of Arizona and Utah. *University of California Department of Science Bulletin* 27: 295–468.

Pearson, P. N. 1992. Walking traces of the giant myriapod Arthropleura from the Strathclyde Group (Lower Carboniferous) of Fife. *Scottish Journal of Geology* 28: 127–33.

Pemberton, S. G., W. A. S. Sarjeant, and H. S. Torrens. 1996. Footsteps before the flood: the first scientific reports of vertebrate footprints. *Ichnos* 4: 321–4. (Contains a reprint of the first newspaper article ever written about fossil footprint discoveries in Scotland.)

Pereda-Suberbiola, J. 1993. Hylaeosaurus, Polacanthus, and the systematics and stratigraphy of Wealden armoured dinosaurs. *Geological Magazine* 130: 767–81.

Pérez-Lorente, F., C. Cuenca-Bescos, M. Aurell, J. I. Canudo, A. I. Soria, and J. I. Ruiz Omenaca. 1997. Las Cerradicas tracksite (Berriasian, Galve, Spain): growing evidence for quadrupedal ornithopods. *Ichnos* 5: 109–20.

Plaziat, J. C. 1964. Pattes d'oiseaux et remaniements synsédimentaires dans le Lutétien du Détroit de Carcassonne (Aude). *Bulletin de la Société Géologique de France* 7: 289–93.

Price, N. 1995. Houses and horses in the Swedish Bronze Age: recent excavation in the Malar Valley. Past. *Newsletter of the Prehistorical Society* 20: 10–2.

Prince, N. K. and M. G. Lockley. 1989. The sedimentology of the Purgatoire Tracksite Region, Morrison Formation of southeastern Colorado. In: D. D. Gillette and M. G. Lockley, eds. *Dinosaur Tracks and Traces*, pp. 155–64. New York: Cambridge University Press.

Quintana, J. 1993. Descripción de un ratro de *Myotragus* e icnitas de *Hypnomys* del yacimento cuaternario de Ses Penyes d'es Perico (Ciutadella de Menorca, Balears). Paleontología i Evolució: Instituto Paleontológico, Dr. M Crusafont, Diputació de Barcelona. No 26-27: 271--279.

Raaf, J. F. M. de, C. Beets, and G. Kortenbout van der Sluijs. 1965. Lower Oligocene bird tracks from northern Spain. *Nature* 207: 146–8.

Raath, M. A. 1972. First record of dinosaur footprints from Rhodesia. *Arnoldia* 5: 1–5.

Radley, J. D. 1994. Stratigraphy, palaeontology and palaeoenvironment of the Wessex Formation (Wealden Group, Lower Cretaceous) at Yaverland, Isle of Wight, southern England. *Proceedings of the Geological Association* 105: 199–208.

Radley, J. D., M. J. Barker, and I. C. Harding. 1998. Paleoenvironment and taphonomy of dinosaur tracks in the Vectis Formation (Lower Cretaceous) of the Wessex sub-basin, southern England. *Cretaceous Research.*

Rainforth, E. C. and M. G. Lockley. 1996. Tracks of diminutive dinosaurs and hopping mammals from the Jurassic of North and South America. In: *The Continental Jurassic (Symposium Volume)*, Bulletin 60, pp. 265–9. Flagstaff, AZ: Museum of Northern Arizona.

Renders, E. 1984. The gait of *Hipparion* sp. from fossil footprints in Laeotoli, Tanzania. *Nature* 308: 179–81.

Roberts, G., S. Gonzalez, and D. Huddart. 1996. Intertidal holocene footprints and their archaeological significance. *Antiquity* 70: 647–51.

Rocek, Z. and J-C. Rage. 1995. The presumed amphibian footprint *Notopus petri* from the Devonian: a probable starfish trace fossil. *Lethaia* 27: 241–4.

Rogers, D. A. 1990. Probable tetrapod tracks rediscovered in the Devonian of Scotland. *Journal of the Geological Society of London* 147: 746–8.

Russell, D. 1989. *An Odyssey in Time: The Dinosaurs of North America.* Minocqua, WI: Northwood Press.

Santamaria, R., G. Lopez, and M. L. Casanovas Cladellas. 1989. Nuevos yacimentos con ichnitas de mamíferos del Oligoceno de los alrededores de Agramunt (Leida, España). Paleontología i Evolució: Instituto Paleontológico, Dr. M Crusafont, Diputació de Barcelona 23: 141–52.

Santos, V. F. dos. 1998. Dinosaur tracksites in Portugal: the Jurassic and Cretaceous record. In: A. M. Galopim de Carvalho et al., eds. *Ist International Meeting on Dinosaur Paleobiology: Museology Program for the Portuguese Dinosaur Tracksites,* pp. 7–16. Lisbon: Museu Nacional de Historia Natural.

Santos, V. F. dos, G. Carvalho, and C. Marques da Silva. 1995. *A jazida de pedreira da ribeira do cavalo (Sesimbra) ou a historia das pegadas de dinosaurio que nunca mais poderemos visitar,* pp. 175–7. Al-madan: Centro de Arquelogía de Almada, Municipality of Sesimbra, II (serie 4).

Santos, V. F. dos, M. G. Lockley, C. A. Meyer, J. Carvalho, A. M. Galopim de Carvalho, and J. J. Moratalla. 1994. A new sauropod tracksite from the Middle Jurassic of Portugal. In: M. G. Lockley et al., eds. *Aspects of Sauropod Paleobiology. Gaia: Revista de Geociencias,* Museu Nacional de Historia Natural, Lisbon, Portugal 10: 5–14.

Santos, V. F. dos, M. G. Lockley, J. J. Moratalla, and A. Galopim, A. 1992. The longest dinosaur trackway in the world? Interpretations of Cretaceous footprints from Carenque, near Lisboa, Portugal. *Gaia: Revista de Geociencas,* Museu Nacional de Historia Natural, Lisbon, Portugal 5: 18–27.

Sarjeant, W. A. S. 1970. Fossil footprints from the Middle Triassic of Nottinghamshire and the Middle Jurassic of Yorkshire. *Mercian Geologist* 3: 269–82.

Sarjeant, W. A. S. 1974. A history and bibliography of the study of fossil vertebrate footprints in the British Isles. *Palaeogeography Palaeoclimatology Palaeoecology* 16: 265–378.

Sarjeant, W. A. S. 1975. A vertebrate footprint from the Stonesfield Slate (Middle Jurassic) of Oxfordshire. *Mercian Geologist* 5: 273–7.

Sarjeant, W. A. S. 1996. A reappraisal of some supposed dinosaur footprints from the Triassic of the English Midlands. *Mercian Geologist* 14: 22–30.

Sarjeant, W. A. S., J. Delair, and M. G. Lockley. 1998. The footprints of *Iguanodon:* a history and taxonomic study. *Ichnos* 6: 183–202.

Scarboro, D. D. and M. E. Tucker. 1995. Amphibian footprints from the mid-Carboniferous of Northumberland, England: sedimentological context, preservation and significance. *Palaeogeography Palaeoclimatology Palaeoecology* 113: 335–49.

Schulp, A. S. and A. W. Brokx. 1999. Maastrichtian sauropod footprints from the Fumanya site, Bergueda, Spain. *Ichnos* 6: 239–50.

Soergel, W. 1925. *Die Fährten der Chirotheria. Eine paläobiologische Studie.* Jena: Gustav Fischer.

Sollas, W. J. 1879. On some three-toed footprints from the Triassic Conglomerate of South Wales. *Quarterly Journal of the Geological Society of Southern Wales* 35: 511–6.

Sternberg, C. M. 1932. Dinosaur tracks from Peace River, British Columbia. Annual Report, National Museum of Canada (for 1930), pp. 59–85.

Stokes, W. L. 1957. Pterodactyl tracks from the Morrison Formation. *Journal of Paleontology* 31: 952–4.

Stössel, I. 1995. The discovery of a new Devonian tetrapod trackway in SW Ireland. *Journal of the Geological Society of London* 152: 407–13.

Tagart, E. 1946. On markings in the Hastings sands near Hastings, supposed to be the footprints of birds. *Quarterly Journal of the Geological Society of London* 2: 267.

Taquet, P. 1998. *Dinosaur Impressions: Postcards from a Paleontologist.* New York: Cambridge University Press.

Thomas, T. H. 1879. Triassic Uniserial Ichnolites in the Trias at Newton Nottage, near Portcawl, Glamorganshire. Cardiff: Cardiff Naturalists' Society, pp. 72–91.

Thompson, W. I. 1981. *The Time Falling Bodies Take to Light.* New York: St. Martin's Press.

Thulborn, R. A. 1990. *Dinosaur Tracks.* London: Chapman & Hall.

Tisljar, J., I. Velic, J. Radovic, and B. Cernokovic. 1983. Upper Jurassic and Cretaceous peritital lagoonal sediments of Istria. In: L. Babic and V. Jelaska, eds. *Contributions to Sedimentology of Some Carbonate and Clastic Units of the Coastal Dinardes. International Association of Sedimentologists Excursion Guidebook,* pp. 13–32.

Tresise, G. 1996. Sex in the footprint bed. *Geology Today* 12: 22–6.

Tresise, G. and W. A. S. Sarjeant. 1997. *The Tracks of Triassic Vertebrates: Fossil Evidence from Northwest England.* London: Her Majesty's Stationery Office.

Tucker, M. E. and T. P. Burchette. 1977. Triassic dinosaur footprints from South Wales: their context and preservation. *Palaeogeography Palaeoclimatology Palaeoecology* 22: 195–208.

Turner, A. and M. Anton. 1997. *The Big Cats and Their Fossil Relatives.* New York: Columbia University Press.

Unwin, D. M. 1989. A predictive method for the identification of vertebrate ichnites and its application to pterosaur tracks. In: D. D. Gillette and M. G. Lockley, eds. *Dinosaur Tracks and Traces,* pp. 259–74. New York: Cambridge University Press.

Unwin, D. M. 1997. Pterosaur tracks and the terrestrial ability of pterosaurs. *Lethaia* 29: 373–86.

Valenzuela, M., J. C. Garcia Ramos, and C. Suarez de Centi. 1986. Los dinosaurios de Asturias a partir de sus huellas. *Asturia* 5: 66–71.

Valenzuela, M., J. C. Garcia Ramos, and C. Suarez de Centi. 1988. Las huellas de dinosaurios del entorno de Ribadessella. Central Lechera Asturiana. Amigos de Ribadessella.

Vallois, H. V. 1927. *Les Empreintes de Pieds Humains du Tuc d'Audoubert, de Caberets et de Ganties,* pp. 328–334. Amsterdam: Institut International d'Anthropologie, III session.

Vallois, H. V. 1931. Les empreintes de pieds humains des grottes préhistoriques du Midi de la France. In: *Paleobiologica*, pp. 79–98. Vienna: Verlag von Emil Haim & Co.

Vialou, D. L'art des grottes en Ariège magdalénienne. XXIIe Supplément à *Gallia Préhistoire*, 28 pl.

Vialov, O. 1972. The classification of the fossil traces of life. Proceedings of the 24th International Congress, Montreal, 1972, section 7. *Paleontology* 639–44.

Viera, L. I. and J. A. Torres. 1992. Sobre dinosaurios coeluridos gregarios en el yacimiento de Valdevajes (La Rioja, Espana). Nota de replica y critica. *Revista Española Paleontología* 7: 93–6.

Walter, H. von and R. Werneburg. 1988. Über Liegespuren (Cubichnia) aquatischer Tetrapoden (Diplocauliden, Nectridea) aus den Rotteroder Schichten (Rotliegendes, Thüringer Wald/DDR). *Freiberger Forschungshefte (Leipzig)* C419: 96–106.

Weidmann, M. and M. Reichel. 1979. Trace des pattes d'oiseaux dans la Molasse Suisse. *Eclogae geologicae Helvetiae* 72: 953–71.

Weishampel, D. B. 1984. Trossingen: E. Frass, F. vo Huence, R. Seemann and the "Schwäbische Lindwurm" Plateosaurus. In: W-E. Reif and F. Westphal, eds. *3rd Symposium. Mesozoic Terrestrial Ecosystems. Short Papers*, pp. 249–53. Tubingen: Attempto Verlag.

Weishampel, D. B., P. Dodson, and H. Osmolska. 1990. *The Dinosauria.* Berkeley: University of California Press.

Wendt, H. 1968. *Before the Deluge.* London: Victor Gollancz Ltd.

Whyte, M. A. and M. Romano. 1993. Footprints of a sauropod from the Middle Jurassic of Yorkshire. *Proceedings of the Geological Association* 104: 195–9.

Whyte, M. A. and M. Romano. 1994. Probable sauropod footprints from the Middle Jurassic of Yorkshire, England. *Gaia* 10: 15–26.

Wills, L. J. and W. A. S. Sarjeant. 1970. Fossil vertebrate and invertebrate tracks from boreholes through the Bunter Sandstone Series (Triassic) of Worcestershire. *Mercian Geologist* 3: 399–414.

Woodhams, K. and J. Hines. 1989. Dinosaur footprints from the Lower Cretaceous of East Sussex, England. In: D. D. Gillette and M. G. Lockley, eds. *Dinosaur Tracks and Traces.* New York: Cambridge University Press.

Woodhams, K. and J. Hines. In press. Footprint evidence for theropod diversity in the Wealden of east Sussex. *Gaia: Revista de Geociencias,* Museu Nacional de Historia Natural, Lisbon, Portugal.

Wright, J. L. 1996. Fossil terrestrial trackways: function, taphonomy and paleoecological significance [Ph.D. thesis]. Department of Geology, University of Bristol, England.

Wright, J. L. In press. Ichnological evidence for the use of the forelimb in iguanodontoid locomotion. *Palaeontology.*

Wright, J. L. and I. J. Sansom. 1998. The earliest known terrestrial tetrapod skin impressions (Upper Carboniferous, Shropshire, UK). *Journal of Vertebrate Paleontology* 18 (Suppl): 88A.

Wright, J. L., D. M. Unwin, M. G. Lockley, and E. C. Rainforth. 1997. Pterosaur tracks from the Purbeck Limestone Formation of Dorset, England. *Proceedings of the Geologists' Association* 108: 39–48.

Yang, S-Y., M. G. Lockley, R. Greben, B. R. Erikson, and S-Y. Lim. 1994. Flamingo and duck-like bird tracks from the Late Cretaceous and Early Tertiary: evidence and implications. *Ichnos* 4: 21–34.

Zhen, S., R. Li, C. Rao, N. Mateer, and M. G. Lockley. 1989. A review of dinosaur footprints in China. In: D. D. Gillette and M. G. Lockley, eds. *Dinosaurs Past and Present,* pp. 187–98. New York: Cambridge University Press.

INDEX